Laboratory Methods in Immunology

Volume II

Editor

Heddy Zola, B.Sc., Ph.D.
Chief Hospital Scientist and Professor of Immunology
Department of Clinical Immunology
Flinders Medical Centre
Bedford Park, South Australia, Australia

CRC Press, Inc.
Boca Raton, Florida

Library of Congress Cataloging in Publication Data

Laboratory methods in immunology / editor, Heddy Zola.
     p. cm.
    Includes bibliographical references.
    ISBN 0-8493-4481-6 (v. 1). — ISBN 0-8493-4482-4 (v. 2)
    1. Immunology—Laboratory manuals. I. Zola, Heddy.
    [DNLM: 1. Immunologic technics. QW 525 L123]
QR183.L34 1990
616.07′9′078—dc20
DNLM/DLC   89-22184
for Library of Congress   CIP

This book represents information obtained from authentic and highly regarded sources. Reprinted material is quoted with permission, and sources are indicated. A wide variety of references are listed. Every reasonable effort has been made to give reliable data and information, but the authors and the publisher cannot assume responsibility for the validity of all materials or for the consequences of their use.

Direct all inquiries to CRC Press, Inc., 2000 Corporate Blvd., N.W., Boca Raton, Florida, 33431.

International Standard Book Number 0-8493-4481-6 (Volume I)
International Standard Book Number 0-8493-4482-4 (Volume II)

s Card Number 89-22184
the United States

PREFACE

Some of the most exciting advances in biology, biotechnology, and medicine in the last 10 years have been in immunology or have used immunological techniques. Biologists, from neuroscientists to plant pathologists, from cancer researchers to toxicologists, need to know how to make and use monoclonal antibodies, how to purify proteins on immunoaffinity columns, and how to stain tissue with specific antibody. While biologists from other disciplines have been forced to embrace immunology (often wearing an expression betraying their distaste for the subject), immunologists have needed to acquire at least the language, and often also the technology, of molecular genetics.

These two volumes were planned to contain cookbook-style recipes for immunological techniques, interspersed with reviews on the strategies required, and the choice of techniques. This volume and its companion contain detailed methods and reviews, written by people who are actually using the methods and written to be used at the bench. As the editor, I have had the opportunity of seeing these chapters early, and I have found several of the chapters useful in my own work already.

Volume 1 contains chapters on tissue culture and hybridoma technology, intended to extend and update my book entitled *Monoclonal Antibodies: A Manual of Techniques,* CRC Press, Inc., Boca Raton, Florida, a series of chapters on lymphokines and functional assays *in vitro;* aspects of molecular biological techniques most clearly related to immunology; and immunochemical techniques. Volume 2 starts with a review of techniques used with small laboratory animals and continues with chapters on more specialized procedures with animals. Antigen detection in cells and tissue is covered in the next section, followed by a section containing chapters on the important area of protein purification using monoclonal antibodies. This section includes chapters by several workers with successful experience of affinity purification on an industrial scale.

The authors were all invited to contribute because they were known by me or by members of the Advisory Board to be successful in using their particular technique. Difficulties in setting up published techniques are the bane of many a biologist's life; authors were asked to include *all* important details and even details which just *might* be important, with the aim of providing techniques which do transfer well from one laboratory to another.

THE EDITOR

Heddy Zola, B.Sc., Ph.D., is Chief Hospital Scientist in the Department of Clinical Immunology, Flinders Medical Centre, and Professor of Immunology in the Flinders University of South Australia. He received a Bachelor of Science degree (with Honours) in Chemistry from the University of Birmingham in England in 1962, followed in 1965 by a Ph.D. degree in Biophysics from the University of Leeds in England.

Following his Ph.D. work, Dr. Zola carried out research in the Department of Biophysics at Leeds University and subsequently in the Departments of Biochemistry and Protein Chemistry at the Wellcome Research Laboratories, Beckenham. At that time, his research was principally on the physical properties of large molecules in solution, but he became interested in immunology and transferred to the Department of Experimental Immunobiology at Wellcome in 1975. In 1978 Dr. Zola came to Australia to take up an appointment as Chief Hospital Scientist in the Department of Clinical Immunology at the Flinders Medical Centre, a teaching hospital and research centre in Adelaide, South Australia, and was subsequently appointed Professor at the Flinders University of South Australia.

In the last few years, Dr. Zola's research has concentrated on the use of monoclonal antibodies, flow cytometry, and cell culture to study the differentiation and function of human B-lymphocytes. He is the author of *Monoclonal Antibodies: A Manual of Techniques,* CRC Press, Inc., Boca Raton, Florida, and over 170 research papers.

ADVISORY BOARD

VOLUME II CONTRIBUTORS

John Aaskov, B.Sc., Ph.D.
Senior Lecturer
Department of Medical Laboratory
  Science
Queensland University of Technology
Brisbane, Queensland, Australia

David J. Allan, B.V.Sc., M.B.BS.,
  Ph.D.
Senior Lecturer
Department of Medical Laboratory
  Science
Queensland University of Technology
Brisbane, Queensland, Australia

Anthony C. Allison
Department of Immunology
Syntex Research
Palo Alto, California

Judith K. Blackshaw, B.Sc., M.A.Ed.,
  Ph.D.
Department of Farm Animal Medicine
  and Production
University of Queensland
Brisbane, Queensland, Australia

John Bradley, M.D.
Sandoz Professor and Chairman
Department of Clinical Immunology
Flinders University and Medical Center
Adelaide, South Australia, Australia

Noelene E. Byars
Department of Immunology
Syntex Research
Palo Alto, California

Raffaela Comacchio, B.Sc.
Research Assistant
Department of Clinical Immunology
Flinders Medical Center
Bedford Park, South Australia, Australia

Edward C. De Fabo, Ph.D.
Associate Research Professor
Department of Dermatology
George Washington University Medical
  Center
Washington, D.C.

Sigbjørn Fossum, M.D.
Assistant Professor
Anatomical Institute
University of Oslo
Oslo, Norway

Arthur W. Hohmann, Ph.D.
Senior Hospital Scientist
Department of Clinical Immunology
Flinders Medical Center
Bedford Park, South Australia, Australia

George W. Jack, Ph.D.
Division of Biotechnology
P. H. L. S./C. A. M. R.
Salisbury, England

Andrew Lyddiatt, Ph.D.
Lecturer and Research Manager
School of Chemical Engineering
Biochemical Recovery Laboratory
University of Birmingham
Birmingham, England

Graham Mayrhofer, B.M. B.Ch.,
  M.A., D.Phil.
Senior Lecturer
Department of Microbiology and
  Immunology
University of Adelaide
Adelaide, South Australia, Australia

Frances P. Noonan, Ph.D.
Associate Research Professor
Department of Dermatology
George Washington University Medical
  Center
Washington, D.C.

Duncan S. Pepper, Ph.D.
Top Grade Biochemist
H. Q. Unit Laboratory
Scottish National Blood Transfusion
  Service
Edinburgh, Scotland

Christopher V. Prowse, D.Phil.
Top Grade Scientist
Blood Transfusion Service
Edinburgh, Scotland

Nigel Quadros, B.Sc.
Department of Immunology
Flinders Medical Center
Bedford Park, South Australia, Australia

Bent Rolstad, Dr.Med.
Anatomical Institute
University of Oslo
Oslo, Norway

Llewellyn D. J. Spargo, B.Sc. (Hons.)
Research Assistant
Department of Microbiology and
    Immunology
University of Adelaide
Adelaide, South Australia, Australia

Heddy Zola, B.Sc., Ph.D.
Chief Hospital Scientist and Professor of
    Immunology
Department of Clinical Immunology
Flinders Medical Center
Bedford Park, South Australia, Australia

ACKNOWLEDGMENTS

This book started from my conviction that there is a need for more books which allow fully detailed descriptions of methods. Books of this type are usually to be found on the laboratory bench rather than on the shelves. Translating this belief into hard copy involved persuading a large number of people to help. They all shared my enthusiasm for this project, at least to a degree, but for all of them to get involved meant giving up a considerable amount of time and putting in a great deal of work. I greatly appreciate the efforts of the authors and the members of the Advisory Board. For the publishers, this book has been developing at a time when their group was undergoing considerable change. The courteous and efficient handling of the manuscript by Janice Morey, Coordinating Editor at CRC Press, and more recently by Carolyn Lea, has made my task easier and more enjoyable than I expected. Mary Brown has as usual provided the efficient secretarial support which is essential to anyone trying to write or edit a book. I would also like to thank two scientists, who, while not members of the Advisory Board, helped me greatly with expert comment on particular subjects — Dr. Keith James of Edinburgh University and Dr. Pam Sykes of Flinders University of South Australia.

LABORATORY METHODS IN IMMUNOLOGY

VOLUME I OUTLINE

VOLUME II TABLE OF CONTENTS

ANIMAL WORK, *IN VIVO* STUDIES

ANTIGEN DETECTION IN TISSUES AND CELLS

IMMUNOAFFINITY PURIFICATION

Animal Work, *In Vivo* Studies

Chapter 1

BASIC TECHNIQUES WITH EXPERIMENTAL ANIMALS

John G. Aaskov, David J. Allan, and Judith K. Blackshaw

TABLE OF CONTENTS

I. INTRODUCTION

Animals have been used in medical research at least since Galen (129—199 A.D.) a Greek medical scientist, used a pig for his experiments. Because of the difficulties associated with obtaining human cadavers in the Middle Ages, many of the early anatomic dissections were also carried out on domestic animals. Fundamental physiological studies have employed animals as diverse as frogs (Galvani, 1791) and dogs (Pavlov, 1900 +). In more recent years, there has been interest by the community at large in the use of experimental animals in research, teaching, and industrial institutions. This has led to the preparation of "guidelines" and, in some instances, legislation to regulate animal experimentation. While this may have been irksome to some, it has helped scientists design safer, less stressful, and, often, more statistically significant experiments. Before using any animals, investigators should familiarize themselves with all institute, state, county or national guidelines and legislation pertaining to the use of experimental animals. Some sources of such information are

1.	Australia:	National Health and Medical Research Council, Canberra
2.	Canada:	Canadian Association for Laboratory Animal Science
3.	Germany:	Gesellschaft für Versuchstierkunde
4.	Japan:	Japan Experimental Animal Research Association
5.	Scandinavian countries:	Scandinavian Federation for Laboratory Animal Science
6.	Spain:	Sociedad Espanola de Experimentacion Animal
7.	UK:	(1) Laboratory Animal Science Association, the representative national scientific society; (2) Research Defence Society, for legal problems related to the use of experimental animals in the UK; (3) Institute of Animal Technicians
8.	U.S.A.:	American Association for Laboratory Animal Science

It should also be remembered that regulations and guidelines pertaining to the use of toxic or infectious agents in a laboratory also apply in an animal house. This is of particular importance where routine animal care is performed by staff who may not be familiar with the organisms or agents being used. CAGES SHOULD ALWAYS BE CLEARLY AND EXPLICITLY LABELED. Again, it is the responsibility of the scientist to obtain all safety clearances before beginning animal experiments.

Once familiar with the guidelines and regulations pertaining to use of experimental animals and having obtained the necessary licenses, ethics approval, etc., the scientist should develop a working understanding of good husbandry. Although most facilities have trained staff to care for animals, it is in a scientist's own interest to ensure animals are not stressed by overcrowding, poor ventilation, lack of, or incorrect, food, lack of, or stale, water, excessive noise, etc. It is also important (mandatory for some granting bodies) that the genetic backgrounds of animals are checked regularly even if the animals are purchased from a commercial supplier.

II. EQUIPMENT

A. GENERAL

While there are many specialized items of equipment which can be purchased to aid handling of animals, the most common procedures (injections and bleeding) in the most frequently used animals (mice, rats, and rabbits) can be performed with four pieces of equipment which can be constructed in the laboratory.

FIGURE 1. Metal cage converted to act as a warming box for mice or rats. Animals are separated from the heat source (a conventional 60-W light bulb) by a metal grid.

1. A warming box (Figure 1)
2. A mouse/rat restrainer (Figure 2)
3. A rabbit restrainer (Figure 3)
4. A mouse/rat bleeding apparatus (Figure 4A and B)

There are a large number of variations in the design of each of these items, and generally all perform the task for which they are intended. However, we have found that the stability and versatility of the mouse/rat restrainer offers a number of advantages over other designs, particularly the mouse "tube", i.e., a plastic tube or syringe sealed with a notched stopper through which the tail is drawn.

B. SYRINGES AND NEEDLES

Most laboratories now use disposable plastic syringes. However, it should be remembered that these usually contain a small amount of lubricant which may interfere with some antigens. Conversely, any agent — particularly, adjuvants containing oil — in which the lubricant is soluble will make the piston on the end of the plunger difficult to move. Some oils also make the rubber piston swell. If the lubricant has been solubilized or the piston

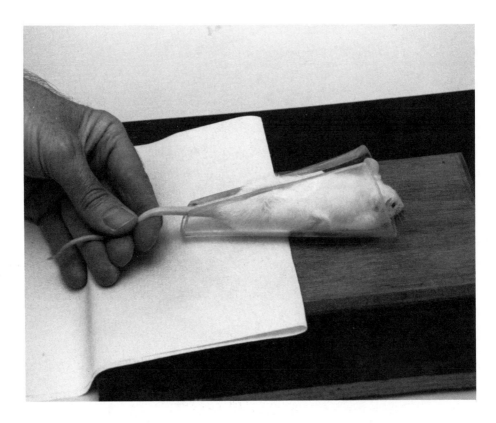

FIGURE 2. A cone-shaped piece of Perspex with a slit at the top is fastened to a wooden base to act as a mouse/rat restrainer. The slot is wide enough to permit the passage of the tail of a mouse or a rat and the cone shape permits the restraint of animals of a variety of sizes. Four 2 × 2-cm squares of soft rubber have been attached to the underside of the wooden base to prevent it sliding on a bench top.

swollen, the plunger will move erratically along the barrel when pressure is applied to it to operate the syringe.

The older style, all-glass syringe is usually used with such oil-based antigens or adjuvants. However, if the fluid is too viscous, there is a tendency for it to flow back around the plunger and out of the top of the syringe, even if the plunger has been adequately lubricated with a suitably inert silicone-based grease.

Before using any glass syringe, ensure the appropriate needles are available. Many of the present disposable needles will not fit the older glass syringes.

Where antigens are very expensive, or available only in small quantities, consideration should be given to using disposable syringes with fixed needles. These syringes can be obtained in a style which has a negligible dead space in the head of the needle and which permits injection of almost all antigen. One such example is the Micro-fine III syringe-needle produced by Becton-Dickinson (Cat. No. 678410 A).

III. HANDLING AND RESTRAINT FOR LABORATORY PROCEDURES

Every person — student, researcher, or technician — who works with laboratory animals should master the techniques of handling and restraint so that animals are not unduly stressed.

A. RAT

If rats are gently handled they become tame. They will not bite unless frightened. A rat can be grasped from above, but if forced into a corner it may bite. An old male rat may

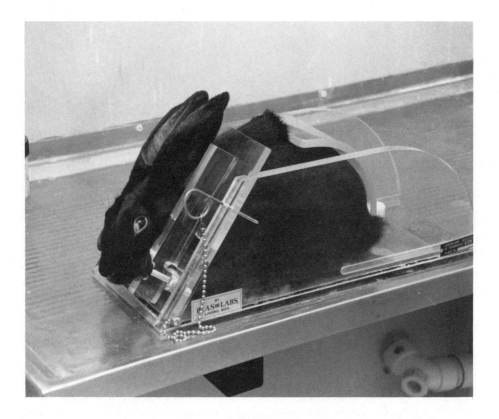

FIGURE 3. A commercially available rabbit restrainer. Similar items can be constructed in the laboratory from wood or Perspex. An alternative design can be found in Reference 17.

become irritable and bite if handled excessively. The tail of a rat is easily stripped of skin if it is pulled roughly. However, the rat can be restrained by holding it gently near the base of the tail for short periods.

An animal in a wire-bottomed cage should not be held by the tail as the animal will grasp the wire. When transferring the rat onto a surface from the cage, a nonslip surface must be provided so the rat has something to hold onto and does not slip or feel insecure. An adult rat can usually be held gently in the hand (Figure 5).

Rats are restrained for injections by placing the hand firmly over the rib cage and using the thumb and forefinger to restrain the head at the jaw (Figure 6). Another method of restraint is in the "rat bag" which can be made of calico or "Chux" cloth and into which rats easily crawl (Figure 7). For intravenous (IV) injections it is usually most convenient to use a "mouse/rat restrainer" (see Figure 2).

B. MOUSE

Mice are more likely than rats to bite and must be immobilized for any injections. If a mouse is placed on a grid or a towel, where it can hold on, it is then easy to grip the large fold of skin at the scruff of the neck (Figure 8A). The tail is then transferred and held against the palm of the first hand by the fourth and fifth fingers (Figure 8B). Mice can be picked up by the tail without damage. When they are between 2 and 3 weeks old, they are very lively and may jump vertically out of an open cage or off a bench.

C. GUINEA PIG

Guinea pigs seldom bite or scratch, and they respond well to frequent handling. They should be picked up with both hands: one hand supporting the hindquarters and the other

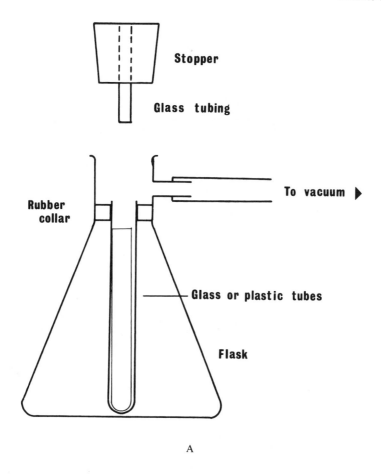

A

FIGURE 4.   Apparatus for obtaining blood from the tail of a rat or mouse. (A) Diagram of the apparatus; (B) apparatus ready for use. The plastic tubing is attached to a vacuum pump or a torricellian vacuum apparatus on a laboratory tap (see Figure 19 and Section VII.C.1.a for use).

either over the shoulders (Figure 9) or under the body (Figure 10). Care must be taken not to damage the thoracic viscera.

D. RABBIT

Handling of rabbits should be firm and gentle. To lift or restrain a rabbit, the scruff of the neck is firmly grasped with one hand while the other hand supports the rear quarters (Figure 11 A and B). This support of the rear is important as a rabbit can flick back its hind legs and either inflict a severe scratch or fracture its lumbar vertebrae. The rabbit can be held for long periods with its head tucked under the arm of the person restraining it. A rabbit should never be held by the ears as they are sensitive and fragile. For injection or other treatment rabbits should be restrained either in a holding box (Figure 3) or by wrapping them firmly in a towel or laboratory coat. It is the experience of the authors that restraint in a coat or towel is less stressful to the rabbit than use of a restraining box.

The rabbit is placed on the coat and while the neck is still held (Figure 12), the coat is first folded firmly over the rear toward the head. This prevents the rabbit drawing away from the operator. The flaps of the coat are then folded over the rabbit, and the animal is held between the arm and chest of an operator while still resting on a bench or similar surface (Figure 13). Rabbits also appear to be more settled if their eyes are covered during injections and bleeding.

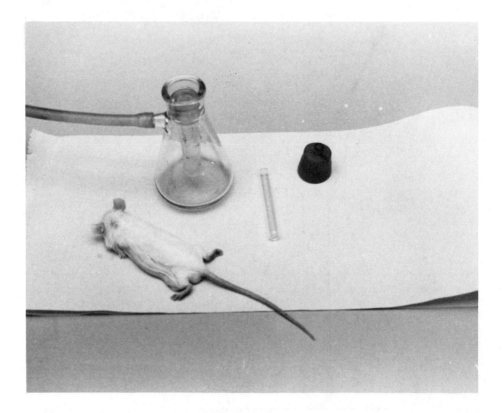

FIGURE 4B.

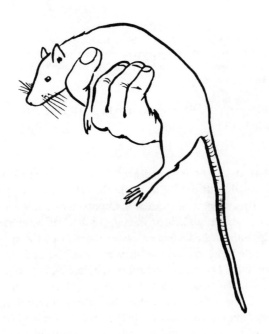

FIGURE 5.   Handling of an adult rat.

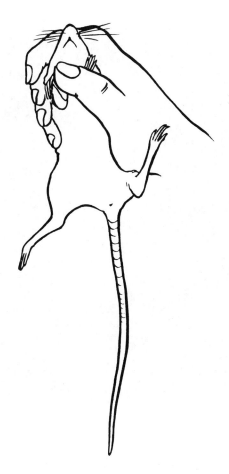

FIGURE 6.   Rat restraint using the hands.

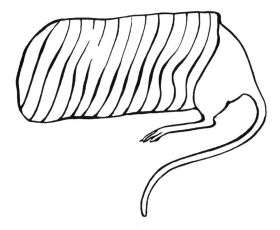

FIGURE 7.   Restraint of a rat in a ''rat bag''.

A

FIGURE 8.   (A) Method for picking up a mouse; (B) method for holding a mouse.

Intramuscular (IM), subcutaneous (SC), and intradermal (ID) injections can be performed by turning the rabbit around so the hindquarters are forward and the flank or a rear leg exposed.

IV. ANESTHESIA AND ANALGESIA

A. INTRODUCTION

Having decided on the scientific component of an animal experiment, it is essential that consideration be given to the need for anesthesia and/or analgesia. Related to this, and part of experimental design, is the need to plan the safe, painless disposal of animals at the conclusion of the study. While there are many simple procedures such as intraperitoneal (IP) or IV injections in mice which can be performed without anesthesia or analgesia, there are others such as cardiac puncture where anesthesia is necessary to reduce stress in both the animal and the operator!

Again, many countries have legislation or guidelines as to which procedures should be performed under anesthesia and/analgesia, and scientists should refer to these.

Anesthesia is a state characterized by a loss of sensation. The anesthetic agent acts by depressing central nervous system functions leading to sedation, lack of pain perception (analgesia), reduction of reflex muscle activity following a noxious stimulus, and, with most agents, loss of consciousness. These useful effects of anesthetics are accompanied by depression of the brain stem centers controlling respiration, and this may lead to cessation of spontaneous breathing in the anesthetized animal. With an increasing dose of an anesthetic, the normal animal will become sedated, then anesthetized, and, with an overdose, die.

FIGURE 8B.

FIGURE 9. Transport of a guinea pig supporting the hindquarters.

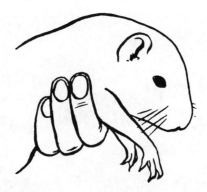

FIGURE 10.   Alternative procedure for handling a guinea pig. Care should be taken with this procedure not to allow the guinea pig to scratch the holder's forearm with its hind legs.

A

FIGURE 11.   (A) Sketch of the technique for holding a rabbit; (B) procedure in use.

The safety margin is an expression of the difference between the dose required to produce the desired effect and the higher dose at which death is probable. A drug with a wide safety margin, e.g., one requiring six times the anesthetic dose to produce death, is preferred to one with a narrow safety margin, e.g., one fifth increase in anesthetic dose to produce death.

The procedures described here will enable the worker inexperienced in anesthetic techniques to carry out basic anesthetization of the rat, mouse, guniea pig, and rabbit, and meet the following criteria:

FIGURE 11B.

1. Simple equipment. Injectable anesthetics are recommended so equipment is limited to syringes and needles.
2. Simple manipulative skills. Injection of drugs by IP or IM routes only.
3. Safety and welfare of the animal. Reliable anesthetic, analgesic, and sedative agents with wide safety margins are recommended.
4. Safety of the worker. Most agents are nonvolatile, nonexplosive, and nontoxic during normal storage and use.

Anesthetic agents are commonly administered by inhalation if volatile (nitrous oxide, diethyl ether, halothane) or intravenously, intraperitoneally, intramuscularly, or into the spinal canal if nonvolatile (barbiturates, ketamine, etc.).

Although intravenous anesthesia has a rapid onset of peak effect and the dose of anesthetic can be readily adjusted to achieve a desired effect, it is technically more difficult to administer and consequently slower when a number of animals are involved. We have, therefore, concentrated on anesthetic agents for IP and IM injection.

B. INJECTABLE ANESTHETIC AGENTS

Trade names for reagents are marked with an asterisk (*) and the name of pharmaceutical companies with a double asterisk (**).

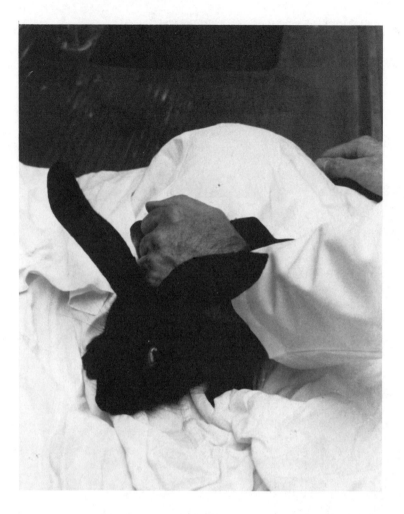

FIGURE 12.   Wrapping a rabbit in a laboratory coat as a method of restraint.

1. Fentanyl and Droperidol Mixture

Innovar-vet injection* is a fixed ratio preparation containing 0.4 mg/ml fentanyl and 20 mg/l droperidol and is marketed by Ethnor**. Fentanyl is a narcotic (sleep inducing) analgesic, similar to, but 100 times more potent than morphine. Fentanyl is rapidly absorbed producing profound depression of the central nervous system with analgesia, sedation, respiratory depression, and bradycardia (slowing of the heart rate). The effects of fentanyl can be reversed by the narcotic antagonists naloxone hydrochloride (Narcon injection*, du Pont**) or nalorphine hydrobromide (Lethidrone*, Wellcome**).

Droperidol is a sedative or tranquilizer which produces a general quiescence and a reduced responsiveness to environmental stimuli. Droperidol potentiates the analgesic effect of fentanyl.

Administration of the mixture IM or IP produces sedation and analgesia in 3 (IP) to 15 (IM) min but not complete unconsciousness (a state called neurolept analgesia). The animals remain sensitive to auditory stimuli so a quiet environment should be maintained. Analgesia lasts 20 to 40 min but sedation lasts for several hours, resulting in slow recovery. Fentanyl-droperidol has a wide safety margin, e.g., anesthetic dose in guinea pig 0.66 ml/kg and lethal dose of 4.40 ml/kg.[1]

Fentanyl-droperidol potentiates or is active with other central nervous system depressants

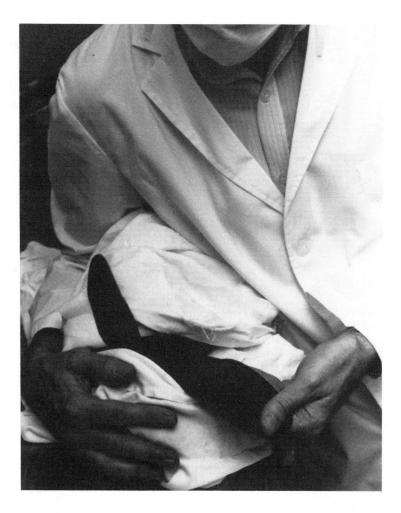

FIGURE 13.   Rabbit restrained in a laboratory coat, with eyes covered, readied for an IV injection or bleeding from the marginal ear vein.

such as barbiturates. When used in combination, the dose of Innovar-vet should be reduced by approximately one half and the barbiturate dosage also reduced to achieve a desired effect.

Advantages (of Innovar-vet)
    Wide safety margin
    Rapid effect
    Reliable analgesia and sedation
    Narcotic effects (analgesia and respiratory depression) reversible by naloxone or nalorphine
Disadvantages
    Slow recovery from droperidol sedation
    Respiratory depression produced by fentanyl (reversible with naloxone or nalorphine)
    Bradycardia produced by fentanyl (preventable by administration of atropine sulfate prior to anesthesia)

a. Dose

Because Innovar-vet is a fixed ratio preparation it is convenient to give the dose rates as milliliters of standard solution containing fentanyl 0.4 mg/ml plus droperidol 20 mg/ml.

1. Sedation for minor procedures, e.g., bleeding. Walden[2] reported the use of Innovar-vet by IM injection with minimum restraint, absence of pain reactions, uneventful recovery in 1 to 2 h, and minimal stress to the animal.

Rat	0.13 ml/kg
Mouse	0.005 ml/kg (to achieve a manageable volume for injection, dilute the standard solution 1 in 10 with normal, 0.9% saline)
Guinea pig	0.08 ml/kg
Rabbit	0.17 ml/kg

2. Surgical anesthesia

Rat	0.3 to 0.4 ml/kg IM, IP[3]
Mouse	0.01 ml/kg IP (dilute 1 in 10 as described above) plus diazepam 5 mg/kg IP before giving Innovar-vet[3]
Guinea pig	0.66 to 0.88 ml/kg IM[1]
	or
	1.3 ml/kg IM or IP with concurrent administration of diazepam 2.5 mg/kg IP to potentiate sedative and improve muscle relaxation[2]
Rabbit	0.3ml/kg IM[3]

2. Ketamine Hydrochloride and Combinations

Ketamine hydrochloride (100 mg/ml: Ketalar Vet*, Parke Davis**; Ketavet 100 [v]*, Delta**) produces unconsciousness, intense somatic analgesia, variable visceral (internal) analgesia, no muscle relaxation, and even an increase in muscle tone. This state is described as catalepsy or dissociative anesthesia. The animal has a waxy rigidity of the limbs but is unaware of the environment.

Advantages
  Does not depress respiratory function
  Wide safety margin
Disadvantages
  Enhances muscle tone; tremors or convulsions may be produced
  Causes a marked increase in salivary secretions which may block the airway
  Wide variation in response between species and between individuals

a. Ketamine Alone

Ketamine used as a sole anesthetic agent is not a satisfactory surgical anesthetic but may be adequate for minor procedures such as cardiac puncture.

Subject	Dose	Effect	Ref.
Rat, rabbit, and guinea pig	40—60 mg/kg IM	Sedation in about 10 min lasting 30 min; variable, usually poor analgesia	4
Mouse	80 mg/kg IM	Produced sedation; up to 400 mg/kg IM failed to produce analgesia even in animals which appeared to be deeply sedated	4

| Rat, mouse, guinea pig, and rabbit | 22 mg/kg IM | Good restraint | 5 |
| | 44 mg/kg IM | Surgical anesthesia induced in all four species in 8—10 min, effective surgical anesthesia for 15—25 min after injection. These results differ greatly from those of Green et al.[4] | 5 |

b. Ketamine + Xylazine

Xylazine hydrochloride (20mg/ml: Rompun*, Bayer**) is a powerful hypnotic (produces sleep) and muscle relaxant which can be mixed with, or given before, ketamine to improve muscle relaxation. Xylazine produces significant disturbances of cardiac rhythm, an effect minimized by administration of atropine before anesthesia.

	Dose[4]	
	Ketamine (mg/kg)	**Xylazine (mg/kg)**
Mouse	80 IP	16 IP
Rat	80 IP	12 IP
Guinea pig	60 IM	8 IM
Rabbit	30 IM	2 IM
		(+ diazepam 5 mg/kg IM)

The peak effect achieved in all species was sedation and relaxation with analgesia being insufficient for major surgery in many cases.

Subject	**Dose**	**Effect**	**Ref.**
Rabbit	Ketamine 35 mg/kg + xylazine 5 mg/kg (single injection)	Produced the most desirable surgical anesthesia	6
Rat	Ketamine 44 mg/kg + xylazine 8 mg/kg IM	Produced surgical anesthesia in 4—8 min and lasted 45—60 min. A second dose given 15 min after the first produced about 90 min of surgical anesthesia.	7
Mouse	IM combination of 0.1 ml ketamine (100 mg/ml) with 0.1 ml xylazine (20 mg/ml) and 0.8 ml of 0.9% saline used at a dose of 0.1 ml for mice averaging 30 g weight	Produced good analgesia and muscle relaxation averaging 12-min duration in outbred mice	7

c. Ketamine + Xylazine + Diazepam

Diazepam (5 mg/ml: Diazepam Injection*, Astra**; Diazepam Injection USP*, David Bull**; 10 mg/2 ml, Valium 10*, Roche**) is a potent tranquilizer, muscle relaxant, and anticonvulsant which can help control the increased muscle tone and tendency to convulsions produced by ketamine. Diazepam potentiates the analgesic effects of ketamine and has minimal effects on the cardiovascular and respiratory systems.

Diazepam must be injected separately as it is not water soluble and cannot be mixed with standard ketamine solution without precipitation.

	Subject	Dose	Ref.
Ketamine + xylazine as above			
Diazepam	Rat	2.5 mg/kg IP	3
	Mouse	5 mg/kg IP	
	Guinea pig	2.5 mg/kg IP (IM)	
	Rabbit	1.0 mg/kg IM (IV)	

3. Pentobarbitone Sodium

(60 mg/ml): Anathal*, VR/Syntex**; Nembutal sterile solution*, BOMAC**; Sagatal*, May and Baker**)

Barbiturate anesthetics, particularly pentobarbitone, have been widely used for laboratory animal anesthesia for effects ranging from restraint to major prolonged surgery. In spite of widespread use, pentobarbitone has several major disadvantages which should lead the laboratory animal worker to consider other anesthetic agents in preference to pentobarbitone. The major disadvantages are

1. Low margin of safety
2. Poor analgesia except at very high doses which depress respiratory and cardiovascular functions
3. Hypothermia (loss of body heat), particularly in rats and mice during and after anesthesia
4. Additional doses to prolong anesthesia result in disproportionately long recovery times

Doses: IP injection (or IM)
Rat, guinea pig, rabbit: 30—50 mg/kg (more suitable by IV in rabbit)
Mouse: 40—90 mg/kg

Young rats and mice are particularly sensitive and need lower doses: 10—20 mg/kg. To extend the period of anesthesia, small doses of 5 mg/kg may be given cautiously to effect.

Peak effect is achieved 5 to 10 min after injection and lasts 10 to 20 min in rabbits (IV); 90 to 120 min in guinea pigs; 20 to 40 min in rats and mice. Full recovery is prolonged in all species taking several hours up to about 24 h in some mice and guinea pigs.

C. ANESTHESIA BY INHALATION
1. Agents
Ether (diethyl ether)
Methoxyflurane
Halothane
Nitrous oxide
Enflurane

One of the most commonly used anesthetic agents in small laboratories is diethyl ether. This agent is highly flammable and explosive and should never by used near flames or any electrical equipment capable of generating a spark. Great care should also be exercised when disposing of animals euthanatized with ether as they may explode in an incinerator.

The other anesthetic agents capable of inhalation require reasonably sophisticated equipment for their administration and are therefore more appropriate to a large laboratory.

Green[8] reviewed the health risks to laboratory personnel from anesthetic gases and made a strong plea for all to be aware of the hazard of anesthetic gas pollution in the laboratory. Green recommended that good ventilation and, preferably, purpose-built scavenging equipment be installed wherever inhalation agents are used.

2. Procedure

Smaller rodents (mice, rats, guinea pigs) can be anesthetized in glass dessicators. Three to 5 ml of diethyl ether is applied to cotton wool, and this is placed in the base of the dessicator with a metal or ceramic tray above it for the animals to rest on. The ceramic tray with holes in it, supplied with most desiccators, is ideal. The animal to be anesthetized is then placed in the desiccator and the lid replaced. The animal will go through a phase of irregular breathing and then resume a steady breathing pattern. It should be removed at this stage.

The period of anesthesia is short, usually only 1 to 3 min but the animal may be reanesthetized. However, it should be remembered that ether (diethyl ether) irritates the mucosa of the mouth and respiratory tract, and this in turn may stimulate sufficient secretions to block the airway, particularly in guinea pigs. Repeated ether anesthesia also predisposes animals to respiratory infection.

Small animals which have been overdosed with ether can sometimes be resuscitated by holding their rear legs and tail and gently swinging them in large circles at arm's length.

Although a restrained rabbit can be anesthetized with ether using a face mask or cotton wool soaked in ether in a plastic bag, the distress to the animal indicates use of an injectable anesthetic.

It should also be remembered that blood recovered from animals anesthetized with diethyl ether will contain significant amounts of the anesthetic, and this may be sufficient to destroy, for example, some enveloped viruses or to interfere with chemical assays.

D. SPINAL ANESTHESIA

Spinal anesthesia is achieved by injecting a local anesthetic agent into the lower part of the spinal canal, which leads to a temporary block of spinal nerve functions. Sensory loss and motor paralysis to lower abdomen, pelvis, and lower limbs is achieved. The undesirable features of general anesthesia are avoided when surgery of lower body region is required. This is a technically demanding procedure and should only be performed after comprehensive training. For further details see Thomasson et al.[9] and Kero et al.[10]

V. MANAGEMENT OF ANESTHESIA

Many factors influence the success of anesthesia. Some of these are addressed below.

A. BEFORE ANESTHESIA

1. Choice of appropriate anesthetic agent

    - Depth and duration of anesthesia required
    - Effect of drug(s) on experimental findings
    - Number and value (monetary and experimental) of animals to be anesthetized
    - Will the animal be destroyed before it recovers consciousness?
    - Experience of worker in anesthetic techniques

2. Planning anesthetic technique

    - Availability of equipment and drugs for anesthesia and any experimental procedures
    - Handling and restraint of animal for injection
    - The time interval between the administration of anesthetic to successive animals must allow for time to carry out manipulation on the anesthetized animal and the time taken for the next animal to reach an effective depth of anesthesia.

- Recording system to document details of drugs, administration, manipulations, and the animal's response. Careful documentation is necessary to monitor anesthetic technique and outcome, and also to aid in the interpretation of experimental results.

3.   Preanesthetic animal manipulation

- Check health status of animal: presence of respiratory disease, in particular, increases the risk of anesthesia.
- Food, but not water, should be withheld for 6 h prior to anesthesia: from rabbits and guinea pigs routinely; from rats and mice when abdominal surgery is planned.
- Weigh animals immediately before anesthesia for accurate calculation of drug doses; in the guinea pig, gastrointestinal contents may account for 20 to 40% of total body weight.
- Premedication: injection of atropine is recommended in rabbits, rats, and mice 30 min before anesthesia. Atropine reduces respiratory tract secretions and minimizes cardiac depression during anesthesia (rabbit 1 to 3 mg/kg IM or SC; rats and mice 0.05 mg/kg SC).

B. DURING ANESTHESIA
1. Determining the Depth of Anesthesia
Light anesthesia or sedation — The animal becomes progressively uncoordinated in movements and will lie down but be able to lift its head without support indicating a mild degree of central nervous depression. The righting reflex will be lost with increasing depth of anesthesia; the animal will not attempt to turn over when placed on its back.

Surgical or deep anesthesia — This is indicated by the following:
(1)Respiratory patterns: breathing is slow and regular and not altered by painful or surgical stimuli, and (2)Reflexes: reflex movement in response to a stimulus diminishes with increasing depth of anesthesia and is absent when surgical anesthesia is reached.

1.   Blink reflex — a blink in response to a gentle tap at the medial end of the eyelids; used for rats, mice, guinea pigs, and rabbits.
2.   Pedal reflex — withdrawal and flexion of the hind limbs when the interdigital skin is pinched; use for rats and mice.
3.   Tail pinch — movement or "jump" following a firm pinch of the tail; use for rats and mice.
4.   Ear pinch — head shaking in response to an ear pinch; used for guinea pigs and rabbits.

Dangerously deep anesthesia (anesthetic overdose) — Changes in respiration and tissue color are the best guide to anesthetic overdose.

Respiratory pattern — Depression of respiration results in excessive slowing of the rate and labored inspiration with very large effort of the abdominal muscles and diaphragm. When the spontaneous breathing rate falls to about one third of the normal rate (unanesthetized), respiratory depression is usually irreversible.[3] Normal respiratory rates are rabbit 30 to 60/min; guinea pig 40 to 100/min; rat 70 to 110/min; mouse 90 to 160/min.[11]

Color changes — extreme pallor (pale) or cyanosis (blue) indicates a major disturbance of the circulation and oxygenation of tissues. Sites to observe tissue color are the ears, feet, and muzzle of rats and mice; the conjunctiva and gums; reflected color of light directed into the eye of rat and rabbit.[3]

2. Maintenance of Adequate Respiration
Three steps should be taken.

- Prevent respiratory center depression by drugs.
- Maintain adequate airway, i.e., the pathway the air follows from nose to lungs. Check for secretions, foreign objects, or the tongue causing obstruction to the upper air passages, check for external pressure or abnormal neck position obstructing the trachea.
- Allow free thoracic and/or abdominal movements. With small animals, particularly, care is necessary to avoid compression of the trunk by the weight of the operator's arm or hand during manipulations or surgery.

3. Management of Inadequate Respiration

- Check above factors and correct where necessary.
- Do not administer more anesthetic.
- Apply artificial respiration by gentle manual compression of the sternum 50 to 100 times per minute, blowing in gentle puffs by plastic tube into the animal's mouth, or gently swinging the animal by its hind limbs.
- Administer nalorphine (2 mg/kg SC) to reverse fentanyl overdose where suspected.
- Administer doxapram (1 mg/kg IP, IM, or SC) to stimulate respiration and circulation; the short duration of action (15 to 20 min) may necessitate repeat doses.

4. Maintenance of Body Temperature

Anesthetized animals are susceptible to body cooling which can be minimized by maintaining a warm, draft-free environment during anesthesia and recovery. Because of small body weight, rats and mice are most sensitive. During prolonged anesthesia, heat loss can be minimized by wrapping the animal in cotton wool or reflecting aluminum foil. The use of a heat source such as hot-water bottle or lamp must be closely monitored to avoid overheating or excessive body water loss leading to dehydration. Isothermic pads are designed specifically to maintain the body temperature of animals and are the safest method of heat application during anesthesia.

C. AFTER ANESTHESIA

Following anesthesia (and surgery), the health and welfare of the animal is safeguarded by careful supervision of the animal until it is fully conscious.

1. Environmental Considerations

- Warmth: 30 to 35°C is recommended particularly for rats and mice.
- To minimize the risk of attack by other animals, single cages for rats and mice are recommended. Animals in the same stage of recovery may be caged together.
- No containers or objects on which the partly conscious, uncoordinated animal may be injured.
- Open mesh floor on the cage lessens the risk of airway obstruction from muzzle compression in the corner of the cage. The mesh allows air to reach the nose.

2. Maintenance of Adequate Respiration

- Respiratory tract obstruction becomes less likely the nearer the animal is to full consciousness as the respiratory tract reflexes are regained.
- Attention to retaining the normal position of the head to neck will prevent tracheal obstruction during recovery.
- Preanesthetic restriction of food minimizes postanesthetic vomiting of gastric contents and their inhalation into the respiratory tract.

- Where recovery to full consciousness takes more than 1 to 2 h, the animal should be turned every 15 to 30 min to prevent blood accumulating (gravitational effect) in the lower-most lung (postural hypostatic congestion).
- Management of inadequate respiration is the same as described for during anesthesia.

3. Hydration and Cardiovascular Function

Water should be offered to the animal as it regains awareness of the environment. With prolonged absence of fluid intake before, during, or after anesthesia and where blood loss has occurred during surgery, body fluid and blood volume may decrease significantly and threaten adequate circulation. Blood loss of up to approximately 20% of blood volume can be adequately treated by replacement with IP (or SC) administration of sterile 0.9% sodium chloride (normal saline) in volumes two to three times the volume of blood loss. Blood volume in the laboratory animal is in the range of 6 to 8% of body weight with the small animals (e.g., mice) having the higher figure i.e., blood volume of a 30-g mouse is approximately 2.4 ml and 20% loss of blood is a loss of approximately 0.5 ml which could be treated by IP injection of 1 to 1.5 ml of 0.9% sodium chloride solution.

Rule of thumb: to minimize the effect of blood loss and/or dehydration, 0.9% sodium chloride can be given IP or SC in a volume equal to about 50% of the normal blood volume of the animal, i.e., 50% of 6 to 8% = 3 to 4% of body weight, without risk of fluid overload.

4. Pain Relief (Analgesia)

The relief of pain is of the utmost importance in the management of laboratory animals, in particular, following surgery. Pain is a complex phenomenon which is difficult to define satisfactorily in humans, and is extremely difficult to recognize and interpret in animals.[12,13] Until further progress is made in assessing the nature of pain in animals, the worker should assume that if a procedure is likely to cause pain in the human, it will produce a similar degree of pain in an animal.

The following behavioral changes often indicate that the animal is experiencing pain.

1. Relative inactivity
2. Complete immobility
3. Movement with abnormal gait
4. Uncharacteristic aggression
5. Unusual restlessness with constant movement and getting up and down
6. Food and water intake severely reduced; dehydration may result
7. Abnormal vocalization; e.g., rats squeak at unusual pitch
8. Abnormal posture; tensing of the muscles of the abdomen and back producing a "tucked-up" appearance

Effective analgesia can be assumed to have been achieved when the abnormal behavior(s) are reduced or disappear following administration of an analgesic.[14]

Flecknell[14] has summarized the data concerning the types of analgesic drugs available and the experimental evidence for their efficacy in laboratory animals. Although special circumstances may indicate the use of a particular analgesic, the most generally useful agent at present was found to be buprenorphine (0.3 mg/ml: Temgesic*, Reckitt and Colman**) which is a potent, long-acting opiate (narcotic)-type analgesic. In animal tests for physical dependence liability, buprenorphine has the least capability of any opioid tested, being lower than codeine. This makes it a much safer drug (for the animal and the worker) than other narcotic analgesics such as morphine and pethidine, which easily induce physical dependence.

Dosage of buprenorphine[14]

Mouse	2.5 mg/kg IP, 6—8 hourly
Rat	0.1—0.5 mg/kg SC, 8—12 hourly
Rabbit	0.02—0.05 mg/kg SC, 8—12 hourly
Guinea pig	No information available

In the guinea pig, buprenorphine could be tried at the same dose as the rat. The recommended doses for other narcotic analgesics are the same in these two species.

For details of efficacy and doses of other analgesic drugs refer to Flecknell.[14]

POISON REGULATIONS: THE PURCHASE AND USE OF MANY OF THE ANESTHETICS AND ANALGESICS LISTED ABOVE ARE CONTROLLED BY LEGISLATION IN MOST COUNTRIES. SCIENTISTS MAY NEED TO TAILOR THEIR ANESTHETIC PROCEDURES TO CONFORM WITH SUCH REGULATIONS.

VI. EUTHANASIA

Euthanasia means a quiet, painless death. The word is derived from the Greek, "eu-", meaning good, and "thanatos", meaning death, i.e., "good death".

Factors to be considered when undertaking euthanasia and in selecting a technique include the following.

A. WELFARE
1. Welfare of the Animal

- Necessity for euthanasia includes relief of suffering; culling unwanted animals; research requirements; management of infectious disease.
- Quick, quiet, painless death of animal is necessary.
- Technique is not to induce fear, apprehension, or panic in the animal.
- Unconsciousness is to be achieved quickly so that the animal is unaware of respiratory and circulatory arrest.
- Other live animals must be protected from the sight, sound, or smell of the technique.

2. Welfare of Personnel

- Procedure must be physically and chemically safe for the operator and staff.
- Procedure must be aesthetically acceptable to the staff involved and not carried out in a public or communal area.

B. PRACTICAL REQUIREMENTS

1. Animal involved: species, size, ease of handling, number to be euthanatized.
2. Skill of operator: simple reliable technique is best; operator experienced in the technique and animal type involved.
3. Requirement for post-mortem material: following euthanasia tissues may be collected for disease diagnosis or research observations, and the euthanasia technique should not physically or chemically interfere with analyses to be carried out.
4. Cost: of materials, apparatus, staff time.
5. Procedure: technique must be approved by ethics committee, funding body, and, if publication is planned, scientific journal.

C. METHODS OF EUTHANASIA

The practical methods of euthanasia produce death either by extreme depression of the respiratory and cardiovascular centers in the brain stem, or by interrupting the cervical spinal cord connection of the respiratory centers in the brain stem to the origins of the phrenic and intercostal nerves which innervate the muscles of respiration. In either case, respiration ceases and is associated with simultaneous or subsequent circulatory collapse.

1. Physical Methods[15,16]

These methods should be used only when (1) the animal is easily handled (or has been anesthetized or tranquilized to achieve control), and (2) the operator has adequate skill to carry out the procedure reliably and efficiently.

a. Cervical Dislocation
i. Mouse

With the mouse on a solid surface hold a strong thin object (e.g., closed shaft of a pair of scissors) over the back of the neck just below the skull. Quickly and firmly pull up and back with the tail and press down with the thin object. The neck dislocates and the spinal cord is severed.

ii. Rat, Guinea Pig, Rabbit

While this procedure can be performed with these larger animals, the potential for pain and suffering when training staff to perform this procedure is so high as to make this method of euthanasia almost unethical.

b. Decapitation

Young rats or mice may be euthanatized efficiently by decapitation with a large pair of scissors. The method is not aesthetically acceptable for many workers.

2. Chemical Methods
a. Pentobarbitone Sodium Injection

An overdose (250 to 300 mg/kg) of pentobarbitone solution is an effective technique for mice, rats, guinea pigs, and rabbits. IP injection is suitable in most cases. Intracardiac or IV injection may be suitable, but SC injection is not suitable because of irregular absorption of the drug. Special high concentrations of pentobarbitone sodium (about 300 mg/ml) are available commercially for specific use in euthanasia.

b. Gas

The Universities Federation for Animal Welfare (UFAW) recommends carbon dioxide as the most suitable chemical from the point of view of both the animal and the operator. The animal(s) is placed either in a special euthanasia cabinet or in a cage which is then placed in a large polyethylene bag. Carbon dioxide from the cylinder is released via a tube into the chamber or bag, and the animal becomes unconscious. Death occurs by depression of the respiratory centers and subsequent respiratory failure.

VII. SPECIFIC LABORATORY PROCEDURES

A. WEIGHING

There is a variety of specialized equipment available for weighing rodents. It varies from a balance, with attached cage which will record individual weights of animals added to the cage as well as calculating mean weight and standard deviation, to a simple one-pan beam balance with a cage on the pan to weigh individual animals. Small rodents can easily be weighed by placing them in a preweighed (or tared) plastic box on a conventional top-loading laboratory balance. Rabbits can be weighed either in a restraining box or confined in a towel (see Figures 3 and 12).

B. INJECTIONS
There are a number of issues to be considered before injecting an animal.

1. Nature of the Antigen
a. Purity
While it is possible to inject cell suspensions intravenously, it may not be possible to inject, for example, a crude tissue homogenate without killing the recipient. At the other extreme some highly purified viruses may induce emboli formation and kill a mouse or rabbit within minutes of being injected intravenously.

b. Solubility
Preparations containing oil or oil-based adjuvants should never be administered IV. Because of the acute peritonitis they cause when given IP, oil-based adjuvants should only be administered IP when all other routes have proved ineffective.

One of the most common adjuvants used to enhance the immunologic response is Freund's (complete, containing heat-killed mycobacteria; incomplete, without the mycobacteria). To work efficiently, the Freund's adjuvant needs to form an emulsion with the antigen preparation. This can be done in a number of ways:

1. By sonication
2. Using a glass syringe and needle to draw up and expel the mixture rapidly
3. Using two glass syringes coupled end to end via a special adaptor[17]

It is important to confirm that the antigen has formed an emulsion in the Freund's adjuvant before injecting the mixture. This is done by placing a drop of the antigen-adjuvant mixture on the surface of some cold water. If the drop disperses immediately, the emulsion requires further mixing.

If the drop remains on the surface for a finite interval, the emulsion is ready for injection. This should be done within 5 to 10 min, before the oil and water phases begin to separate. The most suitable routes for the injection of adjuvants are SC and IM.

NOTE: Having administered Freund's complete adjuvant, only the incomplete (without mycobacteria) form should be used for subsequent injections. Repeated use of Freund's complete adjuvant will lead to a strong delayed-type hypersensitivity reaction to the mycobacterial antigens in the adjuvant.

c. pH
Unless it is essential for the experiment, material for injection should not have a pH outside the range 5.5 to 8.0. Antigens outside the pH range 6.8 to 7.5 should not usually be administered IV.

d. Volume
Where an experiment demands injections of large volumes, these should usually be made by IP injection.

e. Desired Immune Response
If a strong serological response is required, antigen should usually be administered by IV or IP means. However, large doses of antigen given IV may tolerize some animals. The

best cell-mediated responses are usually obtained when antigens are administered by SC or ID injections.

There is no golden rule for the number of injections, the time between multiple injections, the route of injection, or the antigen dose that will produce a peak immune response in all animals with all antigens. All variables have to be evaluated empirically.

2. Filling Syringes

Ideally, syringes should be labeled with a solvent-based ink or a strip of sticking plaster labeled with pencil if alcohol, diethyl ether, or chloroform is to be used. The syringes can then be filled, the sheath replaced over the needle (WITH CARE), and the syringes, ready for use, taken to the animal facility. Where this is not possible, it is desirable to have the antigen in a closed bottle or vial containing a rubber septum in the lid. This permits the syringe to be filled without risk of cross-contamination of animals, operator, or antigen.

Air is removed from the syringe after taking up antigen by gently tapping the vertical barrel with a finger until all air bubbles have risen to the needle end of the syringe. While this is being done, and when the air is expressed, the needle should be encased in sterile cotton wool, sterile cotton gauze, or, preferably, still through the septum into the antigen receptacle. The air is then expressed by pushing the plunger until no more bubbles are visible in the syringe.

Wherever possible, syringes and needles should be discarded, without replacing the sheath, into a rigid waterproof vessel which can be sealed and incinerated. Such "sharps containers" are available from a variety of suppliers. If these are not available, small metal cans with replaceable lids will suffice, provided they are incinerated at a high temperature.

3. Choice of Animal

This is usually determined by price and availability or some specific requirement of a particular experiment. However, there are a number of other factors to bear in mind.

1. While rabbits are probably the animal of choice for the production of large volumes of polyclonal antisera, their sera are often anti-complementary in standard complement-fixation tests.
2. While guinea pigs may be a good source of complement, they are difficult to inject by IV means (in the dorsal vein of the penis for males and the saphenous leg vein in females) and to bleed safely and reliably (cardiac puncture).
3. Rats are relatively easy to inject by all routes and are easy to bleed. However, care must be taken not to strip the skin from the tail.
4. Mice are easy to inject and bleed. However, there are limits to the volume of antigen which can be administered and the volume of blood recoverable.

4. Procedures

a. Intraperitoneal (IP)

Technically, this is the simplest route for injection. However, it may produce pelvic inflammatory disease and subsequent infertility in females.

Equipment: Syringe, 26-gauge needle for mouse, 21-gauge for rat and rabbit.
Volume: Adult mouse - up to 2 ml. Younger mouse - up to 0.7 ml/10 g body weight. Larger rodent - up to 25 ml/k body weight.

For IP injections in rats and mice, the animal should be held on its back with head tilted downward from horizontal (Figure 14). This presents the injection site and allows the abdominal contents to fall away from the injection site. The abdominal skin is slightly stretched and the needle should be inserted about 0.5 to 1 cm (depending on size of animal) anterior to the pubic symphysis (i.e., front of pelvis).

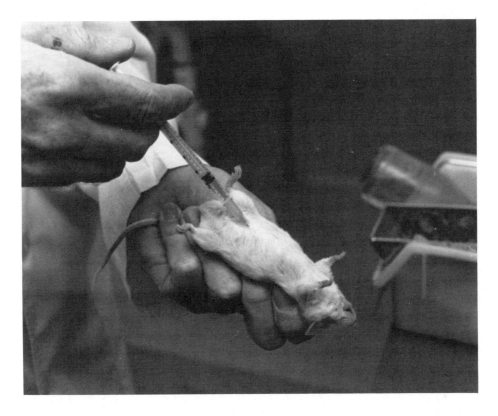

FIGURE 14.   IV injection in mice. Note the head of the animal is lowered so the abdominal contents will fall away from the injection site.

A firm push of the needle pierces the skin and underlying abdominal muscle and reaches the peritoneum. The guinea pig is held in a similar position to rats and mice, but the injection site is 2.5 cm in front of the pubis. A rabbit is held by the hind legs so the gut falls forward, and the point of insertion is 2.5 to 3.5 cm in front of the pubis.

Alternatively, a rabbit may be injected when restrained in the crouching position. A fold of skin from the flank between the ribs and hind leg is held between thumb and forefinger and drawn away from the animal. If a 26-gauge needle is used in place of a 21-gauge, an intraperitoneal injection can be made into the pouch formed behind the stretched skin.

b. Intramuscular (IM)

This is probably the easiest route of injection after IP. While it is possible to restrain a rabbit for an IM injection, it is usually a lot less traumatic if smaller rodents are anesthetized.

Equipment: 26-gauge needle for mouse and 21-gauge for rats and rabbit
Volume: 50 µl in mouse, 100 µl in rat and guinea pig, and up to 500 µl in rabbit

The thigh muscle is the most convenient site for an IM injection in all species. The site is prepared on the posterolateral aspect. The needle is passed through the skin and into the muscle mass, taking care not to hit the femur. An attempt should then be made to draw back the syringe plunger gently. If blood is seen, the needle should be withdrawn and the injection repeated at a different site. If no blood is seen, the injection may be completed. Gently massaging the injection site may also help spread the injected bolus.

FIGURE 15.   SC injection sites in the loose neck skin of a rat.

c. Subcutaneous (SC)

While it is possible to perform SC injections in laboratory animals without the use of an anesthetic, it is much easier to perform these injections in mice which have been anesthetized. Suitable injection sites are the loose abdominal skin, and, in rats and mice, the loose skin of the neck region may also be used (Figure 15). The site of injection in rabbits and guinea pigs should first be shaved.

Equipment: Syringe. 26-gauge needles
Volume: Mouse 50 to 100 μl per site
Rat, guinea pig, and rabbit, up to 500 μl per site.

A fold of skin is taken between the thumb and forefinger and drawn over the forefinger. The needle of the syringe is then inserted immediately in *front* of the forefinger and pushed as close as possible to parallel with the surface of the skin for approximately 0.5 cm (Figure 16 A). A small "blip" should rise under the skin at the tip of the needle as the plunger of the syringe is depressed (Figure 16 B).

If the needle has passed for approximately 0.5 cm subcutaneously, it can be withdrawn slowly without any of the injected material leaking from the injection site along the route of the needle.

d. Intravenous (IV)

While IP and IM injections can be performed by a relative novice, reliable IV injections require some practice.

Equipment: 26-gauge needles for mouse and rabbit. 21-gauge needles can be used in rabbits
             but are rarely required.
Volume: Mouse: up to 1.0 ml if pH, tonicity, etc. are all satisfactory, but 50 to 100
             μl is a safer volume.
             Rat: up to 1.0 ml via the tail vein, but 100 to 500 μl is safer.
             Guinea pig: 500 μl.
             Rabbit: up to 5 ml, but 1 ml is safer and easier to administer.

i. Mouse and Rat

Animals should be warmed for approximately 5 min in a warming box. Alternatively, a wad of cotton wool soaked in warm water (approximately 45°C) can be wrapped around the tail of a restrained animal for 1 to 2 min. These procedures produce vasodilation of the tail vein at the site of injection.

Place the animal in the mouse/rat restrainer by drawing the tail along the slot at the top. Holding the tail between the thumb and second finger, drape it over the forefinger. Rotate the tail 45° to the left or right, and a vein will be seen running the length of the tail.

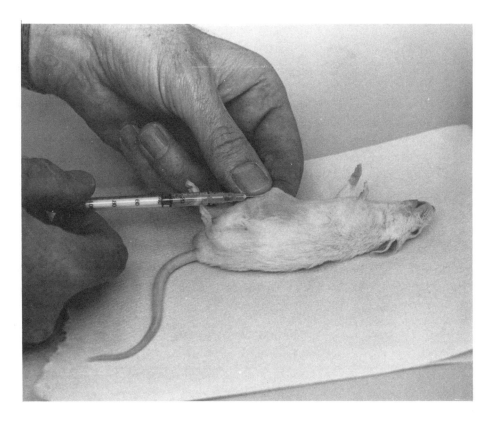

A

FIGURE 16. SC injection in a mouse. (A) Technique for injection; (B) completed injection. The arrow shows the "blip" of antigen at the site of injection.

Insert the needle of the syringe into the vein, bevel up, parallel with the tail, half to one third the distance from the base to the tip of the tail. (Figure 17).

If the needle is in the vein, a small amount of blood may come back into the syringe. Alternatively, the syringe may be pivoted up and down over the forefinger. When the syringe is at its lowest, the bevel of the needle may be seen on the upper surface of the vein, and when the syringe is lowered, the tip of the needle will disappear.

Gently push the plunger on the syringe. A bolus of antigen can often be seen disturbing the uniform red appearance of the vein. There should be no resistance to the movement of the plunger. If there is resistance, or if the skin on the tail blanches, then the needle is not in the vein and should be removed.

A second attempt to insert the needle can be made further toward the base of the tail or in the vein on the opposite side of the tail.

ii. Guinea Pig

The guinea pig should first be anesthetized. It is then turned on its back with a hand under the upper half of the body. The second hand is then placed under the lower half with the thumb resting on the midline anterior to the penis. By pushing down with the thumb the penis is extruded.

The penis can then be grasped with forceps GENTLY or with fingers by a second operator who injects the dorsal vein. This procedure is not recommended unless the operator has been trained to perform it and unless there is absolutely no alternative.

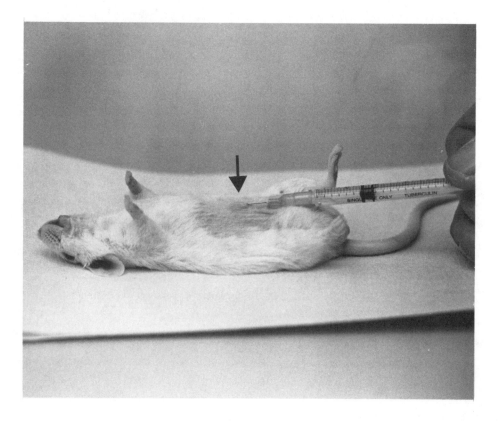

FIGURE 16B.

iii. Rabbit

The rabbit should first be restrained either in a towel or laboratory coat or in a holding box (see Figures 12 and 13). Anesthetic is not necessary. The skin over the marginal ear vein should be swabbed with 70% ethanol in water and shaved. This can be done using a scalpel blade.

Vasodilation of the marginal ear vein can be achieved by warming the ear near an electric bulb, by applying cotton wool soaked in warm water (approximately 45°C), by applying a small amount of xylol to the top of the ear, or by gently flicking the shaved skin over the vein with a finger. If xylol has been used, it should be washed away carefully with alcohol at the end of the procedure.

For injections, the latter procedure is usually adequate. While one operator restrains the rabbit, the second, positioned at the head of the animal, drapes the ear over the fingers of one hand and then holds it still using the thumb (Figure 18). *NEVER PLACE A FINGER ON TOP OF THE EAR IN FRONT OF THE POINT OF INJECTION.* If the rabbit starts, it is very easy to inject the finger.

The needle is then inserted in the vein, bevel uppermost and pointing toward the base of the ear and the rabbit's head. If there is any doubt as to whether the needle is in the vein, repeat the procedure used for the mouse and rat; i.e., move the syringe up and down in a vertical plane using the forefinger as a pivot. The tip of the syringe should appear against the upper surface of the vein and then disappear. If there is any resistance when the plunger is pushed or the skin over the vein blanches, the injection has not been IV. Remove the needle and repeat the injection at a site closer to the base of the ear.

When removing the needle from the ear, place a piece of cotton wool gently over the injection site. When the needle has been completely removed apply firm pressure for 1 to 2 min to prevent any bleeding.

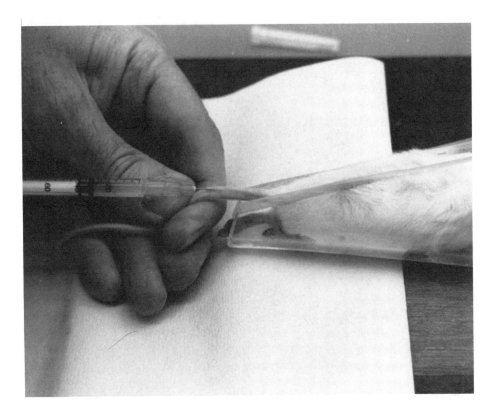

FIGURE 17.   IV injection in the tail vein of a mouse.

C. BLOOD COLLECTION

In a long-term experiment, it is essential to use the same method of blood sampling throughout, both for the control and experimental animals. When planning experiments, the choice of animals may be influenced by the number of times an animal is to be bled and the available routes for bleeding. Rats and mice can be bled from the tail repeatedly as can rabbits from the ear. Guinea pigs can only really be bled by cardiac puncture. There is absolutely no risk to an animal bled from the tail or ear. However, even the most skilled operator occasionally kills an animal when performing a cardiac puncture.

There is little difference in the volume of blood that can be obtained safely from the vein of a rat, mouse, or rabbit and that obtained by cardiac puncture. However, if the animal is to be sacrificed, far larger volumes of blood can be obtained by cardiac puncture than by venipuncture.

Because of the damage done to smaller rodents when teaching untrained staff to obtain blood from the retroorbital plexus, we believe this procedure is unethical. Furthermore, it offers no particular benefit over other procedures, such as tail bleeding.

1. Mice
a. Bleeding from the Tail Vein

Animals should be warmed in a hot box or under a lamp until there is marked vasodilation of vessels in the ears and tail (usually 3 to 5 min).

They are then anesthetized and a small length of tail removed using a large pair of sharp scissors. When removing the tip of the tail approximately 1.0 cm should be taken to ensure a good blood flow. On subsequent occasions 0.2 to 0.5 cm is adequate.

The mouse is then held as for an IP injection and the tail inserted through the stopper of the bleeding jar. The rear legs are restrained using the thumb and forefinger of the second

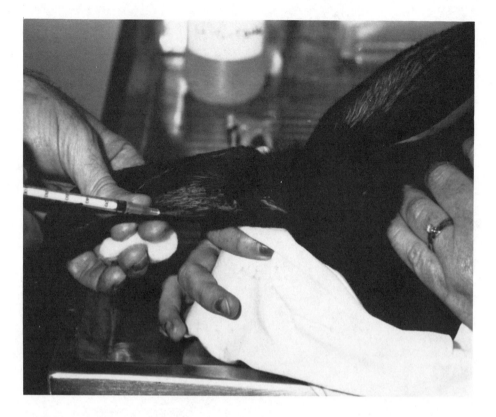

FIGURE 18.   IV injection in the marginal ear vein of a rabbit. Note the wad of cotton wool in the fingers ready to be placed over the injection site when the syringe is removed.

hand (Figure 19). A vacuum just sufficient to draw the tip of a finger onto the stopper is adequate. A high vacuum will prevent bleeding by collapsing blood vessels. Approximately 500 μl of blood can be obtained from a 25-g mouse in this manner. If plasma is required, heparin or EDTA can be added to the tube in the bleeding apparatus.

It is difficult to restrain a struggling mouse or rat on this apparatus, so the animal should be returned to its cage as soon as it starts to recover from the anesthetic or when blood flow slows.

The tip of the tail may be cauterized with the flat edge of a red-hot scalpel blade. However, since the bleeding stops quite readily, this appears an unnecessary discomfort for the mouse.

b. Cardiac Puncture (Exsanguination)

The animal should be anesthetized, preferably with some long-acting anesthetic such as Nembutal* (sodium phenobarbitone). It is then placed on its back and the heart palpated under the fifth and sixth ribs using the index finger. A 1-ml syringe with a 21-gauge needle attached is then inserted at a 45° angle to the long axis of the body. When the needle enters the heart, a small amount of blood will usually appear in the needle or syringe. The plunger of the syringe is then drawn back slowly until no further blood can be withdrawn.

Anticoagulant can be added to the syringe prior to use if unclotted blood is required.

Before disposing of the animal, its cervical spine should be dislocated to ensure it is dead.

2. Rat
a. Bleeding from the Tail Vein

This procedure is the same as for the mouse except that the hole in the stopper of the

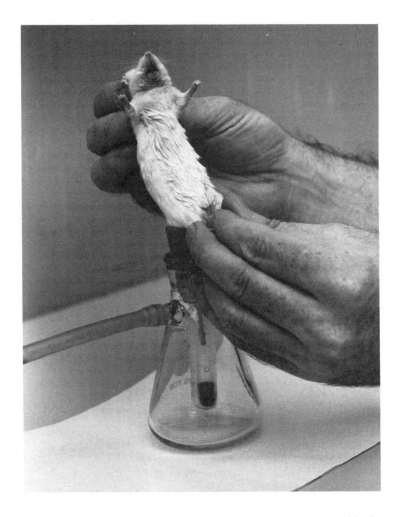

FIGURE 19.   Blood being recovered from the tail of a mouse using a vacuum bleeding apparatus.

bleeding apparatus must be large enough to contain an unbent tail while allowing the rat to rest on the stopper at the top of the bleeding flask.

b. Cardiac Puncture

This is the same as for the mouse. Five milliliters can be recovered from a rat without adverse effects. Larger volumes (10 to 20 ml) can be obtained if the animal is exsanguinated.

3. Guinea Pig
a. Cardiac Puncture

This procedure is the same as for the rat and mouse.

4. Rabbit
a. Bleeding from the Marginal Ear Vein

The rabbit is restrained either in a restraining box or preferably wrapped in a towel or laboratory coat. The area over the marginal vein is swabbed with 70% V/V ethanol/water solution and shaved. A scalpel blade is ideal for shaving. The vein can be dilated by warming, by gently flicking with a finger, or by the addition of a small volume of xylol to the top of the ear. Since xylol, in our hands, gives the best results, it will be described here.

Xylol is added to a pad of cotton wool and wiped over a 1-cm^2 area of both the anterior and posterior surfaces of the tip of the ear. Care must be exercised not to get any xylol near the shaved skin over the marginal vein. After 2 to 3 min all xylol is removed from the ear by wiping with cotton wool or an alcohol-soaked swab. All veins in the ear will have become dilated. The ear of the rabbit is then held as for an IV injection; i.e., fingers under the ear, thumb on top and behind the point at which the needle will be inserted, bevel up, into the marginal vein. The needle of a 19-gauge winged infusion set (Terumo) is then inserted into the vein, pointing toward the base of the ear and the head of the rabbit. The end of the infusion tube containing the attachment for the syringe is placed in a glass bottle or glass centrifuge tube and the vessel lowered as far as the plastic tube will permit. This applies a small hydrostatic pressure which aids blood flow. Some 20 to 30 mls of blood can readily be obtained in this manner. A wad of cotton wool is then placed over the point at which the needle is inserted, the needle is withdrawn, and pressure is applied for 1 to 2 min to stem any bleeding.

Caution: If xylol is allowed to contaminate the area around the injection site, the clotting system is obstructed, and the injection site will continue to bleed for 10 to 15 min or until the xylol is removed.

If winged infusion sets are not available, they can be constructed using a 19-gauge needle and some plastic tubing. Alternatively, a 19-gauge needle can be inserted into the vein and the blood issuing from it collected in a tube.

The procedure of making an incision in the marginal vein with a scalpel blade is stressful to the rabbit (and the scientist) and leads to scarring, which may make sections of the vein unusable. Use of needles avoids this.

Clean all blood from the animals before returning them to their cages.

b. Cardiac Puncture

Restrain the rabbit and administer an anesthetic. Place the animal on its back and draw its forelegs forward beside its head. Swab the area over the rib cage with 70% V/V ethanol/water solution. Palpate the heart with a thumb and forefinger either side of the rib cage and identify the area where the heartbeat is strongest.

Insert the needle of a 50-ml syringe or a winged infusion set between the ribs and toward the strongest heartbeat at an angle of approximately 45° to the vertical and at right angles to the direction of the spine.

After the needle has traveled 2 to 3 cm a definite pulse should be felt. If the needle is inserted another 1 to 2 cm, blood should appear at the top of the needle. If the animal is to be kept alive, 20 ml of blood can safely be collected.

There are two approaches to exsanguination. In the first, syringes are filled with blood and then detached from the needle or infusion set and replaced with a fresh syringe until no further blood can be withdrawn. This usually recovers 50 to 90 ml blood. If, however, an IV drip is placed in a marginal ear vein, 200 to 300 ml of blood plus IV fluid can be obtained. This procedure is most appropriate where a large volume of serum is required and a slight dilution of the serum is acceptable.

D. RECOVERY OF PERITONEAL CELLS

Cells washed from the peritoneal cavity provide a good source of macrophages for immunologic assays. The yield of cells can be improved by injecting stimulants such as thioglycolate medium (Difco) into the peritoneal cavity prior to washing.

The peritoneal washings also contain other cells such as lymphocytes and polymorphonuclear leukocytes. The relative concentration of each cell type will depend on whether or not stimulants are used to induce the exudate and the interval between administration of the stimulant and the harvest of peritoneal cells.

The procedure for all small rodents is similar except that volumes used vary (mice, 5

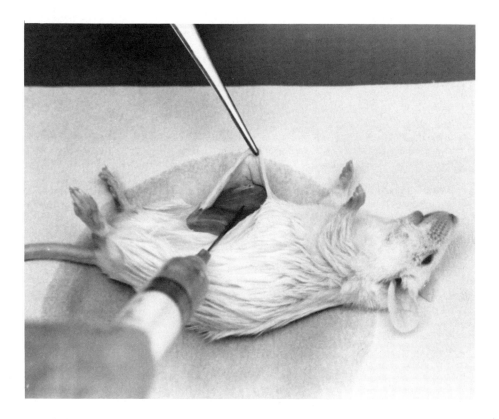

FIGURE 20.   Use of a reservoir formed between the skin and peritoneal wall to collect a peritoneal aspirate.

ml; guinea pigs and rats, 25 to 50 ml). Animals should be anesthetized before rinsing out the peritoneal cavity and euthanatized at the end of the procedure.

The anesthetized animal is placed on its back, the skin over the abdomen raised using forceps and any commercial culture medium (H199, MEM, RPMI 1640) injected by IP means. The animal should then be repeatedly rolled from left to right over an interval of about 5 min. An incision is then made along the midline of the skin of the abdomen (Figure 20) and the skin separated from the peritoneal wall.

A small incision in the peritoneum toward the spine will allow the injected fluid to flow out into a reservoir, formed using the skin flaps. The cell lavage can then be recovered using a syringe. If there is a tendency for cells to clump, this can be prevented by the addition of preservative-free heparin (10 IU/ml) to the culture medium before use.

VIII. FURTHER READING

Three other publications which may be helpful to those working with laboratory animals are

1.   The UFAW Laboratory Handbook[18]
2.   The Handbook of Experimental Immunology[19]
3.   Principles of Laboratory Animal Management[20]

REFERENCES

1. **Rubright, W. C. and Thayer, C. B.**, The use of Innovar-vet (R) as a surgical anaesthetic for the guinea pig, *Lab. Anim. Care,* 20, 989, 1970.
2. **Walden, N. B.**, Effective sedation of rabbits, guinea pigs, rats and mice with a mixture of fentanyl and droperidol, *Aust. Vet. J.,* 54, 538, 1978.
3. **Green, C. J.**, Animal anaesthesia, in *Laboratory Animal Handbook 8,* Laboratory Animals Ltd., London, 1979.
4. **Green, C. J., Knight, J., Precious, S., and Simpkin, S.**, Ketamine alone and combined with diazepam or xylazine in laboratory animals: a ten year experience, *Lab. Anim.,* 15, 163, 1981.
5. **Weisbroth, S. H. and Fudens, J. H.**, Use of ketamine hydrochloride as an anaesthetic in laboratory rabbits, rats, mice and guinea pigs, *Lab. Anim. Sci.,* 22, 904, 1972.
6. **White, G. L. and Holmes, D. D.**, A comparison of ketamine and the combination ketamine-xylazine for effective surgical anaesthesia in the rabbit, *Lab. Anim. Sci.,* 26, 804, 1976.
7. **Bain, S.**, Anaesthesia of the laboratory mouse and rat, in *Aust. Soc. Lab. Anim. Sci. Newsl.,* August, p. 4, 1984.
8. **Green, C. J.**, Anaesthetic gases and health risks to laboratory personnel: a review, *Lab. Anim.,* 15, 397, 1981.
9. **Thomasson, B., Ruuskanen, O., and Merikanto, J.**, Spinal anaesthesia in the guinea pig, *Lab. Anim.,* 8, 241, 1974.
10. **Kero, P., Thomasson, B., and Soppi, A. M.**, Spinal anaesthesia in the rabbit, *Lab. Anim.,* 15, 347, 1981.
11. **Harkness, J. E. and Wagner, J. E.**, *The Biology and Medicine of Rabbits and Rodents,* Lea & Febiger, Philadelphia, 1983.
12. **Erickson, H. H. and Kitchell, R. L.**, Pain perception and alleviation in animals, *Fed. Proc.,* 43, 1307, 1984.
13. **Morton, D. B. and Griffiths, P. H. M.**, Guidelines on the recognition of pain, distress and discomfort in experimental animals and an hypothesis for assessment, *Vet. Rec.,* 116, 431, 1985.
14. **Flecknell, P. A.**, The relief of pain in laboratory animals, *Lab. Anim.,* 18, 147, 1984.
15. **Buckland, M. D., Hall, L., Mowlen, A., and Whatley, B. F.**, *A Guide to Laboratory Animal Technology,* William Heinemann Medical Books, London, 1981.
16. **Scott, W. N. and Ray, P. M.**, Euthanasia, in *UFAW Handbook on the Care and Management of Laboratory Animals,* Churchill Livingstone, Edinburgh, 1976, 160.
17. **Dresser, D. W.**, Immunization of experimental animals, in *Handbook of Experimental Immunology,* Weir, D. M., Ed., Blackwell Scientific, Oxford, 1986, 8.1.
18. Universities Federation for Animal Welfare, *Handbook on the Care and Management of Laboratory Animals,* Churchill Livingstone, Edinburgh, 1987.
19. **Weir, D. M., Ed.**, *Handbook of Experimental Immunology,* Blackwell Scientific, Oxford, 1986.
20. **Blackshaw, J. K. and Allan, D. J., Eds.**, *Principles of Laboratory Animal Management,* Australian Society for the Study of Animal Behaviour, Blacktown, Australia, 1987.

Chapter 2

IMMUNOLOGIC ADJUVANTS: GENERAL PROPERTIES, ADVANTAGES, AND LIMITATIONS

Noelene E. Byars and Anthony C. Allison

TABLE OF CONTENTS

I. INTRODUCTION

An adjuvant can be defined as a substance which acts to increase the specific immune response of an animal to an antigen. Factors affecting the choice of an adjuvant include (1) the type of immune response desired or, conversely, the type of response to be avoided, (2) the species to be vaccinated, (3) the route of administration, and (4) the tolerance of the target animals for side effects induced by the adjuvant. Adjuvants can be divided into several broad classes: (1) aluminum salts, (2) surface-active agents, (3) bacterial derivatives, and (4) vehicles and slow-release materials.

II. ALUMINUM SALTS

In 1926, Glenny et al.[1] discovered that a suspension of alum-precipitated diphtheria toxoid was more antigenic than the toxoid from which it was derived. Since then, the aluminum salts, principally those of phosphate or hydroxide, have been widely used to immunize humans as well as laboratory and farm animals. The salts of other metals including calcium, zirconium, beryllium, and iron have also been used to adsorb antigens, but these have not received the widespread usage of the aluminum salts. Alum has relatively few side effects, although injection-site granulomas have been observed, particularly following s.c. or i.d. rather than i.m. injection.[2-4] Similar granulomas have also been observed following injection of zirconium or beryllium.[3,5] Alum is most useful in situations in which increased antibody synthesis is desired and cell-mediated immunity is relatively unimportant. In mice, the IgG1 subclass is increased the most following immunization.[6,7] Alum can also increase the formation of IgE in rabbits and rodents.[8]

Alum has been extensively used in veterinary vaccines, principally for large animals. However, there is wide variation in the relative amounts of alum and antigen used in commercially available products, even with a well-characterized antigen such as tetanus toxoid.[9] This fact suggests that optimal augmentation of immunogenicity has not been achieved for such livestock vaccines, and further research is warranted in this area. Aluminum salts can be difficult to manufacture in a reproducible fashion; the physicochemical properties of the various salts differ and can change with the age of the preparations. Hem and White[10] discuss factors which affect the structure of aluminum hydroxide (e.g., pH, nature of anions present), the effect of morphology on surface area, and the methods for determining both surface area and charge. They also discuss mechanisms by which proteins are adsorbed to aluminum hydroxide, as well as how some of these properties might affect the adjuvant activity of aluminum hydroxide.

III. SURFACE-ACTIVE AGENTS

Saponins, including Quil A, which is a purified saponin, have been used in a number of experimental and veterinary vaccines (e.g., foot and mouth disease virus, FMDV).[11] Recent studies with saponin in mice have shown it to be an excellent adjuvant when used with glycoproteins of *Trypanosoma cruzi*.[6] Vaccines prepared with saponin protected mice against challenge with live parasites, while vaccines prepared with *Corynebacterium parvum*, *N*-acetyl muramyl-L-alanyl-D-isoglutamine (MDP), and MDP analogs in either saline or Freund's incomplete adjuvant (FIA), alum (alhydrogel), dimethyl dioctadecyl ammonium (DDA), or Pfizer CP20,961 (avridine) offered no protection. Among the adjuvants tested, saponin increased IgG2a and IgG2b antibodies as well as IgG1 antibodies, while the others, with the exception of a squalene emulsion, augmented IgG1 only. However, unlike saponin, the squalene emulsion did not induce delayed hypersensitivity nor did it protect against challenge.

However, saponin has a number of undesirable side effects: it is irritating and proinflammatory, binds to cholesterol, and lyses red blood cells.[11,12] Hence, ways to retain the adjuvant activity but diminish the side effects have been sought. One approach has been the production of "immune-stimulating complexes", ISCOMs, which were originally described by Morein et al.[13] as particles consisting of Quil A and membrane proteins. Use of ISCOMs allowed reduction in Quil A concentrations. Later studies by Lövgren and Morein[14] showed that lipids are also essential for the formation of the cage-like structures characteristic of ISCOMs, whereas preparations lacking lipids tend to form aggregates or micelles.

ISCOM preparations have been reported to increase antibody responses to a number of viral membrane proteins, compared to responses obtained with aqueous or liposome preparations of the antigens.[13] Proteins from bovine herpesvirus type 1, cytomegalovirus, hepatitis B virus, Epstein-Barr virus, and canine distempter virus are among those which have been tested with ISCOMs.[15-19] ISCOMs have also been used to improve an equine influenza vaccine,[20] and a commercial ISCOM-influenza vaccine is now licensed for use in Sweden.[21] Rhesus monkeys were vaccinated with inactivated Simian immunodeficiency virus (SIV) either in ISCOMs or in SAF-1 (Syntex Adjuvant Formula) (see below). The group vaccinated using ISCOMs developed good anti-envelope titers but variable anti-core titers, while the SAF-1 group had strong responses to both envelope and core antigens. However, despite the presence of antibodies, both groups became viremic following challenge with live SIV.[22]

DDA bromide and avridine (CP20,961) are quaternary amines with adjuvant activity.[23,24] Avridine is insoluble in water, which could present formulation problems, especially for large-scale commercial vaccine preparation. Various strategies to incorporate avridine in vaccines have included use of ethanol solutions in oil emulsions, incorporation into liposomes, suspension in oil or saline suspensions, and mixing with alum-precipitated antigens. Despite these difficulties, avridine has been used in a number of species.[23]

Anderson and Reynolds[24] found avridine to be an effective adjuvant for induction of neutralizing antibodies and protection of mice against challenge with Venezuelan equine encephalitis (VEE) virus. Hilgers et al.[25] found avridine and DDA had similar efficacy in the induction of delayed-type hypersensitivity (DTH) in mice following immunization with azobenzene-arsonate coupled to phosphatidylethanolamine. In guinea pigs, DDA was shown to stimulate DTH to DNP_{22}-BSA, but little or no antibody to either the hapten or carrier was produced.[26] In mice, the route of administration appears to influence the type of immune response obtained. Gordon et al.[27] found both increased antibody and DTH responses following i.p. administration, whereas the intracutaneous or i.m. routes appear to favor DTH.[28,29]

It has been proposed that for DDA or CP20,961 to act as adjuvants, the antigens must bind the adjuvant.[25] If dextran-SO_4 was added to DDA, or CP20,961, the adjuvant activity was markedly reduced, and dextran-SO_4 inhibited the binding of labeled antigen to DDA at the same dose levels at which inhibition of adjuvant activity occurred. Dextran did not inhibit the adjuvant activity, suggesting that the negatively charged SO_4 groups of dextran-SO_4 are important for its inhibitory action. These data are interesting since dextran-SO_4 has been reported to act as an adjuvant with sheep red blood cells (SRBC) and with *Listeria monocytogenes*, although there is also a report of suppression of cell-mediated immunity (CMI) by dextran SO_4.[30] Dextran-SO_4 has significant side effects, unlike DDA or CP20,961 which appear to be nontoxic at levels required for adjuvant activity.

It has been suggested that induction of interferon (IFN-α or IFN-β) may be the underlying mechanism of action of DDA and other adjuvants which induce DTH. However, in a comparison of DDA and polyinosinic polycytidylic acid (poly-I:poly-C) as IFN inducers and adjuvants, it was found that DDA was a much more effective adjuvant than poly-I:poly-C, but the converse was true for IFN induction.[31] When the authors compared different mouse strains for DTH responses to lysozyme or Semliki Forest Virus (SFV) with DDA, marked differences were found in response to these vaccines. C_3H/HeJ mice had negligible responses to the vaccines, while BALB/c mice responded strongly. However, both strains produced similar amounts of IFN in response to DDA. Thus, it appears that the adjuvant activity of DDA in mice is not directly related to IFN induction.

IV. BACTERIAL DERIVATIVES

Bacteria and their derivatives have long been recognized as adjuvants. The classic bacterial adjuvant, Freund's complete adjuvant (FCA) consisting of killed mycobacteria in mineral oil and arlacel A, stimulates both cell-mediated and humoral responses. FIA, which lacks the mycobacteria, stimulates only humoral responses. In guinea pigs, FCA stimulates antibodies of both the IgG1 and IgG2 classes, whereas FIA stimulates only IgG1.[32,33] In BALB/c mice immunized with human serum albumin (HSA) in FCA, IgG1 antibodies were increased to a greater extent than those of the IgG2a or IgG2b subclasses.[34] The ratios of the different murine subclass responses depends on the strain of mouse as well as the adjuvant used. For example, Natsuume-Sakai et al.[35] reported that BALB/c and C3H/He mice had greater increases in IgG1 antibodies against bovine gamma globulin than did C57BL/6 mice, following immunization using FCA. However, the C57BL/6 mice showed much greater increases in the IgG2b isotype than either the BALB/c or C3H/He mice. These changes reflect the pattern of predominant isotypes in the sera of these strains: IgG1 was found to be the predominant serum isotype for BALB/c and C3H/He strains, while IgG2b was predominant isotype for C57BL/6.

The use of FCA is restricted to experimental animals, since the resulting granuloma formation at the injection site and tuberculin sensitivity make this adjuvant unsuitable for either human or veterinary vaccines. There is additional concern about the carcinogenic potential of both the mineral oil and the arlacel A emulsifier in FCA.[36,37] Another factor which should be considered when using either FCA or FIA is that the process of producing the antigen-containing emulsion is likely, at least partially, to denature the antigen. Animals immunized with HSA in FCA make antibodies to epitopes found on denatured HSA as well as to native epitopes;[34] with some other adjuvant formulations (including SAF) antibodies are directed predominantly against determinants in the native molecule. Thus, FCA is not the best adjuvant to use with unstable antigens.

Considerable progress in understanding the mechanism of action of FCA has been made since Ellouz et al.[38] published their report describing MDP as the adjuvant active component of the mycobacterial cell wall. MDP and its many analogues stimulate both cell-mediated and humoral immune responses. MDP is frequently used in FIA, or some other oil emulsion, although there are reports of aqueous vaccines with MDP as adjuvant.[39,40] However, such aqueous vaccines are often less effective than emulsion-based vaccines.[7,41] Like FCA, MDP stimulates both IgG1 and IgG2 antibodies in guinea pigs.[32] Using aqueous MDP vaccines in CBA/H or NMRI mice, Bomford[7] and Heymer et al.[42] found IgG1 responses to BSA to be increased, while Ohkuni et al.[43] found both IgG and IgE levels were increased in A/J mice. However, using the threonyl analogue of MDP in Syntex Adjuvant Formulation in BALB/c mice, we have found that antigen-specific IgG2a antibodies as well as IgG1 antibodies were significantly increased while smaller increases in IgM and Ig2b were observed. No IgE was found, and there was little or no increase in IgG3 or serum IgA. The relative increases in IgG2a and IgG1 antibodies were found to vary with the antigen used, e.g., HSA,[34] influenza B,[44] or HBsAg.[86]

The parent MDP (i.e., [ala¹]-MDP) is pyrogenic, induces uveitis in rabbits and monkeys, and induces guinea pig distress syndrome in guinea pigs.[45,46] A number of MDP analogues, especially α-aminobutyryl derivatives and lipophilic MDP analogues, have similar side effects. Such MDPs are, therefore, unsuited for use in vaccines for veterinary or human vaccines and are far from ideal for experimental laboratory animals. It seems reasonable and humane to use adjuvants which produce the best stimulation of specific immune responses with the least distress to the immunized animals.

We have found that *N*-acetyl muramyl-L-threonyl-D-isoglutamine (threonyl MDP or [thr^1]-MDP) is an excellent adjuvant and is neither pyrogenic nor uveogenic at doses needed for adjuvant activity. Only when enormous doses which are more than 300-fold higher than the adjuvant-active dose are given are such side effects observed.[45] Chedid et al.[47] have also described a nonpyrogenic MDP, namely murabutide. Murabutide has a somewhat broader spectrum of activity than [thr^1]-MDP since it is also active as an antiinfective agent,[47] whereas [thr^1]-MDP has no activity as a stimulator of nonspecific immune stimulation.[48]

A number of liphophilic MDP derivatives have been synthesized, and while they are often good adjuvants, some also tend to show greater toxicity than their hydrophilic counterparts.[45] In some cases incorporation of such lipophilic MDPs into liposomes reduced the side effects, but in other experiments liposomes provided no protection. For example, *N*-acetyl muramyl-L-alanyl-D-isoglutaminyl-L-alanine-phosphatidylethanolamine (MTP-PE) given i.v. to rabbits or dogs at a dose of 0.1 mg/kg/d resulted in the development of carditis, synovitis, and arthritis, whereas liposome-encapsulated MTP-PE did not elicit any manifestations of generalized inflammatory responses, even after 3 months of dosing.[49] However, incorporation of a lipophilic MDP derivative into liposomes did not protect rabbits against the uveogenic properties of the compound.[87]

Threonyl-MDP was originally tested for adjuvant activity using FIA as the vehicle. It was found to have considerably more potent adjuvant activity than the parent MDP.[86] However, FIA has not been approved for use in human vaccines and has drawbacks when used in veterinary vaccines. Sterile abscesses at the injection site in meat-producing animals can lead to carcass condemnation and consequent economic losses. Furthermore, abscesses or granulomas at vaccination sites are poorly accepted by pet owners. Thus, a new vehicle was clearly needed. We developed Syntex Adjuvant Formulation (SAF) to meet this need.[50] It consists of an oil-in-water type emulsion of Pluronic L121®, squalene, and Tween 80 in phosphate-buffered saline, to which appropriate concentrations of threonyl MDP are added. To prepare the final vaccine, equal volumes of the preformed SAF emulsion and antigen solution or suspension are mixed gently for 2 or 3 min. This method increases the likelihood that the native structure of the antigen will be retained, unlike antigens in Freund's-type emulsions, where the emulsification process denatures antigens. SAF has been found to be efficacious with a wide variety of antigens including ovalbumin,[50] bovine serum albumin, influenza A and B virus hemagglutinins,[44] hepatitis B virus surface antigen, gp 340 of Epstein-Barr virus, formalin-inactivated SIV,[22] inactivated FeLV, inactivated type D retrovirus,[51] gp120 of HIV,[52] and human IgM. Animal species which have been successfully vaccinated using SAF-based vaccines include mice, rats, guinea pigs, rabbits, cats, dogs, rhesus and cynomolgus monkeys, cotton-top tamarins, and horses.

Figures 1 and 2 show examples of data obtained using SAF as adjuvant for influenza B virus hemagglutinin (HA) and hepatitis B virus surface antigen (HBsAg), respectively. In both cases, the SAF-adjuvant vaccines are compared to the currently available vaccines. Antibody responses to influenza vaccines currently used are inconsistent, especially in people aged 65 years and older, who are particularly susceptible to the disease. The influenza vaccines now in use for humans are simply aqueous solutions of HA with no added adjuvant, while HBsAg vaccines are adjuvanted with alum. We found that SAF-adjuvant vaccines induced significantly higher antibody titers to both these antigens, in mice as well as guinea pigs. Furthermore, the responses were more consistent, particularly when young immunologically immature mice (3 weeks old) or old mice (13$\frac{1}{2}$ months) were immunized with influenza B HA (B/USSR/100/83) in SAF. For example, all of the old mice (11/11) in the group immunized with 1 μg of HA in SAF responded, while only 2/10 mice in the group given 1 μg of HA in saline produced detectable anti-HA antibodies. Thus, we believe that SAF is safe and efficacious with a broad range of antigens, and it is free of the deleterious side effects of many other adjuvants.

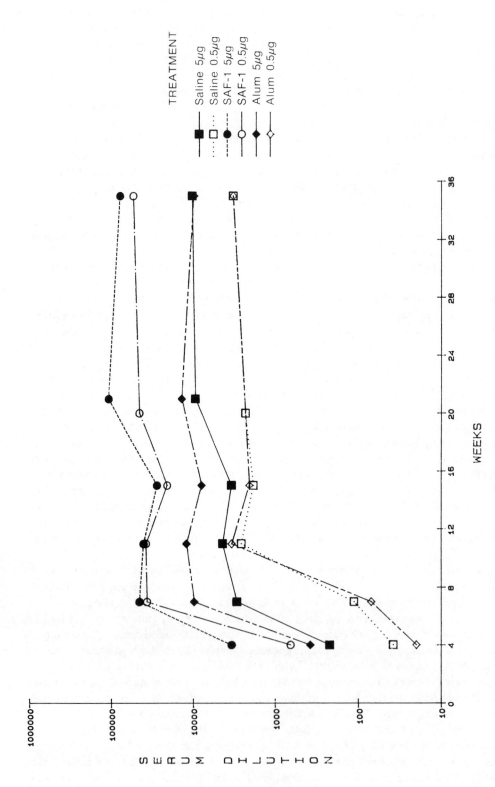

FIGURE 1. Time course of anti-HBsAg titers in pooled sera of guinea pigs following immunization with HBsAg in SAF-1, saline, or alum. Antibody titers were determined by ELISA.

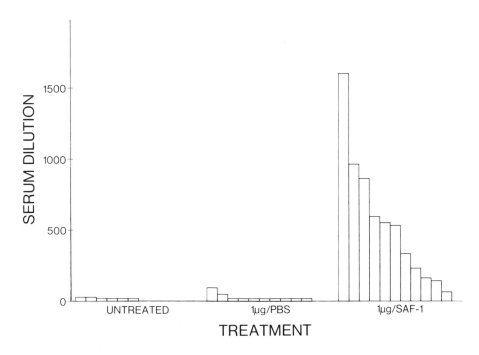

FIGURE 2.   Anti-HA titers of individual mice 19 weeks after immunization of old mice with 1 μg of influenza B virus HA in PBS or SAF-1. Titers were determined by ELISA.

Gram-negative bacteria such as *Escherichia, Salmonella,* and *Pseudomonas* have endotoxins which induce fever, hypotension, changes in leukocyte count, shock, and uveitis.[53] The endotoxins are lipopolysaccharides (LPS), consisting of a hydrophilic polysaccharide covalently linked to the hydrophobic lipid A component. LPS has potent immunologic adjuvant activity, as well as causing macrophage activation and nonspecific stimulation of immune responses. The structures of lipid A from *Escherichia coli* and *Salmonella* have been elucidated and are quite similar.[53-55] Synthetic lipid A and analogues have been produced and data obtained using the synthetic compounds confirms that the biological activity of LPS resides in the lipid A portion of the molecule.

Chemical modifications of the lipid A structure have been made in attempt to reduce the toxicity without reducing the adjuvant activity of the molecule. Monophosphoryl lipid (MPL) of *E. coli* has been synthesized by Imoto et al.,[56,57] while Johnson et al.[58] have chemically removed the phosphate moiety from the C-1 position of the toxic diphosphoryl lipid A of *Salmonella*.[58] The MPL compounds are much less toxic than the diphosphoryl molecule, but retain much of the adjuvant and mitogenic activity of lipid A. MPL also increases nonspecific protection against bacterial infections.[58] The efficacy of MPL as an adjuvant is enhanced by the addition of bacterial cell wall derivatives and trehalose dimycolate (TDM).[59] TDM, originally known as cord factor, appears to be responsible for much of the toxicity attributed to the mycobacteria in FCA.[60] Toxic side effects in mice including inflammation, granuloma formation, pulmonary hemorrhage, weight loss, and death have been reported.[61,62] Naturally occurring TDM and analogues having different fatty acids have been synthesized. The synthetic analogues were much less toxic than the parent TDM.[62] The physical form of TDM also affects the toxicity and adjuvant activity: monolayers of TDM retain both activities while aqueous suspensions of TDM micelles were biologically inert.[61]

Bordetella pertussis bacteria have been tested as adjuvants for proteins, SRBC, and parasites with varying degrees of success. For example, Ruiz et al.[63] reported that mice immunized with a lyophilized fraction of trypomastigotes of *T. cruzi* mixed with *B. pertussis* bacteria were much better protected against live parasite challenge than mice given only the parasites. All the mice in the former group survived, whereas only 66% of the latter group did. McColm et al.[64] compared the efficacy of a number of adjuvants in induction of protective immunity to *Plasmodium yoelii* in mice. Vaccination with killed parasites and *B. pertussis* resulted in significant reduction in parasitemia following challenge, compared to controls vaccinated with parasites alone. *Corynebacterium parvum*, FCA, and alum were also efficacious, although somewhat less so than saponin. Bomford[7] found that *B. pertussis* was an ineffective adjuvant in mice for humoral responses to BSA, but increased secondary antibody responses to SRBC. *C. parvum* was ineffective with either antigen for antibody production, but augmented CMI to SRBC in mice and guinea pigs, especially if it was combined with either FIA or alum.[65] *C. parvum* was effective in a vaccine with irradiated M4 tumor cells, whereas *B. pertussis* added to tumor cells provided no protection against challenge with live tumor cells.[65] However, *B. pertussis* or fractions thereof induce IgE synthesis.[66,67] Thus, there is the potential for undesirable allergic reactions following use of a vaccine containing *B. pertussis*. The adjuvant effects of *B. pertussis* as well as neurotoxicity are attributed to pertussis toxin, which makes it unsuitable as an adjuvant for human use.

Many other bacteria and their derivatives have been tested for adjuvant activity. Early experiments by Degrand and Raynaud[68] had shown that in guinea pigs, good delayed hypersensitivity to azobenzenearsonate-*N*-acetyl-L-tyrosine could be obtained if *C. granulosum* bacteria or cell walls were used with the antigen in FIA. A more recent example is that of the exopolymer of *Cytophaga* species described by Usinger and Mishell.[69] The adjuvant-active fraction consists of amino-sugar polymers ranging in size from 5 to 500 × 10^6 relative molecular mass. The polymer augments antibody responses to cytochrome C and also induces synthesis of IL-1 and CSF by murine macrophages *in vitro*. It appears to be well tolerated by the mice, unlike many of the other bacterial products.

V. VEHICLES AND SLOW-RELEASE MATERIALS

Another group of adjuvants should perhaps more properly be called vehicles. Included in this group are liposomes, oil emulsions such as FIA, Adjuvant 65,[70] and a metabolizable, lipid-based adjuvant[71] now marketed as Lipovant. A considerable amount of work has been done with these vehicles both for veterinary vaccines and, in some cases, for human use.

FIA is, of course, the classic example of an oil-emulsion vehicle. Extensive clinical trials were done with influenza vaccines in FIA for humans, but final approval from the regulatory authorities has not been forthcoming.[72] Oil emulsions have also been used in a number of veterinary vaccines, for example, rabies, canine distemper, and foot-and-mouth disease.[73] Woodhour et al.[70] developed Adjuvant 65, consisting of peanut oil, aluminum monostearate stabilizer, and Arlacel A. This adjuvant also underwent extensive clinical trials for use with influenza vaccines.[74] Reynolds et al.[71] prepared a metabolizable lipid emulsion (Lipovant) consisting of peanut oil, glycerol, and lecithin. Vaccines were prepared by mixing the lipid emulsion with aqueous suspensions of inactivated Venezuelan or Western equine encephalomyelitis virus or with Rift Valley fever virus. Good responses to these antigens were obtained in mice, hamsters, sheep, and monkeys.[71] However, Harrington et al.[75] found no augmentation of the immune responses of sheep to inactivated Rift Valley fever virus, using either the lipid emulsion, or poly-I:poly-C. Further, Brugh et al.[76] failed to observe a positive response in chickens following vaccination with Newcastle disease virus in the lipid emulsion. They did, however, observe adjuvant effects of nine different emulsions which contained mineral oil, with the emulsifiers polysorbate 80 (Tween 80) and sorbitan monooleate (Arlacel 80). Hence, the use of oil emulsions as vaccine vehicles is not straightforward, particularly with emulsions based on metabolizable oils.

Liposomes have promise as vehicles in vaccines, but the lipid composition, the method of preparation, and the location of the antigen (entrapped or surface associated) differ widely among experiments reported by workers in this field. It may be necessary to customize liposome preparations for each antigen to be tested. Readers are referred to recent reviews for discussions of these variables.[77,78]

VI. CYTOKINES

Since the induction of cytokine synthesis has been proposed as a mechanism of action of adjuvants, some workers have attempted to use the cytokines themselves as adjuvants. Staruch and Wood[79] have described the adjuvant activity of IL-1 in mice, using BSA as the antigen. IL-2 has been tested by several workers using whale myoglobin in mice, herpes simplex 2 in guinea pigs, or *Haemophilus pleuropneumoniae* (HPP) in swine.[80-83] However, in some of these experiments multiple doses of IL-2 were required, and in the experiment with HPP in swine some toxicity was observed in the animals given the higher doses of IL-2. Interferon-γ (IFN-γ) has also been tested as an adjuvant. Playfair and De Souza[84] augmented resistance of mice to lethal challenge with *Plasmodium yoelii* by incorporating IFN-γ in a vaccine prepared from the schizont stage of the parasite. Yilma et al.[85] reported an augmentation of the humoral response of cattle to the G glycoprotein of vesicular stomatitis virus by administration of bovine IFN-γ. While these experiments are academically interesting, practical application of cytokines as adjuvants presents several difficulties. For example, multiple doses may be required, and the optimal time of dosing may not be at the time of antigen administration. Further, if a cytokine from a different species is used, it may be immunogenic in the target species.

VII. DISCUSSION

This chapter outlines many of the immunologic adjuvants that are available, including their main properties, advantages, and limitations. It is a general account of the properties of adjuvants and guide to their use rather than a list of recipes for each one. Such lists are not very helpful. In the case of commercially available adjuvants the vendors provide recommendations for use. These must be taken with reservation; for example, the optimal conditions for adsorption to aluminum hydroxide vary from antigen to antigen, and for some antigens, such as influenza hemagglutinin, aluminum salts are ineffective adjuvants. Evidence is accumulating that the composition of liposomes must be varied for optimal use with different antigens. In some cases immunogenicity is greater when antigens are covalently bound to the surface of liposomes, in other cases when the antigens are entrapped within the liposomes. ISCOMs are more useful for membrane proteins than soluble proteins. Thus, adjuvants must be selected according to the antigen and need.

Needs are now becoming better defined. It is no longer sufficient to measure the levels of antibodies elicited by a convenient method, e.g., ELISA or hemagglutination. The affinities of the antibodies for antigens in solution and their isotypes can also be important. Often antibodies of certain isotypes (e.g., IgG2a in the mouse and IgG1 in humans) activate complement and act synergistically with antibody-dependent effector cells more effectively than antibodies of other isotypes. For most vaccines IgE antibodies are undesirable: they are not required for protection and pose the risk of hypersensitivity. Adjuvants clearly affect the isotypes of antibodies formed. Aluminum salts, LPS, and MPL favor the formation of IgG1 and IgE antibodies in the mouse whereas SAF favors IgG2a antibodies. However, antigens themselves and the mouse strains responding also influence isotypes, so all these factors have to be taken into consideration.

Adjuvants should stimulate T-cells as well as B-cells. Antibody responses usually require T-cell help, and cell-mediated immunity is required for protection against some pathogenic organisms. Some adjuvants clearly elicit cell-mediated immunity better than others. This is shown not only by delayed hypersensitivity, but also by proliferative responses of lympho-cytes to antigens and production of mediators such as IL-2 and by genetically restricted cell-mediated cytotoxicity. Some adjuvants, e.g., FCA and SAF, elicit cell-mediated immune responses, whereas others do not. It is notable that the adjuvants eliciting cell-mediated immunity also favor the formation of IgG2a antibodies in the mouse. An explanation can now be provided: strongly responding T-cells would release IFN-γ, which favors the production of IgG2a antibodies.

Trends in adjuvant use are discernible. FCA has been a mainstay of laboratory immunology for 4 decades. There is now growing realization that other adjuvants, such as SAF and ISCOMs, are as efficacious as Freund's and more humane to use. Procedures such as repeated intraperitoneal injection of antigens with Freund's adjuvant to raise monoclonal antibodies are unnecessary. Already in some large research establishments the use of FCA is restricted, and by the end of the century it seems likely that Freund's adjuvant will be, like some other immunologic procedures, largely of historical interest.

Some of the new adjuvants are already in commercial veterinary vaccines, e.g., ISCOMs for equine influenza. This trend is likely to continue. With the production of viral and other microbial subunits by recombinant technology and with the development of peptide vaccines, a new generation of vaccines is possible. However, that exciting possibility will be exploited only if safe and efficacious adjuvants are developed and authorized for use in human and veterinary vaccines.

REFERENCES

1. **Glenny, A. T., Pope, C. G., Waddington, H., and Wallace, V.,** Antigenic value of toxoid precipitated by potassium alum, *J. Pathol. Bacteriol.,* 29, 38, 1926.
2. **Butler, N. R., Voyce, M. A., Burland, W. L., and Hilton, M. L.,** Advantages of aluminum hydroxide adsorbed diphtheria, tetanus and pertussis vaccines for the immunization of infants, *Br. Med. J.,* 1, 663, 1969.
3. **Turk, J. L. and Parker, D.,** Granuloma formation in normal guinea pigs injected intradermally with aluminum and zirconium compounds, *J. Invest. Dermatol.,* 68, 336, 1977.
4. **Straw, B. E., MacLachlan, N. J., Corbett, W. T., Carter, P. B., and Schey, H. M.,** Comparison of tissue reactions produced by *Haemophilus pleuropneumoniae* vaccines made with six different adjuvants in swine, *Can. J. Comp. Med.,* 49, 149, 1985.
5. **Salvaggio, J. E., Flax, M. H., and Leskowitz, S.,** Studies in immunization. III. The use of beryllium as a granuloma-producing agent in complete Freund's adjuvant, *J. Immunol.,* 95, 846, 1965.
6. **Scott, M. T., Bahr, G., Modabber, F., Afchain, D., and Chedid, L.,** Adjuvant requirements for protective immunization of mice using a *Trypanosoma cruzi* 90K cell surface glycoprotein, *Int. Arch. Allergy Appl. Immunol.,* 74, 373, 1984.
7. **Bomford, R.,** The comparative selectivity of adjuvants for humoral and cell-mediated immunity. I. Effect on the antibody response to bovine serum albumin and sheep red blood cells of Freund's incomplete and complete adjuvants, alhydrogel, *Corynebacterium parvum, Bordetella pertussis,* muramyl dipeptide and saponin, *Clin. Exp. Immunol.,* 39, 426, 1980.
8. **Yin, J.-Z., Furusawa, S., Hirano, T., and Ovary, Z.,** Studies on immunity in hybridoma-bearing mice. I. Immune response to antigens, *Int. Arch. Allergy Appl. Immunol.,* 83, 414, 1987.
9. **Bunn, T. O., Nervig, R. R., and Pemberton, J. R.,** Metallic salts as adjuvants for veterinary biologics, in *Advances in Carriers and Adjuvants for Veterinary Biologics,* Nervig, R. M., Gough, P. M., Kaeberle, M. L., and Whetstone, C. A., Eds., Iowa State University Press, Ames, 1986, 105.
10. **Hem, S. L. and White, J. L.,** Characterization of aluminum hydroxide for use as an adjuvant in parenteral vaccines, *J. Parenter. Sci. Technol.,* 38, 2, 1984.
11. **Dalsgaard, K.,** Adjuvants, *Vet. Immunol. Immunopathol.,* 17, 145, 1987.
12. **Bomford, R. H. R.,** The differential adjuvant activity of Al(OH)$_3$ and saponin, in *Immunopharmacology of Infectious Diseases: Vaccine Adjuvants and Modulators of Non-Specific Resistance,* Madje, J., Ed., Alan R. Liss, New York, 1987, 165.

13. **Morein, B., Sundquist, B., Höglund, S., Dalsgaard, K., and Osterhaus, A.,** ISCOM, a novel structure for antigenic presentation of membrane proteins from enveloped viruses, *Nature (London),* 308, 457, 1984.
14. **Lövgren, K. and Morein, B.,** The requirement for lipids for the formation of immuno-stimulating complexes (ISCOMS), *Biotech. Appl. Biochem.,* 10, 161, 1988.
15. **Trudel, M., Nadon, F., Sequin, C., Boulay, G., and Lussier, G.,** Vaccination of rabbits with a bovine herpes virus type 1 subunit vaccine: adjuvant effect of ISCOMS, *Vaccine,* 5, 239, 1987.
16. **Howard, C. R., Sundquist, B., Allen, J., Brown, S. E., Chen, S. H., and Morein, B. J.,** Preparation and properties of immune-stimulating complexes containing hepatitis B virus surface antigen, *J. Gen. Virol.,* 68, 2281, 1987.
17. **Wahren, B., Nordlund, S., Akesson, A., Sundquist, V. A., and Morein, B.,** Monocyte and ISCOM enhancement of cell-mediated response to cytomegalovirus, *Med. Microbiol. Immunol.,* 176, 13, 1987.
18. **Morgan, A. J., Finerty, S., Longren, K., Scullion, F. T., and Morein, B.,** Prevention of Epstein-Barr (EB) virus-induced lymphoma in cotton-top tamarins by vaccination with the EB virus envelope glycoprotein gp 340 incorporated into immunostimulating complexes, *J. Gen. Virol.,* 69, 2093, 1988.
19. **De Vries, P., Uytdehaag, F. G., and Osterhaus, A. D.,** Canine distemper virus (CDV) immune stimulating complexes (ISCOMS) but not measles virus iscoms, protect dogs against CDV infection, *J. Gen. Virol.,* 69, 2071, 1988.
20. **Sundquist, B., Lövgren, K., and Morein, B.,** Influenza virus ISCOMS: antibody response in animals, *Vaccine,* 6, 49, 1988b.
21. **Morein, B.,** The ISCOM antigen-presenting system, *Nature (London),* 332, 287, 1988.
22. **Letvin, N. L., Daniel, M. D., King, N. W., Kannagi, M., Chalifoux, L. V., Sehgal, P. K., Desrosiers, R. C., Arthur, L. O., and Allison, A. C.,** AIDS-like disease in macaque monkey induced by simian immunodeficienty virus: a vaccine trial, in *Vaccines 87,* Chanock, R. M., Lerner, R. A., Brown, F., and Ginsberg, H., Eds., Cold Spring Harbor Laboratory, Cold Spring Harbor, NY, 1987, 209.
23. **Jensen, K. E.,** Synthetic adjuvants: avridine and other interferon inducers, in *Advances in Carriers and Adjuvants for Veterinary Biologics,* Nervig, R. M., Gough, P. M., Kaeberle, M. L., and Whetstone, C. A., Eds., Iowa State University Press, Ames, 1986, 79.
24. **Anderson, A. O. and Reynolds, J. A.,** Adjuvant effects of lipid amine CP20,961, *J. Reticoloendothel. Soc.,* 526, 667, 1979.
25. **Hilgers, L. A. T., Snippe, H., van Vliet, K. E., Jansze, M., and Willers, J. M. N.,** Suppression of the cellular adjuvanticity of lipophilic amines by a polyanion, *Int. Arch. Allergy Appl. Immunol.,* 80, 320, 1986.
26. **Snippe, H., De Reuver, M. J., Kamperdijk, E. W. A., van den Berg, M., and Willers, J. M. N.,** Adjuvanticity of dimethyl dioctadecylammonium bromide in guinea pigs. I. Skin test reactions, *Int. Arch. Allergy Appl. Immunol.,* 68, 201, 1982.
27. **Gordon, W. C., Praeger, M. D., and Carroll, M. C.,** The enhancement of humoral and cellular responses by dimethyl dioctadecylammonium bromide, *Cell Immunol.,* 49, 329, 1980.
28. **Snippe, H., Belder, M., and Willers, J. M. N.,** Dimethyl dioctadecyl ammonium bromide as an adjuvant for delayed hypersensitivity in mice, *Immunology,* 33, 931, 1977.
29. **Smith, R. H. and Viola, B.,** Cyclophosphamide and dimethyl dioctadecyl ammonium bromide immunopotentiate the delayed-type hypersensitivity response to inactivated enveloped viruses, *Immunology,* 58, 245, 1986.
30. **Babcock, G. F. and McCarthy, R. E.,** Suppression of cell mediated immune responses by dextran sulphate, *Immunology,* 33, 925, 1977.
31. **Kraaijeveld, C. A., Kamphuis, W., Benaissa-Trouw, B. J., van Harlem, H., Harmsen, M., and Snippe, H.,** Potentiation of the immune response by adjuvants: a limited role for adjuvant induced interferon, *Int. Arch. Allergy Appl. Immunol.,* 81, 148, 1986.
32. **Souvannavang, V., Adam, A., and Lederer, E.,** Kinetics of the humoral and cellular response of guinea pigs after injection of the synthetic adjuvant N-acetyl muramyl-L-alanyl-D-isoglutamine: comparison with Freund's complete adjuvant, *Infect. Immun.,* 19, 966, 1978.
33. **White, R. G.,** Concepts of the mechanism of action of adjuvants, *Front. Biol.,* 24, 112, 1972.
34. **Kenney, J. S., Hughes, B. W., Masada, M. P., and Allison, A. C.,** Influence of adjuvants on the quantity, affinity, isotype and epitope specificity of murine antibodies, *J. Immunol. Methods,* 121, 157, 1989.
35. **Natsuume-Sakai, S., Montonishi, K., and Migita, S.,** Quantitative estimations of five classes of immunoglobulin in inbred mouse strains, *Immunology,* 32, 861, 1977.
36. **Murray, R., Cohen, P., and Hardegree, M. C.,** Mineral oil adjuvants: biological and chemical studies, *Ann. Allergy,* 30, 146, 1972.
37. **Potter, M. and Boyce, C. R.,** Induction of plasma cell neoplasms in BALB/c strain mice with mineral oil and mineral oil adjuvants, *Nature (London),* 193, 1986, 1962.
38. **Ellouz, F., Adam, A., Ciorbaru, R., and Lederer, E.,** Minimal structural requirements for adjuvant activity of bacterial peptidoglycans, *Biochem. Biophys. Res. Commun.,* 59, 1317, 1974.

39. **Audibert, F., Chedid, L., Lefrancier, P., and Choay, J.,** Distinctive adjuvanticity of synthetic analogs of mycobacterial water-soluble components, *Cell. Immunol.,* 21, 243, 1976.

40. **Jolivet, M., Audibert, F., Beachey, E. H., Tartar, A., Gran-Masse, H., and Chedid, L.,** Epitope specific immunity elicited by a synthetic streptococcal antigen without carrier or adjuvant, *Biochem. Biophys. Res. Commun.,* 117, 359, 1983.

41. **Jolivet, M., Sache, E., and Audibert, F.,** Biological derivatives of lipophilic MDP-derivatives incorporated in liposomes, *Immunol. Commun.,* 10, 511, 1981.

42. **Heymer, B., Finger, H., and Wirsing, C. H.,** Immunoadjuvant effects of the synthetic muramyl dipeptide (MDP) *N*-acetylmuramyl-L-alanyl-D-isoglutamine, *Z. Immunitaetsforsch. Klin. Immunol.,* 155, 87, 1978.

43. **Ohkuni, H., Norose, Y., Ohta, M., Hayama, M., Kimura, Y., Tsujimoto, M., Kotani, S., Shiba, T., Kusomoto, S., Yokoyama, K., and Kawata, S.,** Adjuvant activities in production of reagenic antibody by bacterial cell wall peptidoglycan or synthetic *N*-acetylmuramyl dipeptides in mice, *Infect. Immun.,* 24, 313, 1979.

44. **Byars, N. E., Allison, A. C., Harmon, M. W., and Kendal, A. P.,** Enhancement of antibody responses to influenza B virus hemagglutinin by use of a new adjuvant formulation, *Vaccine,* in press.

45. **Waters, R. V., Terrell, T. G., and Jones, G. H.,** Uveitis induction in the rabbit by muramyl dipeptide, *Infect. Immun.,* 51, 816, 1986.

46. **Byars, N. E.,** Two adjuvant active muramyl dipeptide analogues induce differential production of lymphocyte activating factor and a factor causing distress in guinea pigs, *Infect. Immunol.,* 44, 344, 1986.

47. **Chedid, L., Parant, M. A., Audibert, F. M., Riveau, G. J., Parant, F. J., Lederer, E., Choay, J. P., and Lefrancier, P. L.,** Biological activity of a new synthetic muramyl dipeptide adjuvant devoid of pyrogenicity, *Infect. Immun.,* 35, 417, 1982.

48. **Fraser-Smith, E. B., Waters, R. V., and Matthews, T. R.,** Correlation between *in vivo* anti-*Pseudomonas* and anti-*Candida* activitites and clearance of carbon by the reticuloendothelial system for various muramyl dipeptide analogs, using normal and immunosuppressed mice, *Infect. Immunol.,* 35, 105, 1982.

49. **Dukor, P. and Schumann, G.,** Modulation of non-specific resistance by MTP-PE, in *Immunopharmacology of Infectious Diseases: Vaccine Adjuvants and Modulators of Non-Specific Resistance,* Madje, J. A., Ed., Alan R. Liss, New York, 1987, 255.

50. **Byars, N. E. and Allison, A. C.,** Adjuvant formulation for use in vaccines to elicit both cell-mediated and humoral immunity, *Vaccine,* 5, 223, 1987.

51. **Marx, P. A., Pedersen, N. C., Lerche, N. W., Osborn, K. G., Lowenstine, L. J., Lackner, A. A., Maul, D. H., Kwang, H.-S., Kluge, J. D., Zaiss, C. P., Sharpe, V., Spinner, A. P., Allison, A. C., and Gardner, M. B.,** Prevention of simian acquired immune deficiency syndrome with a formalin-inactivated type D retrovirus vaccine, *J. Virol.,* 60, 431, 1986.

52. **Robey, W. G., Arthur, L. O., Matthews, T. J., Langlois, A., Copeland, T. D., Lerche, N. W., Oroszolan, S., Bolognesi, D., Gilden, R. V., and Fischinger, P. J.,** Prospect for prevention of human immunodeficiency virus infection: purified 120-kDa envelope glycoprotein induces neutralizing antibody, *PNAS,* 83, 7023, 1986.

53. **Rietschel, E. Th., Brade, L., Schade, U., Zahringer, U., and Brade, H.,** Bacterial endotoxins: relation of chemical structure to biological activity, in *Immunopharmacology of Infectious Diseases: Vaccine Adjuvants and Modulators of Non-specific Resistance,* Madje, J. A., Ed., Alan R. Liss, New York, 1987, 75.

54. **Imoto, M., Kusomoto, S., Shiba, T., Rietschel, E. T., Galanos, C., and Lüderitz, O.,** Chemical structure of *Escherichia coli* lipid A., *Tetrahedron Lett.,* 26, 907, 1985.

55. **Takayama, K., Qureshi, N., Ribi, E., and Cantrell, J. L.,** Separation and characterization of toxic and non-toxic forms of lipid A, *Rev. Infect. Dis.,* 6, 439, 1984,

56. **Imoto, M., Yoshimura, H., Sakaguchi, N., Kusomoto, S., and Shiba, T.,** Total synthesis of *Escherichia coli* lipid A, *Tetrahedron Lett.,* 26, 1545, 1985.

57. **Kotani, S., Takada, H., Tsujimoto, M., Ogawa, T., Takahashi, I., Ikeda, T., Otsuka, K., Shimauchi, H., Kasai, N., Mashimo, J., Nagao, S., Tanaka, A., Tanaka, S., Harada, K., Nagaki, K., Kitamura, H., Shiba, T., Kusomoto, S., Imoto, M., and Yoshimura, H.,** Synthetic lipid A with endotoxic and related biological activities comparable to those of a natural lipid A from an *Escherichia coli* re-mutant, *Infect. Immun.,* 49, 225, 1985.

58. **Johnson, A. G., Tomai, M., Solem, L., Beck, L., and Ribi, E.,** Characterization of non-toxic monophosphoryl lipid A, *Rev. Infect. Dis.,* 9, S512, 1987.

59. **Ribi, E., Ulrich, J. T., and Masihi, K. N.,** Immunopotentiating activities of monophosphoryl lipid A, in *Immunopharmacology of Infectious Diseases: Vaccine Adjuvants and Modulators of Non-specific Resistance,* Madje, J. A., Ed., Alan R. Liss, New York, 1987, 101.

60. **Bloch, H.,** Virulence of mycobacteria, *Adv. Tuberc. Res.,* 6, 49, 1955.

61. **Retzinger, G. S., Meredith, S. C., Takayama, K., Hunter, R. L., and Kezdy, F. J.,** The role of surface in the biological activities of trehalose-6,6′-dimycolate. I. Surface properties and development of a model system, *J. Biol. Chem.,* 256, 8208, 1981.

62. **Numata, F., Nishimura, K., Ishida, H., Ukei, S., Tone, Y., Ishihara, C., Saiki, I., Seikikawa, I., and Azuma, I.,** Lethal and adjuvant properties of cord factor (trehalose-6,6′-dimycolate) and synthetic analogs in mice, *Chem. Pharm. Bull.,* 33, 4544, 1985.

63. **Ruiz, A. M., Estera, M., Riarte, A., Subias, E., and Segura, E. L.,** Immunoprotection of mice against *Trypanosoma cruzi* with a lyophilized flagellar fraction of the parasite plus adjuvant, *Immunol. Lett.,* 12, 1, 1986.

64. **McColm, A. A., Bomford, R., and Dalton, L.,** A comparison of saponin with other adjuvants for the potentiation of protective immunity by a killed *Plasmodium yoelii* vaccine in the mouse, *Parasite Immunol.,* 4, 337, 1982.

65. **Bomford, R.,** The comparative selectivity of adjuvants for humoral and cell-mediated immunity. II. Effect on delayed-type hypersensitivity in the mouse and guinea pig and celll-mediated immunity to tumour antigens in the mouse of Freund's incomplete and complete adjuvants, alhydrogel, *Corynebacterium parvum, Bordetella pertussis,* muramyl dipeptide and saponin, *Clin. Exp. Immunol.,* 39, 435, 1980.

66. **Suko, M., Ogita, T., Okudaira, H., and Horiuchi, Y.,** Preferential enhancement of IgE antibody formation by *Bordetella pertussis, Int. Arch. Allergy Appl. Immunol.,* 54, 329, 1977.

67. **Hirashima, M., Yodoi, J., and Ishizaka, K.,** Formation of IgE-binding factors by rat T-lymphocytes. II. Mechanisms of selective formation of IgE-potentiating factors by treatment with *Bordetella pertussis* vaccine, *J. Immunol.,* 127, 1804, 1981.

68. **Degrand, F. and Raynaud, M.,** Quantitative effect of *Mycobacteria* and of anaerobic *Corynebacteria* upon the immunogenic activity of azobenzenearsonate-*N*-acetyl-L-tyrosine in guinea pigs, *Eur. J. Immunol.,* 3, 660, 1973.

69. **Usinger, W. R. and Mishell, R. I.,** Chemical characterization and immunoenhancing activities of purified *Cytophaga* exopolymer, in *Immunopharmacology of Infectious Diseases: Vaccine Adjuvants and Modulators of Non-Specific Resistance,* Madje, J. A., Ed., Alan R. Liss, New York, 1987, 125.

70. **Woodhour, A. F., Metzgar, D. P., Stim, T. B., Tytell, A. A., and Hillemen, M. R.,** New metabolizable immunological adjuvant for human use. I. Development and animal immune response, *Proc. Soc. Exp. Biol. Med.,* 116, 510, 1964.

71. **Reynolds, J. A., Harrington, D. G., Crabbs, C. L., Peters, C. J., and DiLuzio, N. R.,** Adjuvant activity of a novel metabolizable lipid emulsion with inactivated viral vaccines, *Infect. Immun.,* 28, 937, 1980.

72. **Davenport, F. M.,** Seventeen years experience with mineral oil adjuvant influenza virus vaccines, *Ann. Allergy,* 26, 288, 1968.

73. **McKercher, P. D.,** Oil adjuvants: their use in veterinary biologics, in *Advances in Carriers and Adjuvants for Veterinary Biologics,* Nervig, R. M., Gough, P. M., Kaeberle, M. L., and Whetstone, C. A., Eds., Iowa State University Press, Ames, 1986, 115.

74. **Weibel, R. E., McLean, A., Woodhour, A. F., Friedman, A., and Hilleman, M. R.,** Ten-year followup study for safety of Adjuvant 65 influenza vaccine in man, *Proc. Soc. Exp. Biol. Med.,* 143, 1053, 1973.

75. **Harrington, D. G., Lupton, H. W., Crabbs, C. L., Peters, C. J., Reynolds, J. A., and Slone, T. W.,** Evaluation of a formalin-inactivated Rift Valley fever vaccine in sheep, *Am. J. Vet. Res.,* 41, 1559, 1980.

76. **Brugh, M., Stone, H. D., and Lupton, H. W.,** Comparison of inactivated Newcastle disease viral vaccines containing different emulsion adjuvants, *Am. J. Vet. Res.,* 44, 72, 1983.

77. **Rouse, B. T., Turtinen, L., and Correa-Freire, M.,** Use of liposomes in immunology, in *Advances in Carriers and Adjuvants for Veterinary Biologics,* Nervig, R. M., Gough, P. M., Kaeberle, M. L., and Whetstone, C. A., Eds., Iowa State University Press, Ames, 1986, 121.

78. **Eppstein, D. A., Byars, N. E., and Allison, A. C.,** New adjuvants for vaccines containing purified protein antigens, in *Advanced Drug Delivery Reviews,* Poste, G., Ed., in press.

79. **Staruch, M. J. and Wood, D. D.,** The adjuvanticity of interleukin 1 *in vivo, J. Immunol.,* 130, 2191, 1983.

80. **Kawamura, H., Rosenberg, S. A., and Berzofsky, J. A.,** Immunization with antigen and interleukin 2 *in vivo* overcomes Ir gene low responsiveness, *J. Exp. Med.,* 162, 381, 1985.

81. **Perrin, P., Joffret, M. L., Leclerc, C., Oth, D., Sureau, P., and Thibodeau, L.,** Interleukin 2 increases protection against experimental rabies, *Immunobiology,* 177, 199, 1988.

82. **Anderson, G., Urban, O., Fedorka-Cray, P., Newell, A., Nunberg, J., and Doyle, M.,** Interleukin 2 and protective immunity in *Haemophilus pleuropneumoniae:* preliminary studies, in *Vaccines 87,* Chanock, R. M., Lerner, R. A., Brown, F., and Ginsberg, H., Eds., Cold Spring Harbor Laboratory, Cold Spring Harbor, NY, 1987, 22.

83. **Weinberg, A. and Merrigan, T. C.,** Recombinant interleukin 2 as an adjuvant for vaccine-induced protection. Immunization of guinea pigs with herpes simplex virus subunit vaccines, *J. Immunol.,* 140, 294, 1988.

84. **Playfair, J. H. L. and De Souza, J. B.,** Recombinant gamma interferon is a potent adjuvant for a malaria vaccine in mice, *Clin. Exp. Immunol.,* 67, 5, 1987.

85. **Yilma, T., Anderson, K., Brechling, K., and Moss, B.,** Expression of an adjuvant gene (interferon-γ) in infectious vaccinia virus recombinants, in *Vaccine 87,* Chanock, R. M., Lerner, R. A., Brown, F., and Ginsberg, H., Eds., Cold Spring Harbor Laboratory, Cold Spring Harbor, NY, 1987, 393.
86. **Byars, N.,** unpublished.
87. **Terrell, T.,** Syntex, unpublished.

Chapter 3

LABELING OF LYMPHOCYTES FOR *IN VIVO* TRACING

Sigbjørn Fossum and Bent Rolstad

TABLE OF CONTENTS

I. INTRODUCTION

A. WHY STUDY LYMPHOCYTE MIGRATION?

The discovery that lymphocytes recirculate[1] not only finally explained the enigmatic fate of the thoracic duct cells continuously poured into the blood stream, but also opened a new field of research. With the subsequent realization that lymphocytes comprise different subsets, the stage was set for exploring paths and steps along the migratory routes. A wealth of information has emerged and continues to materialize from these studies, with fundamental contributions to our comprehension of the immune system.

Knowledge of the many highly ordered migratory patterns of the various lymphocyte subsets is a prerequisite for understanding immune responsiveness *in vivo*. Unlike cells of other organ systems, the cells of the immune system are disperesed throughout the body, enjoying only transient contact with each other. Despite this dispersal, the immune system functions as a unit, in spite of the incessant redistribution of its cells.

However, there is also a second, wider perspective to migration studies. Partly because they are so readily available, partly because they have become so well characterized, lymphocytes have become the favorite tools for investigating general principles of cellular biology. During the last decade the emphasis of research on lymphocyte migration gradually shifted from descriptions of migratory patterns to work on molecular mechanisms. In this context, the ability of lymphocytes to discriminate with high specificity between different endothelia may be regarded as an example of more universal mechanisms for cell-cell recognition. Lymphatic tissues preserve a well-defined architecture despite rapid exchange of their cell populations. In most other tissues the forces that direct the component cells to their proper localization act only once during early fetal life. In lymphatic tissues they continue to operate throughout the life of the individual. Understanding the factors controlling these migratory streams is, therefore, not only an important object in itself; it may also provide clues to unraveling basic morphogenetic mechanisms.

B. CONSIDERATIONS ABOUT SPECIES

The chief contributions to our knowledge about lymphocyte migration stem from two separate animal sources: large species such as the sheep, cow, pig, and human, and rodents such as the rat and mouse. A major advantage with the larger animals is that it is possible to cannulate the afferent and efferent lymphatics from single lymph nodes. Thus, the influx to the lymph nodes of unlabeled, unmodified lymph-borne cells can be directly compared with the efflux of the same cells from the nodes. Such studies made in sheep have yielded invaluable information.[2,3] (Pigs represent an interesting, rather special case. They belong to the peculiar group of species that possess so-called inverted lymph nodes that release their effluent lymphocytes to the bloodstream rather than to the efferent lymphatics.)[4] In larger animals it is also possible to employ extracorporeal perfusion of organs such as the thymus and the gut.[5] By this technique most or all lymphocytes from these organs can be labeled in their normal environment in a more controlled manner than is yet possible in the rodents.

A disadvantage with the larger species is that, with the exception of the human, there has been limited access to monoclonal antibodies against differentiation markers of lymphocyte subsets. Another drawback is the lack of inbred strains. This prohibits the transfer of lymphocytes between individuals and thereby restricts the possibilities for experimental designs (see discussion of ^{51}Cr, Section III.B).

In contrast, numerous inbred rat and mouse strains have been raised, and a large number of monoclonal antibodies toward markers for lymphocyte subsets have been developed. Owing to the greater ease with which the thoracic duct of the rat, as compared to the mouse, is cannulated, the rat has become the species most commonly used for lymphocyte migration studies.

II. PRINCIPLES OF LABELING

A. TYPES OF MARKERS

Much information about the migratory behavior of lymphocytes can be obtained with unmodified cells. In larger animals the cells may simply be collected from the lymphatics leading to and from single lymph nodes. The nodes may be unstimulated, challenged with antigen, or influenced by drugs, hormones, or other biologic response modifiers. Even in rodents several aspects of lymphocyte migration can be studied without making use of cell labeling. Simply depleting the animal of recirculating cells by prolonged drainage of the thoracic duct has given valuable information about the tempo of lymphocyte recirculation.[6] Actually, these early results proved in the long run more reliable than the data obtained later with more sophisticated techniques utilizing labeled lymphocytes (see Section III.B.2). The segregation of lymphocyte subsets into separate compartments within the lymphoid tissues was first suggested on the basis of observations from simple histological examination of light microscopic sections from lymphoid tissues. The tissues were from T-cell-deficient individuals, i.e., rats and mice thymectomized at birth[7,8] and congenitally athymic children and mice.[9,10] The findings were corroborated by immunocytochemical studies of lymphoid tissues from euthymic animals.[11,12] However, the decisive proof that lymphocytes migrate into defined areas of lymph nodes and spleen[13] and that recirculating T and B cells localize in different areas within these lymphoid tissues[14] was made with TDL radiolabeled *in vitro* and injected intravenously and their migration patterns followed by autoradiography. Such studies also revealed that the T and the B cells enter the lymphoid tissues by a common gate.[15] For these as well as for many other purposes, the method of tracing a sample of labeled cells proved particularly suitable.

The cells to be traced are marked by applying an extrinsic label or by exploiting an intrinsic marker (Table 1). The extrinsic labels are incorporated into or adsorbed to cellular macromolecules (DNA, RNA, or proteins) either by incubating the cell sample with the label *in vitro* or by injecting the label *in vivo*. For reasons discussed below, it would be preferable to label cells in their natural environment, i.e., *in vivo*. A major obstacle is to restrict the label to the selected sample. In some cases this is not a problem, e.g., labeled thymidine or analogues may be used to mark the cells *in vivo* in order to follow the fate of all lymphocytes that are in the S-phase at the time of labeling. Also, samples of nondividing lymphocytes may be selectively labeled *in vivo*, e.g., thymocytes may be marked simply by injecting solutions containing the label directly into the thymus. Alternatively, extracor-

TABLE 1
Markers Currently in Use for Tracing Lymphocytes

Extrinsic labels:	Marked molecules	Radioactive labels	Nonradioactive labels
Label incorporated into intracellular macromolecules	DNA	^{3}H-TdR ^{14}C-TdR ^{125}IUdr	BrUdR ^{3}H-UdR
	RNA	^{14}C-UdR	
	Proteins	^{3}H-leu ^{14}C-leu	
Label adsorbed or coupled to intracellular macromolecules		^{51}Cr ^{111}In $^{99\text{-}m}$Tc	FITC TRITC

Intrinsic markers:
  Chromosomal
  Alloantigens
  Nuclear morphology (the chicken/quail system)

poreal circulation may be used to label all lymphoyctes or lymphocytes in S-phase in selected organs. This technique requires access to an extracorporeal perfusion system, considerable experience, and is, as stated above, currently limited to use on larger experimental animals.[5] In practice, for most purposes the sample of lymphocytes is, therefore, purified and labeled *in vitro* before reinjection into a host. Unfortunately, the migratory faculty of lymphocytes has turned out to be a delicate function, easily disturbed by isolating the cells from their natural microenvironment and holding them *in vitro*.

B. *IN VITRO* LABELING OF LYMPHOCYTES
1. The Source of Cells
The choice of cell source — lymph, blood, lymphoid, or nonlymphoid tissues — depends on the questions posed in the experiment, but other factors should also be considered carefully.

a. Central Lymph
For many reasons central lymph is preferred as the cell source. Usually, the labeled cells are returned to a host by intravenous injection. Reinjecting lymphocytes collected from the thoracic duct, therefore, comes close to the physiological situation. With some practice, the rat thoracic duct can be cannulated without much effort.[16] An overnight collection of lymphocytes from the thoracic duct (thoracic duct cells or TDL) of an adult rat usually contains 3 to 6 × 10^8 recirculating lymphocytes, of which 90 to 95% are small, nondividing cells. In the collection of the first night the T to B ratio varies between 4:1 and 3:1, but falls on prolonged drainage due to the more rapid depletion of recirculating T cells. Although the great majority of the TDL are accredited recirculators, 5 to 10% of them are lymphoblasts. Many of the blasts produce IgA and are destined to end up in the gut lamina propria.

b. Blood
In the human, the lymphocytes to be labeled are most conveniently isolated from the blood. It should be noted that even though the blood lymphocytes are continuously replaced by thoracic duct cells, these cell populations are not identical. Compared with the latter population, those in the blood will be enriched for subsets that are slow to leave the blood and also will contain nonrecirculating lymphocyte-like cells, some of which are natural killer (NK) cells.[17] Before being labeled and reinjected, lymphocytes must be separated from the

other blood cells. The commonly used methods are based on differences in density, e.g., by centrifugation on Isopaque-Ficoll (IF), as described by Bøyum.[18] This highly efficient method consists of only a single step and is regarded as relatively innocuous, but nevertheless introduces a source of trauma. Furthermore, B cells have a slightly higher average density than T cells and tend to be underrepresented.[19] The standard preparations with a density of 1.077, adapted for separation of human mononuclear cells, are for this reason unsuitable in the rat. In this species a large proportion of the TDL, in particular, B cells, will pass through the density gradient. A somewhat higher IF density (approximately 1.090) is therefore required for a good cell yield in this species.[19] In other species the densities should be similarly adjusted. Also, the osmolarity of the suspending medium is of importance for the results and should be adapted to the leukocytes from the species used. This is of particular importance when using several different density gradients to separate certain subpopulations of lymphoid cells, e.g., NK cells.[20]

c. Lymphoid Tissues

When lymphoid tissues are the source of lymphocytes, the following points should be considered.

1. The cell suspension contains many activated lymphoyctes. A high proportion of such blasts complicates the analysis, because: (1) they take up much more label per cell when, e.g.,[51] Cr is used[14], and (2) their migratory patterns differ from those of resting lymphocytes.[21,22] The altered migratory behavior results at least in part from modulation of the surface molecules mediating adherence to endothelia.[23]
2. The cell suspension is contaminated by many nonlymphocytic cells.
3. Many of the lymphocytes within the lymphoid tissues have recently entered these tissues from the blood. There is evidence that their migratory capacity is transiently reduced by the process of crossing the vascular endothelium.[24]
4. Some lymphocytes in lymphoid organs have left the migratory pool, e.g., the splenic marginal zone B cells do not recirculate,[25] or are by nature nonrecirculating cells, e.g., lymphoblasts.

2. The Effect of Keeping Lymphocytes *In Vitro*

In a series of papers Ford and co-workers[24,26,27] showed that the physical environment experienced by lymphocytes *in vitro* profoundly affects their ability to enter lymph nodes, but not the spleen. The authors compared rat TDL obtained by "standard" collection conditions with TDL obtained by so-called "optimal" conditions. In brief, the former consisted of the conventional method of collecting TDL overnight (16 ± 2 h) in 100-ml flasks placed in an ice bucket. Five milliliters of phosphate-buffered saline solution with Ca^{++} and Mg^{++} ions added (Dulbecco's solution A + B) and containing 20 U of heparin per milliliter had been added to the flasks before beginning the collection. The latter method employed an intermediate host into which lymphocytes collected from a primary donor under standard conditions and labeled *in vitro* were injected and then recollected after 24 h. The idea behind the intermediate host was to give the lymphocytes handled *in vitro* an opportunity to recover from the trauma. In contrast to the standard collection conditions, the duration of collection of thoracic duct cells from the intermediate host was only 1 to 2 h, was performed at room temperature, and required no further handling before reinjection into the final recipient.

The results of these comparisons revealed that the traditional method of collecting and labeling lymphocytes led to a dramatic reduction in the early localization of both B and T cells into lymph nodes and Peyer's patches (Table 2). For example, 60 min after reinjection "standard" collected B cells showed a 20-fold (!) lower localization in cervical lymph nodes

TABLE 2
The Effect of Conditions of Collection on the Organ Distribution of *In Vitro* Labeled T and B Cells

Conditions of collection	Cervical LN		Mesenteric LN		Peyer's patches	
	T[a]	B[b]	T[a]	B[b]	T[a]	B[b]
"Standard"[c] 16 h O°C nonpassaged	7.0[d]	1.1	7.9	3.9	6.8	3.8
"Optimal"[e] 1—2 h, RT passaged	57.9	22.0	52.8	13.3	36.9	27.2

[a]   T cells - unfractionated thoracic duct cells from euthymic rat; first night collection containing 70 to 90% T cells.

[b]   B cells - unfractionated thoracic duct cells from congenitally athymic, nude rats; first night collection containing 93% small B cells.

[c]   Standard collection refers to the commonly used collection overnight of the thoracic duct cells into flasks containing a few milliliters of phosphate buffered saline with heparin, the vials immersed in crushed ice.

[d]   Percentage of injected dose (^{51}Cr-labeled cells) per gram of tissue at 60 min after intravenous injection.

[e]   Optimal collection - cells that have been labeled *in vitro* after standard collection of a primary donor, injected by i.v., and recollected from an intermediate host for 1-2 h at room temperature ("passaged" cells).[24,26]

compared with "optimally" labeled B cells. The explanation for these discrepancies was not that passaging of the TDL selected for a subset of fast recirculators. When the TDL was isolated from the blood rather than from the thoracic duct of the intermediate host, the same effects were observed. To identify the environmental factors causing these adverse effects, passaged cells were subjected to different conditions before reinfusion into the final host.[28] In summary, the two most important factors turned out to be the duration of keeping the cells *in vitro* and the exposure of the cells to heparin. The reversibility of the harmful influence of "standard" collection was corroborated by the comparison with "optimally" collected cells. By 24 h after reinjection into the final host the differences in migratory rates between the two groups had almost vanished.

These findings are noteworthy mainly because of their practical importance for the design of lymphocyte traffic experiments. However, as pointed out by Ford et al.,[28] the findings may also have a deeper significance than merely revealing an artifact produced by handling the lymphocytes *in vitro*. Conditions that reduce the expression of microvilli have been shown to impair entry into lymph nodes.[29] *In vitro* treatment reduces the number of microvilli exposed on the lymphocyte surface and induces formation of blebs. However, according to the results of Ford et al.[28] the correlation between formation of microvilli and entry into lymph nodes turned out to be poor when the optimally collected cells were subjected to various treatments. Whereas the expression of microvilli was most vulnerable to low temperature, the entry into lymph nodes was most affected by the duration of holding TDL *in vitro*.

In contrast, the adverse effects of *in vitro* handling more or less duplicated the effects of light trypsinization of lymphocytes, as reported by Woodruff and Gesner in 1968.[30] They observed that light trypsinization strongly reduced the entry of the treated lymphocytes into lymph nodes without impairing their entry into the spleen. This similarity suggests that the adverse effects of *in vitro* treatment stem from exposure to released proteolytic enzymes. Interestingly, the Manchester group also found that TDL that had recently entered the lymph nodes or the spleen of an intermediate recipient showed a selectively impaired lymph node entry.[24,26] Their conclusion was that lymphocytes undergo cyclical changes in their migratory functions according to their situation within the recirculation pathway. The behavioral changes are probably elicited by the lymphocytes modulating the expression of their adhesion molecules for high endothelial cells.

TABLE 3
**Characteristics of Some of the Isotopes Currently Used to Label
Lymphocytes *In Vitro***

Compound	Isotope	Labeled molecule	Principal emission	Toxicity	Reutilization
$Na_2{}^{51}CrO_4$	^{51}Cr	Protein	γ	+	+
Indium oxine	^{111}In	Protein	γ	+ +	±
^3H-leucine	^3H	Protein	β	+	+ + +
^{14}C-leucine	^{14}C	Protein	β	+	+ + +
^3H-uridine	^3H	RNA	β	+	+ +
^{14}C-uridine	^{14}C	RNA	β	+	+
^3H-thymidine	^3H	DNA	β	+ +	+ +
^{14}C-thymidine	^{14}C	DNA	β	+ +	+
Iododeoxy uridine	^{125}I	DNA	γ	+ + +	±

III. RADIOACTIVE LABELS

A. GENERAL CONSIDERATIONS

The ideal leukocyte label should fulfill the following conditions:

1. It should be nontoxic to the cells.
2. It should label all cells uniformly.
3. It should not interfere with any of the cell functions under study.
4. It should be firmly attached to the cells during their whole lifetime.
5. The small amount of label released due to metabolic turnover or label released from dying cells should be quickly excreted and not reutilized by other cells.

None of the presently available extrinsic markers satisfy all these requirements, and the following examples should clarify why this is so. First, all radionuclides that are incorporated into organic molecules that are intermediates of cell metabolism may be reutilized by other cells. Elution and reutilization poses a major problem in studies of lymphocyte migration into nonlymphoid tissue, where the density of the injected, labeled lymphocytes is small.[31] This is a problem of considerable practical importance with radioactive uridine and leucine, especially with the tritiated compounds. When the cell sample is heterogeneous, such as TDL, which comprise both T and B cells, the label may be treated differently by the subsets. This is the case with ^3H-uridine, which, for unknown reasons, labels T cells about ten times more intensely than B cells in the rat.[14,15] Second, toxicity is a common problem with all extrinsic markers, especially radiotoxicity, examples of which will be given later. Although some of the markers come closer to fulfilling the ideal requirements than others, the choice of marker depends heavily on the type of recording needed for the particular kind of study. Even though compounds like ^3H-uridine or ^3H-leucine are not the markers of choice for studying lymphocyte migration into nonlymphoid tissues by scintillation counting, they are well suited for studying lymphocyte migration at the individual cell level in autoradiographic tissue sections because of their low energy β-emission.[14,15] Even individual cell death within the tissue can be traced with ^3H-uridine labeled cells.[32] In the following we will try to review advantages and pitfalls when using some of the more commonly used markers, the essence of which are listed in Table 3.

B. RADIONUCLIDES THAT BIND TO OR ARE INCORPORATED INTO CELLULAR PROTEIN
1. γ-Emitters

Of the γ-emitters, ^{51}Cr and ^{111}In are the most frequently used. Because they label lymphocytes easily and the detection systems are simple, they are used in several different contexts in studies on the migration of lymphocytes and other leukocytes *in vivo*. The energies of γ-emission of both isotopes permit whole-organ scintillation counting, and no sectioning or processing of tissues is necessary.[33]

a. ¹¹¹In

¹¹¹In has the advantage over ^{51}Cr in that it emits γ-rays of an energy suitable for extracorporeal gamma-camera scanning.[34] Its short half-life (2.8 d) makes it superior to any other available markers for *in vivo* studies of human lymphocyte migration.[35] ¹¹¹In is taken up efficiently by cells only when chelated with a lipophilic compound, such as oxine (hydroxyquinolone). Furthermore, when lymphocytes are incubated with ¹¹¹In-oxine in protein-free medium, 80 to 90% of the radioactivity in the medium is rapidly taken up by the cells after only short-term incubation *in vitro*. The elution of ¹¹¹In from intact cells is very slow, both *in vivo* and *in vitro*.[34,36-38] The main disadvantage with ¹¹¹In is its radiotoxicity. Thus, the low-energy Auger electrons are very damaging to lymphocytes exposed to the isotope over some period of time. Lymphocytes labeled with ¹¹¹In-oxine at 0.37 or 0.74 MBq/ml showed unimpaired capacity to recirculate from blood to lymph over 24 h.[17,34,37] However, 7 d later, overt signs of destruction of the labeled cells were observed in the spleen.[37] Nevertheless, ¹¹¹In has remained one of the most popular γ-emitters for short-term lymphocyte migration studies *in vivo* and is available commercially as ¹¹¹In-oxine. (Amersham, Amersham, England, and MediPhysics Inc., Emeryville, CA.)

b. ⁵¹Cr

Of all the γ-emitters, ^{51}Cr (as $Na_2{}^{51}CrO_4$) has gained the widest popularity in animal studies due to its ease of handling, low toxicity, low cost, and relatively long half-life (28 d). The long half-life makes it ideally suited for longer-term migration studies in animals (at least up to 1 week), but probably less suited for human studies. Although the internal conversion effect can be exploited for autoradiography in cell smears,[39] autoradiographic detection of ^{51}Cr-labeled cells in tissue sections is not possible for unknown reasons.[33] ^{51}Cr binds nonspecifically to cellular proteins. Thus, large lymphocytes will label more heavily than small lymphocytes,[14] and erythrocytes will label at least as well as lymphocytes. This stresses the importance of clearing the cell suspension of unwanted cells, such as erythrocytes, before labeling. Erythrocytes are efficiently removed by centrifuging the cells on a single-step density gradient of IF, as described by Bøyum.[18]

Chromium is reduced from 6^+ to 3^+ when it is taken up by the cell. It has been claimed that Cr released from cells in the reduced form will not easily be reutilized by other cells.[40,41] This may well be true for cells in the circulation, but when labeled cells were destroyed by irradiation[37] or destroyed by nonadaptive cytotoxic mechanisms within the spleen,[42,43] much of the radioactivity associated with the injected cells was transferred to nonlymphocytic cells. In these circumstances the determination of lymphocyte migration into the spleen solely by whole-organ scintillation counting will give misleadingly high estimates of the number of living labeled cells present in the spleen and stresses the importance of supplementary autoradiographic studies.

^{51}Cr is probably the least toxic of all the γ-emitters. Its toxicity is due to radiation and not to chemical damage, at least within the dose ranges commonly used to label lymphocytes.[44] Furthermore, the radiation damage is dose dependent. When we tested an immunologic function, such as the graft-vs-host (GVH) reactivity of ^{51}Cr-labeled T cells, we found that a labeling dose of 3.7 MBq (100 μCi) of ^{51}Cr per milliliter of cells almost abolished their ability to induce a GVH-reaction, whereas a labeling dose of 0.37 MBq (10 μCi) per milliliter did not affect the GVH reactivity of the cells.[44] Since a dose of 0.37 to 0.74 MBq (10 to 20 μCi)/ml is sufficient to label the cells for detection by organ scintillation counting, we recommend that ^{51}Cr doses in excess of this should not be used for migration studies.

Since ^{51}Cr released from dying cells is mainly excreted in the urine and reutilized less by other cells than, e.g., ^{3}H-uridine, it may also be exploited to study cell death *in vivo*. In fact, the use of ^{51}Cr has revealed some essential features behind the phenomenon of allogeneic lymphocyte cytotoxicity (ALC). ALC refers to the rapid rejection of allogeneic lymphocytes that takes place in nonimmune recipients in certain strain combinations of rats and mice.[43,45-47] In ALC there is histological evidence that most of the damage to the donor cells occurs within the first 24 h after cell injection (see later). It was, therefore, possible to detect increased amounts of ^{51}Cr in the body fluids as a consequence of release from damaged cells. Indeed, when ^{51}Cr-labeled lymphocytes were injected into an allogeneic recipient with an indwelling thoracic duct cannula, almost none of the labeled cells recirculated through to the thoracic duct, but instead a high level of radioactivity was found in cell-free lymph. When the injected cells and the recipient were syngeneic, the reverse was the case, i.e., much of the activity was associated with the TDL, and little free activity was found in the lymph.[42] In the allogeneic transfers, most of the excess of ^{51}Cr in cell-free lymph was present in the first 2 h after cell injection, showing that the elimination of the allogeneic lymphocytes took place as soon as they had entered the tissues.[48] The analogy between this experimental system and *in vitro* cytotoxic assays should be obvious.

In studies where the radioactivity has been monitored in several different organs after injection of ^{51}Cr-labeled cells, certain patterns of distribution of the injected cells have emerged that are either associated with cell death or cell survival.[45,49,50] Thus, a high lymph node uptake of the isotope is associated with cell survival, whereas high liver, spleen, kidney, and blood plasma contents of the isotope are signs of cell death. This statement is corroborated by autoradiographic demonstration of destroyed cells within the lymphoid tissue, in situations where the lymph node uptake of the isotope is low[32] (see also Figure 1). When the radioactivity in these tissues was monitored in several different allogeneic strain combinations of rats, a pattern emerged, which showed that ALC takes place only in selected strain combinations of rats differing with respect to the major histocompatibility complex (MHC).[50]

Finally, the difference in the decay of γ-emission between ^{51}Cr and ^{111}In can be exploited in double-labeling studies.[51]

c. Other γ-Emitters

The potential usefulness of radioactive markers to trace lymphocyte migration patterns in humans under physiological or pathological conditions has stimulated the search to find alternatives to ^{51}Cr. Two other isotopes that also label proteins, ^{99m}Tc[31,34] and ^{75}Se-methionine,[52] initially showed great promise. However, interest in these isotopes has faded, either because of high radiotoxicity (^{99m}Tc)[34,53] or due to extensive reutilization (^{75}Se).[31]

d. Labeling Protocols

In the following we shall go through the labeling protocol for ^{51}Cr and ^{111}In. Unless otherwise stated, the procedure is detailed with reference to the rat.

i. Chromium

Isotope in the form of $Na_2^{51}CrO_4$ is purchased as a sterile and isotonic salt solution at a specific activity of 10 to 20 GBq/mg Cr. from Amersham, England or New England Nuclear (NEN), Boston, MA. The chromium should never be allowed more than 1 month of decay (i.e., one half-life) before being used.

The cells should have a high viability, preferably TDL, blood lymphocytes, or lymph node cells and be freed of erythrocytes by centrifugation on IF. Rat cells need slightly higher density IF (specific gravity 1.090) than mouse or human cells (1.077). The procedure for collection of TDL has been detailed elsewhere.[16]

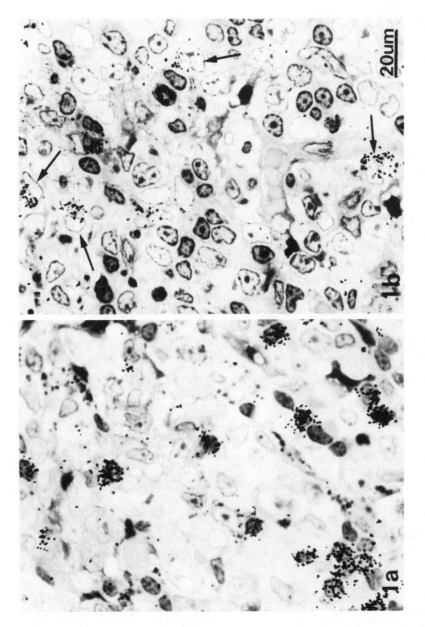

FIGURE 1. Light microscopic (LM) autoradiograms of paracortical areas from nude rat cervical lymph nodes 6 h after intravenous injection of ³H-uridine labeled thoracic duct lymphocytes. (a) Syngeneic (PVG) donor cells: heavily labeled cells scattered throughout the paracortex, all the cells look intact. (b) Allogeneic (AO) donor cells: in this field not a single intact donor cell is seen; however, several large pale host cells (arrows) show radioactive labeling over a cytoplasm containing scattered nuclear fragments, bearing evidence that the majority of allogeneic cells that had entered this area were rapidly phagocytosed and broken down. (From Fossum, S. and Rolstad, B., *Eur. J. Immunol.*, 16, 440, 1986. With permission.)

Label the cells in RPMI 1640 plus 10% fetal calf serum (FCS) at 10 to 50 million cells per milliliter. Add 0.37 MBq (10 μCi) of $Na_2{}^{51}CrO_4$ per milliliter cells and incubate at 37°C with occasional resuspension of the cells.

Wash the cells twice in a large volume (50 ml) of phosphate-buffered saline (PBS) plus 2% FCS. Centrifuge for 10 min at 300 *g*.

Resuspend the cells in 50 ml of PBS-FCS and leave the cells at room temperature for at least 1/2 h. This maneuver will allow the cells to elute most of their loosely bound chromium and will thus lower the "background" level of extracellular chromium in the recipient.

Centrifuge the cells and resuspend them to the desired concentration in PBS or RPMI for injection into the recipient. Fifty million cells per recipient injected in 1 or 2 ml of suspension medium to each recipient is often used in rats. As lymphocyte uptake into various tissues shows no signs of saturation up to at least 100 million cells injected in this species, any cell number between 10 and 100 million cells injected should give the same distribution pattern.

Remove a 50- or 100-μl sample of the cells for determination of the total injected radioactivity.

Inject the cells *slowly* into the lateral tail vein or dorsal penile vein of rats, 8 to 12 weeks of age, with a 2-ml syringe to which is coupled a 23^{11} gauge needle. Adult rats tolerate a 2-ml injection sample well.

ii. Indium

The cells are labeled in protein-free medium, e.g., RPMI or PBS at 10 to 100 million cells per milliliter with 0.37 mBq (10 μCi) of ^{111}In Oxine (MediPhysics Inc., Emeryville, CA, or INS.1, Radiochemicals, Amersham, England) for 10 min at room temperature.[17,34,38]

The procedure for cell sampling and washing is the same as for chromium labeling, except that cell "sweating" between the second and third washing is not necessary. Since the small amount of ^{111}In not absorbed by the cells during labeling is avidly bound to protein, it may be advisable to increase the serum concentration in the washing medium to 5 to 10%, or spin the cells through a layer of protein before resuspension for injection.

2. β-Emitters

Leucine, in the form of ^{3}H-leucine or ^{14}C-leucine, enters the general cellular metabolic pool of amino acids, some of which will be incorporated into proteins. As expected the isotopes are rather quickly released from the cells and reutilized by other cells,[31] ^{3}H more so than ^{14}C. These isotopes, therefore, have their main applicability in short-term migration studies[15] and are better suited for autoradiography than for scintillation counting (see later). The cell labeling procedures are identical to those of ^{3}H-iridine and ^{14}C-uridine, respectively, except that the labeling concentration of ^{3}H-leucine is 0.37 μCi/ml (see next section for details of the labeling procedure).

C. RAIOLABELS THAT ARE INCORPORATED INTO NUCLEIC ACIDS
1. β-Emitters
a. Radiolabels Selectively Incorporated into RNA

^{3}H- and ^{14}C-uridine are both frequently used for lymphocyte labeling, especially when autoradiography is needed. The difference in β-energy emission between the two isotopes makes it possible to follow two different cell populations in the same recipient.[54] However, β particle counting of tissues is laborious, since the tissue first must be dissolved and an appropriate scintillator added. Furthermore, color quenching complicates the counting procedure. ^{3}H and ^{14}C are coupled to different parts of the uridine molecule that are metabolized differently. Thus, the release of ^{3}H from labeled cells is quicker than ^{14}C, which must be taken into account in double labeling studies.[55] Finally, small lymphocytes do not label uniformly with radioactive uridine. In the rat[14] and in man[56] T cells are more intensely labeled than B cells. In order to detect uridine-labeled B cells in the tissues by autoradiography, either fairly intense labeling of the cells is required[15] or the autoradiograms have to be exposed for several weeks, or even several months (see Section IV.D).

³H-uridine can also be applied for migration studies of lymphoid cells other than T or B cells. In the rat, most of the NK cell activity is associated with a morphologically distinguishable cell population, termed large granular lymphocytes (LGL). Unlike T or B cells, these cells do not recirculate from blood to lymph, but are found mainly in the blood, spleen, and liver.[17,57] These cells can be separated from blood to more than 90% purity and incorporate ³H-uridine fairly uniformly *in vitro*. In autoradiographic tissue sections of lymphoid tissue it could be shown that these cells had a migration pattern that was radically different from T or B cells in that they localized in the splenic red pulp and did not enter the lymph nodes.[17]

b. Protocol for Labeling Lymphocytes with Radioactive Uridine

Cell sampling is the same as for ^{51}Cr.

Label the cells in RPMI 1640 + 10% FCS (Gibco), at 50 million cells per milliliter in 50-ml Falcon centrifuge tubes.

5-³H-uridine (Amersham), or 5,6-³H-uridine (NEN) is added to give a concentration of 0.18 to 0.37 MBq (5 to 10 μCi) per milliliter cells. For B cell labeling, up to 0.9 MBq/ml may be needed. U-^{14}C-uridine (Amersham or NEN) is added at 0.037 to 0.074MBq/ml.

Incubate the cells for 1 h at 37°C with occasional gentle shaking.

Wash the cells twice in ≅50 ml of PBS-FCS before resuspension and injection.

Immediately after labeling most of the isotope within the cells is in a low-molecular-weight, acid-soluble material. Approximately half of the label is lost from the cells within the first 12 h after labeling, and the remaining label is then found within acid-insoluble RNA. After the initial rapid loss of label, the radioactivity is lost much more slowly, and this slow release is probably mainly due to metabolic turnover of RNA.

c. Radiolabels That Are Incorporated into DNA

³H-thymidine and ^{14}C-thymidine label cells in the S-phase. Since small lymphocytes in the thoracic duct are nondividing cells, none of them will be labeled by these isotopes after short-term *in vitro* incubation. However, 5 to 10% of TDL are large, rapidly dividing cells. The S-phase in these cells may approach 50% of the cell cycle; around 30% of them were labeled after incubation with ³H-thymidine for 1 h *in vitro*.[13] The technique of *in vitro* labeling, combined with autoradiography, has enabled us to study the turnover of LGL.[58] In contrast to small lymphocytes, a significant proportion (3 to 5%) of LGL was in the S-phase in the blood, spleen, and liver.

S-phase cells can also be labeled *in vivo* by pulse labeling. The animals are injected by i.v. injection with 37 kBq of the isotope per gram of body weight, and killed 1 to 2 h later. However, ³H-thymidine has more frequently been applied to studies on the rate of lymphocyte formation from precursors and to studies of lymphocyte longevity *in vivo*. When animals are continuously infused with ³H-thymidine, the cells that remain *un*labeled cannot have divided during the period of infusion. Such experiments have shown that some lymphocytes have a very long life span. In rats continuously infused with ³H-thymidine for 220 d, about 10% of blood lymphocytes remained unlabeled, which showed that these lymphocytes had not divided for the whole period of labeling.[59] In their comprehensive studies on lymphocyte turnover in rats, Everett and Tyler[60] found that the rate at which labeled lymphocytes accumulate in the blood is highest during the first week of infusion (6 to 10% per day) and then sharply falls off to 1 to 2% per day. The interpretation was that a subpopulation of lymphocytes have a much shorter life span than the average.

Of the different cell classes, lymphocytes and, in particular, T lymphocytes are the only cells with a known long life span.[61,62] In continuous labeling experiments with ³H-thymidine *in vivo*, the rate of disappearance of unlabeled cells showed that the half-life of dendritic cells in peripheral lymph was on the order of 4 d.[63] Similarly, the half-life of LGL in blood and spleen was about 1 week.[64]

The turnover rate of the cell population can thus be measured by the rate of accumulation of labeled lymphocytes, or disappearance of unlabeled cells, in a compartment as a function of time of continuous infusion of lymphocytes. Alternatively, it can be measured as the rate of disappearance of the labeled cells after stopping the infusion. However, there are some fallacies in the interpretation of such data. Disappearance of labeled cells may well indicate cell death, but other possibilities that must be considered are transformation of the labeled cell into a cell with a different morphology, migration out of the compartment, or repeated cell divisions that dilute the label to a point below the level of detection. Cell longevity in this context is, therefore, a technical term and gives no information about the final destiny of the cells.

d. Procedures for Labeling Lymphocytes with ³H-Thymidine

The *in vitro* labeling procedure for ³H-thymidine (methyl-³H-thymidine, Amersham) is the same as for ³H-uridine except that a concentration of 3.7 to 37 kBq/mℓ is sufficient to get good visualization of S-phase cells by autoradiography.

For *in vivo* labeling rats are given 1 μCi (37 kBq) per gram body weight per day, either as a continuous infusion or divided in two to four doses, given by i.p. injection. Four daily injections were as effective as continuous infusion,[60] but even with two daily injections almost 100% of bone-marrow cells were labeled after 7 d.[64] If it is sufficient to label most, but not necessarily all, dividing cells, two daily injections are acceptable.

2. γ-Emitters

5-¹²⁵I-2′-deoxyuridine (¹²⁵IUdR) substitutes for thymidine during DNA-synthesis. It is a useful marker in flash labeling experiments for determining the intensity of DNA-synthesis in whole organs, such as the bone marrow or spleen, when the tracing of individual S-phase cells is not necessary. It can also be used for following the fate of newly formed lymphocytes for limited periods of time.[27] However, the use of this isotope for tracing the fate of S-phase cells over several days is not possible because of its radiotoxicity.

IV. IDENTIFICATION OF RADIOACTIVE LYMPHOCYTES IN THE RECIPIENT

A. DEFINITIONS AND BASIC CONCEPTS

Before going into details with the sampling procedure, we first want to define some of the terms frequently used in lymphocyte migration studies:

Lymphocyte localization — This simply refers to the number of lymphocytes from the injection sample present in the organ at sampling.

Lymphocyte migration — This implies some form of active movement or transport of the cells from one compartment into another, but does not tell anything about the mechanisms.

Lymphocyte recirculation — This should be used only for those cells that have completed one full round of migration from the blood through the lymphoid tissue and back to the blood again. Recirculating cells are not synonymous with TDL: some large lymphocytes among TDL will localize within the lamina propria of the gut and leave the recirculating lymphocyte pool permanently.

Lymphocyte homing — This implies migration to a final destination, but is often misused in the context of lymphocyte recirculation. With this interpretation, "lymphoblast homing to the lamina propria" is correct use of the term, whereas "TDL homing to lymph nodes" is not.

Lymphocyte trapping — This indicates excessive, nonspecific accumulation of recirculating lymphocytes within, e.g., a lymph node as a consequence of antigenic stimulation. Increased influx of cells as well as retention of the cells to be released into the efferent lymph may both contribute to trapping.

Lymphocyte recruitment — This can be both nonspecific or antigen specific. In the latter case it means the selective retention of antigen-specific cells in a lymphoid tissue after antigen stimulation. This occurs through the process of lymphocyte *selection*, whereby specific antigen reactive cells are temporarily removed from the recirculating pool of lymphocytes by antigen.

Some basic concepts of lymphocyte recirculation to be kept in mind when designing lymphocyte recirculation experiments using the tracer sample principle are as follows:

1. The cells usually enter the tissues from the blood. The number of lymphocytes that have entered the tissue during the time between cell injection and organ harvest will, therefore, depend on the average blood concentration of the labeled cells during that period. The blood concentration reflects the number of labeled cells retained or excluded by other organs. This stresses the necessity of making a complete balance sheet, where all the organs belonging to the main traffic areas of the injected cells are measured, including the blood itself.
2. Autoradiographic studies or other histological techniques should confirm that the scintillation data reflect the organ uptake of viable cells.
3. Since the circulation kinetics of the injected cells may differ from one organ to another, several time intervals between $^1/_2$ and 24 h of injection of the cells should be measured. For example, the mean transit time for T cells through the spleen is only about 5 h, compared with a transit time through lymph nodes of about 12 h. The consequence of this is a net redistribution of injected T cells from the spleen to the lymph nodes between $^1/_2$ h and 24 h, which will only be apparent in kinetic studies.[35]

B. SAMPLING PROCEDURES

The sampling routine for studying the recirculation of TDL in the rat is as follows:

The recipients are killed at various time intervals to determine the organ uptake of the isotope. Preferably several intervals between $^1/_2$ h and 24 h should be monitored. The $^1/_2$ h is important, because it shows the rate of initial migration into the organ undisturbed by cells that have already finished their first round of recirculation and started their second. The 24 h is equally important because at this stage the injected cells are in close to equilibrium with the host's own pool of recirculating lymphocytes, and the organ distribution of radioactivity gives an impression of the fraction of the pool of recirculating lymphocytes present in the various organs.

The organs usually examined for the presence of the injected labeled cells are cervical lymph nodes, mesenteric lymph nodes, Peyer's patches, spleen, liver, lungs with bronchi removed, gut less Peyer's patches, kidneys, one tibia representing approximately 4% of the total bone marrow, a sample of white blood cells separated from a standard volume of heparinized blood with IF, and a standard volume of blood plasma. Note that both cervical and mesenteric lymph nodes should be sampled, the former representing somatopleuric or somatic lymph nodes, the latter splanchnopleuric or visceral nodes. We often include other somatic lymph nodes as well, such as popliteal, superficial inguinal, brachial, and axillary nodes. The coecal lymph node represents a special case, as it drains the liver.[63] Note also that in some situations the position of the lymph nodes within a chain "peripheral" vs.

"central" nodes) may be of relevance. At sacrifice, the animal should be thoroughly exsanguinated and, preferably, perfused with a physiological salt solution to minimize counts due to intravascular lymphocytes. This is particularly important when studying lymphocyte migration into nonlymphoid tissues.

C. DETECTION OF RADIOLABEL BY γ-COUNTING

γ-Counting of tissues should be without problems when using [111]In- or ^{51}Cr-labeled cells, as long as the window of the γ-counter is adjusted for the energy of these isotopes. The results are usually presented as the percentage of the injected radioactivity present in the organ, either per whole organ or per gram of tissue. The latter value also makes it possible to compare the densities of the injected cells in different organs. For example, although a large fraction of the injected cells usually localize in the liver, the density of the injected cells is usually much higher within the lymph nodes and spleen.

D. DETECTION BY β-SCINTILLATION COUNTING

β-Scintillation counting of tissues requires extensive processing of the tissue. In this laboratory, we digest samples of tissue in Soluene-350 (Packard) overnight, bleach them with H_2O_2, and finally add Dimilume-30 (Packard) as a scintillator.[49] Since the tissues vary in their chemical composition, color as well as chemical quenching is a problem, especially when tissues with a high content of hemoglobin, such as the spleen and blood, are counted. Therefore, elaborate quench compensation programs adjusted for each tissue have to be generated. The different energy of β particle emission of ^3H vs. ^{14}C has been exploited for double labeling studies.[49,54,55] Again, elaborate quench compensation and computer programs are needed to determine the content of ^3H and ^{14}C simultaneously in the tissue.[55]

E. DETECTION BY AUTORADIOGRAPHY

For autoradiographic detection, ^3H-labeling is usually employed, but it is possible to score ^3H-labeled and ^{14}C-labeled cells separately on smears or in sections. It should be remembered that the weak β-rays of ^3H have a maximal range of only a few microns within tissues. Self-absorption correspondingly limits the useful thickness of tissue sections. For tissues we prefer epoxy resin-embedded, 1- to 2-μm sections. In addition to reducing self-absorption, much finer details can be discerned with these sections than with the traditional paraffin sections. This is of particular importance when higher resolution is needed, see Figure 1, but less important if only the overall localization patterns of the lymphoid cells are studied.

The techniques of autoradiography have been described in detail by Rogers.[66] Cytospins or smears of lymphoid cells are fixed in methanol and washed in water to remove low molecular weight compounds, and tissue sections are fixed, processed, and embedded by routine methods. As we often combine light microscopic autoradiography of 1- to 2-μm-thick section with electron microscopy of neighboring 50-nm-thick sections, we prefer to use perfusion fixation followed by immersion fixation in order to preserve fine cellular detail (Figures 1 and 2). A plastic tube is inserted through the left ventricle into the ascending aorta and tied in place with a ligature. We first perfuse briefly with warm (37°C) phosphate buffered saline and thereafter with phosphate buffer (0.1 *M*) containing glutaraldehyde (2.5%) and formaldehyde (2 to 3%). Immersion fixation is done in the same solution or in cacodylate buffer containing the same fixatives, followed by immersion fixation by OsO_4. The formaldehyde/glutaraldehyde solutions should be freshly made by dissolving paraformaldehyde in the buffer (heating is needed) and then adding glutaraldehyde (Taab, EM-grade) just prior to use.

The photographic emulsion we use in Kodak nuclear track emulsion NTB-2, but Ilford nuclear research emulsion G5 or K5 gives grains about the same size. The NTB-2 emulsion

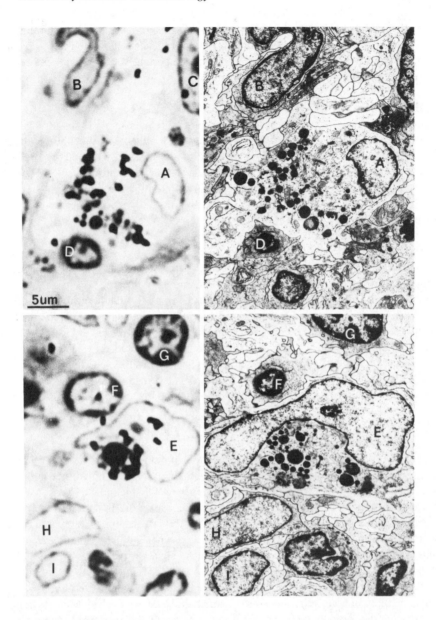

FIGURE 2.   Details from the same nodes as in Figure 1 b. Here we have combined light microscopic autoradigrams (left) with neighboring electron microscopic (EM) sections (right). The EM sections facilitate identification of the phagocytic cells as interdigitating cells. Direct EM autoradiography could have been performed, but would have had two drawbacks. First, the emulsion reduces contrast and quality of the EM preparations. Second, the EM sections are about 30-fold thinner than the LM sections, which means that the exposure time for the autoradiograms would have to be increased manyfold. For the LM sections it was already 6 weeks for allogeneic T cells and 5 months (!) for allogeneic B donor cells. The long exposure time needed for B cells reflects the much lower labeling intensity for B than for T cells (see text). (From Fossum, S. and Rolstad, B., *Eur. J. Immunol.,* 16, 440, 1986. With permission.)

is diluted with an equal volume of distilled water and warmed to 42°C in a water bath. To allow time for air bubbles to escape, the emulsion is left in the water bath for at least 10 min before dipping. Exposure times are usually from 2 to 4 weeks, somewhat longer for tissue sections than for cytospins. However, much longer exposure times may be needed in special cases. First, when we followed the fate of ^{3}H-uridine labeled B cells (which are only weakly labeled by this compound, see above) that were injected into allogeneic hosts, exposure times up to 6 *months* were necessary.[32] Second, when protective films are used as a diffusion barrier between prestained smears or sections and the emulsion (in order to minimize chemography, see below), the exposure time must be increased. The exposure time can be reduced by the use of scintillating fluids,[66] but we have had limited success with this technique.

Light microscopic autoradiographic sections are usually stained after the emulsion has been applied and developed (poststaining techniques). The problems with poststaining include: (1) loss of silver grains or damage to the emulsion by the solutions used for staining or for differentiation of the stains, (2) absorption of the stain by the gelatin-based emulsion, and (3) alteration of the cells chemically by the emulsion or the developing agents, so that they will not bind properly. On the other hand, prestaining, i.e., staining before application of the emulsion, may (1) lead to positive or negative chemography or (2) the stain may be removed during the development of the autoradiogram. These problems can be avoided by the proper choice of stains or staining conditions or by the use of a thin membrane interposed between the stained smear or section and the film (see below).

As for choice of stains, we routinely use a combination of prestaining with p-phenylenediamine (pPd) and poststaining with toluidine blue for our epon-embedded sections. The advantage of using pPd is that it provides a counterstain to the nuclear strains like toluidine blue for use on epon-embedded sections. A 1% solution of pPd (p-phenylene-diamindihydrochloride, Fluka Ag, A 56 663) with a 1:1 mixture (V/V) isopropanol/ethanol as solvent is used for block staining of the tissue samples before embedding into epon. Note that the tissue must have been fixed in OsO_4 for pPd to bind. The point with block staining is that it does not give positive chemography, which is a great problem when sections are prestained with pPd.[67]

Immunocytochemistry combined with autoradiography is a most useful method to follow the selective localization of subsets of radiolabeled heterogeneous lymphocyte samples such as TDL or to aid in the identification of cell types when, e.g., ^{3}H-thymidine labeling *in vivo* is performed (Figure 3). However, immunocytochemistry cannot be performed after application of the photographic emulsion, and e.g., the benzidine deposits of immunoperoxidase cytochemistry give positive chemography. The problems of immunoperoxidase-prestaining may be obviated by the use of a protective film interposed between the section and the emulsion. We use a polyvinylidenechloride (PVC) film (a copolymer of vinylidenechloride and vinylchloride) made by dipping the sections into a PVC solution followed by very slow withdrawal[66,68] (a specially designed apparatus has been made by our mechanical workshop for this purpose). It is also most useful for other types of cytochemistry.[69] We have recently exploited it for autoradiographic studies of dividing large granular lymphocytes, identified by Giemsa staining of the granules.[64] For this study the protective film had to be used because the Giemsa staining gave strong positive chemography when applied before the emulsion and did not stain the granules when applied after the emulsion.

V. NONRADIOACTIVE LABELS

A. DIRECT LABELING OF CELLS WITH FLUOROCHROMES

A drawback with radioactive labeling methods is that individual labeled cells cannot be identified while still alive. By contrast, with fluorescent labels attached to the cells, live marked cells may be reisolated by fluorescence-activated cell sorting. Fluorescence-labeled cells may also be rapidly examined in large numbers by flow cytofluorometry, and they are easily detected within the tissues by fluorescence microscopy. An obvious advantage with the latter method is that the tissue may be examined the same day as it is sampled, without the need for lengthy incubation periods and with much simpler preparation procedures than

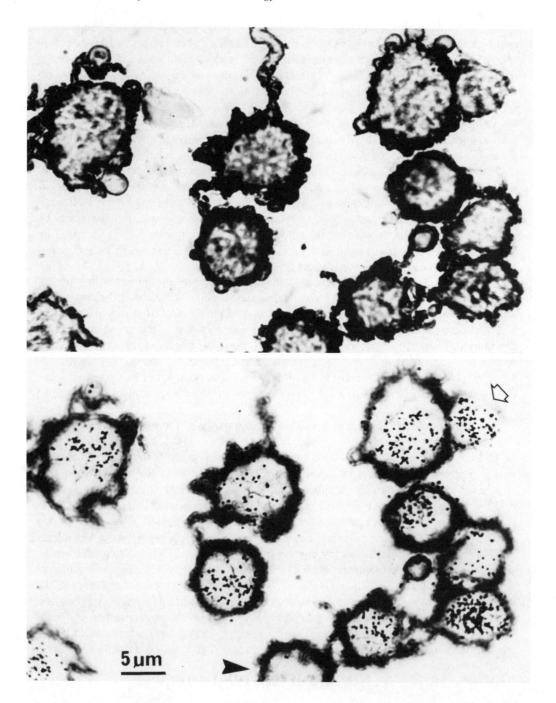

FIGURE 3.   Autoradiogram of cytosmear of dendritic cells collected from rat peripheral lymph after repeated injections
of ^{3}H-thymidine for 4 d. Before application of the photographic emulsion the cells were labeled with a monoclonal antibody
against rat MHC class II molecules followed by the indirect peroxidate-antiperoxidase method. The dark outline of the
cells shows that these cells are all strongly MHC class II positive. The arrowhead points to a single unlabeled dendritic
cell, the arrow to a single weakly MHC class II positive lymphoblast. To avoid positive chemography a thin protecting
PVC film was applied before the sections were dipped in the photographic emulsion, see Section IV.D. (From Fossum,
S., *Scand. J. Immunol.*, 27, 97, 1988. With permission.)

for autoradiography. The fluorescent tag is attached to the sample in two different ways: either by fluorochrome-conjugated antibodies directed against cell surface alloantigens or by labeling cell samples directly with the fluorochromes. The former method is outlined in the section below on intrinsic markers. The latter method involves incubating the lymphocyte sample for 15 min at 37°C in solutions containing fluorescein isothiocyanate (FITC) or tetramethylrhodamine isothiocyanate (TRITC). Lymphocytes may also be labeled *in vivo* by injecting these solutions directly into organs such as the thymus or by whole organ perfusion with the solutions. The concentrations used vary between 10 to 50 μg/ml for FITC and 0.4 to 4.0 μg/ml for TRITC. Details of the labeling procedure are given in Reference 33 which also considers problems such as uneven labeling and cytotoxicity.

B. LABELING OF S-PHASE CELLS WITH BROMODEOXYURIDINE

As an alternative to radiolabeled thymidine, 5-bromodeoxyuridine (BrUdR) may be used to label dividing lymphocytes. As with IUdR, BrUdR is stereochemically similar to thymidine and therefore converted to BrUdR-triphosphate through the salvage pathway and incorporated into DNA during DNA synthesis. The method has certain advantages over the method based on radiolabeled thymidine. First, as there is no need for autoradiographic processing, the method is quicker. Second, it can be combined with immunostaining of differentiation markers without the need for protective films as discussed above. And third, it is claimed to be more sensitive than the autoradiographic method.

We use a method, based on a recipe developed in Mason's laboratory, in which the BrUdR-epitopes are demonstrated with a monoclonal antibody:[71]

1. 5-Bromo-2'-deoxyuridine (Sigma, product number B5002) is stored at −20°C with desiccant.
2. *In vivo* cells are labeled by intravenous injection of 10 mg/kg weight of BrUdR in phosphate buffered saline.
3. *In vitro* cells may be labeled by incubation in medium containing 10 μg/ml of BrUdR at 37°C for 1 h.
4. Cryostat sections or cytospins are stored unfixed at 4°C with desiccant.
5. Before staining, the sections and smears are fixed in acetone at 4°C for 10 min and then air-dried.
6. In order to expose the BrUdR determinants the DNA must be denatured. This is done by treating the sections with preheated deionized formamide. The formamide solution is made by mixing 60 ml deionized formamide with 1 ml concentrated sodium citrate buffer (175 g NaCl, 88.2 g trisodium citrate $2H_2O$ to pH 7.0, distilled water to 1 l) and 2 ml of distilled H_2O. The mixture is poured into a glass-jar and heated to 67°C in a water bath with a lid on the jar. The sections are then put into the preheated formamide for 35 min.
7. The sections are washed three times in phosphate buffered saline at 4°C. All subsequent steps should be performed at 4°C to prevent reannealing of DNA.
8. The sections may then be stained by anti-BrUdR antibody and secondary enzyme-linked antibodies, as with conventional immunocytochemistry.
9. An alternative to denaturing DNA with formamide is digestion with a strong acid or base or digestion with nuclease. A complete "cell proliferation kit," containing BrUdR and also the fluorinated analogue FUdR, nuclease and monoclonal antibodies against the thymidine analogues, secondary antibodies, and substrate, is now available from Amersham (code nr. RPN 20).

VI. INTRINSIC MARKERS

The limitations of the tracer techniques based on extrinsic markers have been described in the previous sections. Intrinsic markers, such as morphological markers, chromosome markers, or proteins that are actively synthesized by the cell during most of its life-time, offer obvious advantages in that they are nontoxic and do not disappear from the cell over longer observation periods.

An elegant example of the use morphological markers is the chicken/quail system. In this system cells or organ rudiments are transferred between the two species *in ovo*. Because the morphological features of the cell nuclei are markedly different in the two species, the cells are easily distinguished in histological sections. The method has yielded invaluable information on cellular migration patterns in early ontogeny of the chicken and within the lymphoid tissues,[70] but is restricted to these avian species.

Chromosome markers have the advantage of being nonantigenic, so that cells can be transferred without evoking rejection mechanisms. A major drawback is that identification of the marked cells can only be performed on mitotically active cells. This requires laborious examination of mitotic spreads and cannot be used for detection of cells *in situ*. These techniques are therefore used only for very special purposes.

In earlier days alloantigens coded for by genes within the MHC were popular for the study of leukocyte differentiation. MHC class I antigens are well represented on most nucleated cells of the body and are easily detected by specific alloantisera. It has been possible to produce long-term radiation chimeras by the reconstitution of lethally irradiated rats with semiallogeneic MHC-incompatible bone-marrow cells (see, for example, Reference 61). However, the introduction of a strong foreign alloantigen may lead to prompt rejection of the injected cells by the ALC-mechanism. As detailed in Section III.B.1, such rejection may occur even in irradiated animals, or in animals that are deficient in alloreactive T cells, such as ''B'' rats or congenitally athymic rats.

Three other alloantigen marker systems have been developed in the rat and mouse and have proved to be useful in cell migration studies as well as in the examination of lymphocyte maturation sequences. These are the leukocyte common antigen (LC-A) family of molecules, various Ig allotypes, and the two allotypes of the Thy-1 antigen in the mouse. Several congenic mouse and rat strains are now available which differ only in one heavy- or light-chain allotype, or with respect to the different allogenic variants of the LC-A or Thy-1 antigens alone. Furthermore, specific monoclonal antibodies toward these markers have also been developed and are available. Combined with indirect immunofluorescence or immunoperoxidase staining, cells positive for the relevant antigens can be identified both in cell suspensions or in tissue sections. For details of the different marker systems, see, e.g., Reference 34.

VII. CONCLUSIONS

Of the various marker systems used to trace lymphocyte migration *in vivo*, radioactive labels, although far from fulfilling the requirements of ideal markers, have remained the most popular ones. When applied critically, the radiolabeling techniques should continue to provide new important information about intricate patterns of lymphocyte migration and interactions *in vivo*. An important supplementary labeling system has emerged in recent years through the fluorescent markers, but one should be aware of the problems of toxicity and the danger of keeping lymphocytes *in vitro*. Alloantigenic markers and chromosome markers can be applied to minimize or avoid these problems and are useful for selected problems in the studies of lymphocyte migration *in vivo*.

REFERENCES

1. **Gowans, J. L.,** The recirculation of lymphocytes from blood to lymph in the rat, *J. Physiol.,* 140, 54, 1959.
2. **Smith, J. B., McIntosh, G. H., and Morris, B.,** The traffic of cells through tissues: a study of peripheral lymph in sheep, *J. Anat.,* 107, 87, 1970.
3. **Frost, H., Cahill, R. N. P., and Trnka, Z.,** The migration of recirculating autologous and allogeneic lymphocytes through single lymph nodes, *Eur. J. Immunol.,* 5, 839, 1975.
4. **Binns, R. M.,** Organisation on the lymphoreticular system and lymphocyte markers in the pig, *Vet. Immunol. Immunopathol.,* 3, 95, 1982.
5. **Reynolds, J. D. and Pabst, R.,** The emigration of lymphocytes from Peyer's patches in sheep, *Eur. J. Immunol.,* 14, 7, 1984.
6. **McGregor, D. D. and Gowans, J. L.,** The antibody response of rats depleted of lymphocytes by chronic drainage from the thoracic duct, *J. Exp. Med.,* 117, 303, 1963.
7. **Waksman, B. H., Arnason, B. G., and Jankovic, B. D.,** Role of the thymus in immune reactions in rats. III. Changes in the lymphoid organs of thymectomized rats, *J. Exp. Med.,* 116, 187, 1962.
8. **Parrott, D. V., de Sousa, M. A., and East, J.,** Thymus-dependent areas in the lymphoid organs of neonatally thymectomized mice, *J. Exp. Med.,* 123, 191, 1966.
9. **Cleveland, W. W., Fogel, B. J., Brown, W. T., and Kay, H. E. M.,** Foetal thymic transplant in a case of Digeorge's syndrome, *Lancet,* 2, 1211, 1968.
10. **de Sousa, M. A. B., Parrott, D. M. V., and Pantelouris, E. M.,** The lymphoid tissues in mice with congenital aplasia of the thymus, *Clin. Exp. Immunol.,* 4, 637, 1969.
11. **Goldschneider, I. and McGregor, D. D.,** Anatomical distribution of T and B lymphocytes in the rat, *J. Exp. Med.,* 138, 1443, 1973.
12. **Fossum, S. and Ford, W. L.,** The organization of cell populations within lymph nodes: their origin, life history and functional relationships. A review for "histopathology", *Histopathology,* 9, 469, 1985.
13. **Gowans, J. L. and Knight, E. J.,** The route of recirculation of lymphocytes in the rat, *Proc. R. Soc. Biol. London,* 159, 257, 1964.
14. **Howard, J. C., Hunt, S. V., and Gowans, J. L.,** Identification of marrow-derived and thymus-derived small lymphocytes in the lymphoid tissue and thoracic duct lymph of normal rats, *J. Exp. Med.,* 135, 200, 1972.
15. **Nieuwenhuis, P. and Ford, W. L.,** Comparative migration of B and T lymphocytes in the rat spleen and lymph nodes, *Cell. Immunol.,* 23, 254, 1976.
16. **Ford, W. L. and Hunt, S. V.,** The preparation and labelling of lymphocytes, in *Handbook of Experimental Immunology,* Vol. 2, Weir, D. M., Ed., Blackwell Scientific, Oxford, 1973, p. 23.1.
17. **Rolstad, B., Herberman, R. B., and Reynolds, C. W.,** Natural killer cell activity in the rat. V. The circulation patterns and tissue localization of peripheral blood large granular lymphocytes (LGL), *J. Immunol.,* 136, 2800, 1986.
18. **Bøyum, A.,** Separation of leucocytes from blood and bone marrow, *Scand. J. Clin. Lab. Invest.,* 21, (Suppl. 97), 77, 1968.
19. **Parish, C. R. and Hayward, J. A.,** The lymphocyte surface. II. Separation of Fc receptor, C'3 receptor and surface immunoglobulin-bearing lymphocytes, *Proc. R. Soc. London,* 187, 65, 1974.
20. **Timonen, T., Reynolds, C. W., Ortaldo, J. R., and Herberman, R. B.,** Isolation of human and rat natural killer cells, *J. Immunol. Methods* 51, 269, 1982.
21. **Hall, J. G., Parry, D. M., and Smith, M. E.,** The distribution and differentiation of lymph-borne immunoblasts after intravenous injection into syngeneic recipients, *Cell Tissue Kinet.* 5, 269, 1971.
22. **Smith, M. E., Martin, A. F., and Ford, W. L.,** Migration of lymphoblasts in the rat. Preferential localization of DNA-synthesizing lymphocytes in particular lymph nodes and other sites, *Monogr. Allergy,* 16, 203, 1980.
23. **Butcher, E. C., Scollay, R. G., and Weissman, I. L.,** Organ specific lymphocyte migration: mediation by highly selective lymphocyte interaction with organ-specific determinants on high endothelial venules, *Eur. J. Immunol.,* 10, 556, 1980.
24. **Smith, M. E. and Ford, W. L.,** The migration of lymphocytes across specialized vascular endothelium. VI. The migratory behaviour of thoracic duct lymphocytes retransferred from the lymph nodes, spleen, blood or lymph of a primary recipient, *Cell. Immunol.,* 78, 161, 1983.
25. **McLennan, I. C., Gray, D, Kumararatne, D. S., and Bazin, H.,** The lymphocytes of splenic marginal zone: a distinct B cell lineage, *Immunol. Today,* 3, 305, 1982.
26. **Smith, M. E. and Ford, W. L.,** The recirculating lymphocyte pool of the rat: a systematic description of the migratory behaviour of recirculating lymphocytes, *Immunology,* 49, 83, 1983.
27. **Fossum, S., Smith, M. E., and Ford, W. L.,** The migration of lymphocytes across specialized vascular endothelium. VII. The migration of T and B lymphocytes from the blood of the athymic, nude rat, *Scand. J. Immunol.,* 17, 539, 1983.

28. **Ford, W. L., Allen, T. D., Pitt, M. A., Smith, M. E., and Stoddart, R. W.,** The migration of lymphocytes across specialized vascular endothelium. VIII. Physical and chemical conditions influencing the surface morphology of lymphocytes and their ability to enter lymph nodes, *Am. J. Anat.,* 170, 377, 1984.

29. **van Ewijk, W., Brons, N. H. C, and Rozing, J.,** Scanning electron microscopy of homing and recirculating lymphocyte populations, *Cell. Immunol.,* 19, 245, 1975.

30. **Woodruff, J. J. and Gesner, B. M.,** Lymphocytes: circulation altered by trypsin, *Science,* 161, 176, 1968.

31. **Rannie, G. H. and Donald, K. J.,** The migration of thoracic duct lymphocytes to non-lymphoid tissues. A comparison of the distribution of radioactivity at intervals following i.v. transfusion of cells labelled with ^{3}H, ^{14}C, ^{75}Se, ^{99m}Tc, ^{125}I and ^{51}Cr in the rat, *Cell Tissue Kinet.,* 10, 523, 1977.

32. **Fossum, S. and Rolstad, B.,** The roles of interdigitating cells (IDC) and natural killer (NK) cells in the rapid rejection of allogeneic lymphocytes, *Eur. J. Immunol.,* 16, 440, 1986.

33. **Butcher, E. C. and Ford, W. L.,** Following cellular traffic: methods of labelling lymphocytes and other cells to trace their migration in vitro, in *Handbook of Experimental Immunology,* 4th ed., Weir, D. M., Herzenberg, L. A., Blackwell, C. C., and Herzenberg, L. A., Eds., Blackwell Scientific, Oxford, 1985, chap. 57.

34. **Rannie, G. H., Thakur, M. L., and Ford, W. L.,** An experimental comparison of radioactive labels with potential application to lymphocyte migration studies in patients, *Clin. Exp. Immunol.,* 29, 509, 1977.

35. **Wagstaff, J., Gibson, C., Thatcher, N., and Crowther, D.,** A method for studying the dynamics of the primary migration of human lymphocytes using indium-111 oxine cell labelling, *Adv. Exp. Med. Biol.,* 149, 153, 1982.

36. **Frost, P., Wiltrout, R., Maciorowski, Z., and Rose, N. R.,** An isotope release cytotoxicity assay applicable to human tumours: the use of ^{111}Indium, *Oncology,* 34, 102, 1977.

37. **Sparshott, S. M., Sharma, M., Kelly, J. I., and Ford, W. L.,** Factors influencing the fate of ^{111}Indium-labelled lymphocytes after transfer to syngeneic rats, *J. Immunol. Methods,* 41, 303, 1981.

38. **Wiltrout, R., Gorelik, H. E., Brunda, M. J., Holden, H. T., and Herberman, R. B.,** Assessment of in vivo natural antitumor resistance and lymphocyte migration in mice: comparison of ^{125}I-iododeoxyuridine with ^{111}indium-oxine and ^{51}chromium as cell labels, *Cancer Immunol. Immunother.,* 14, 172, 1983.

39. **Ronai, P. M.,** High resolution autoradiography with ^{51}Cr, *Int. J. Appl. Radiat. Isot.,* 20, 471, 1969.

40. **Bunting, W. L., Kiely, J. M., and Owen, C. A., Jr.,** Radiochromium-labeled lymphocytes in the rat, *Proc. Soc. Exp. Biol. (N),* 113, 370, 1963.

41. **Ronai, P. M.,** The elution of ^{51}chromium from labelled leukocytes - a new theory, *Blood,* 33, 408, 1969.

42. **Tønnesen, B. and Rolstad, B.,** In vivo elimination of allogeneic lymphocytes in normal and T cell deficient rats. Elimination does not require T cells, *Scand. J. Immunol.,* 17, 303, 1983.

43. **Rolstad, B. and Ford, W. L.,** The rapid elimination of allogeneic lymphocytes: relationship to established mechanisms of immunity and to lymphocyte traffic, *Immunol. Rev.,* 73, 87, 1983.

44. **Rolstad, B. and Toogood, E.,** Toxicity of NA$_2$ ^{51}CrO$_4$ when used to label rat lymphocytes, *J. Immunol. Methods,* 21, 271, 1978.

45. **Bainbridge, D. R., Brent, L., and Gowland, G.,** Distribution of allogeneic ^{51}Cr-labelled lymph-node cells in mice. *Transplantation,* 4, 138, 1966.

46. **Bainbridge, D. R.,** Elimination of allogeneic lymphocytes by mice, *Immunol. Rev.,* 73, 5, 1983.

47. **Heslop, B. F. and McNeilage, L. J.,** Natural cytotoxicity: early killing of allogeneic lymphocytes in rats, *Immunol. Rev.,* 73, 35, 1983.

48. **Rolstad, B., Fossum, S., Bazin, H., Kimber, I., Marshall, J., Sparshott, S. M., and Ford, W. L.,** The rapid rejection of allogeneic lymphocytes by a non-adaptive, cell-mediated mechanism ("NK" activity), *Immunology,* 54, 127, 1985.

49. **Rolstad, B.,** The influence of strong transplantation antigens (Ag-B) on lymphocyte migration in vivo, *Cell. Immunol.,* 45, 389, 1979.

50. **McNeilage, L. J., Heslop, B. F., Heyworth, M. R., and Gutman, G. A.,** Natural cytotoxicity in rats: strain distribution and genetics. *Cell. Immunol.,* 72, 340, 1982.

51. **Issekutz, T. B., Chin, G. W., and May, J. B.,** Lymphocyte traffic through chronic inflammatory lesions: differential migration versus differential retention, *Clin. Exp. Immunol.,* 45, 604, 1981.

52. **Rose, M. L. and Micklem, H. S.,** ^{75}SeL-selenomethionine: a new isotopic marker for lymphocyte localisation studies, *J. Immunol. Methods,* 9, 281, 1976.

53. **Benestad, H. B., Sundrehagen, E., Rolstad, B., and Spretting, A.,** Scintigraphy after infusion of $^{99-m}$Tc-labelled syngeneic lymphocytes. Biological and theoretical models predict unphysiological distribution of the cells, in preparation.

54. **Ford, W. L. and Simmonds, S. J.,** The tempo of lymphocyte recirculation from blood to lymph in the rat, *Cell Tissue Kinet.,* 5, 175, 1972.

55. **Atkins, R. C. and Ford. W. L.,** Early cellular events in a systemic graft-versus-host reaction. I. The migration of donor lymphocytes after intravenous injection, *J. Exp. Med.,* 141, 664, 1975.

56. **Scott, D. W. and Josephs, S. H.,** Uridine labelling of human lymphocytes: differential uptake by T and B cells, *Cell. Immunol.,* 20, 64, 1975.
57. **Reynolds, C. W., Timonen, T., and Herberman, R. B.,** Natural killer (NK) cell activity in the rat. I. Isolation and characterization of the effector cells, *J. Immunol.,* 127, 282, 1981.
58. **Vaage, J. T., Fossum, S., and Rolstad, B.,** Proliferation of large granular lymphocytes in rats: the effect of interferon inducers, *Adv. Exp. Med. Biol.,* 237, 457, 1988.
59. **Robinson, S. H., Brecher, G., Lourie, S. I., and Haley, J. E.,** Leukocyte labelling in rats during and after continuous infusion of tritiated thymidine: implications for lymphocyte longevity and DNA reutilisation, *Blood,* 26, 281, 1965.
60. **Everett, N. B. and Tyler, R. W.,** Lymphopoiesis in the thymus and other tissues: functional implications, *Int. Rev. Cytol.,* 22, 205, 1967.
61. **Howard, J. C.,** The life-span and recirculation of marrow-derived small lymphocytes from the rat thoracic duct, *J. Exp. Med.,* 135, 185, 1972.
62. **Sprent, J. and Basten, A.,** Circulating T and B lymphocytes of the mouse. II. Lifespan, *Cell. Immunol.,* 7, 40, 1973.
63. **Fossum, S.,** The life history of dendritic leukocytes, *Curr. Top. Pathol.,* 79, 101, 1989.
64. **Vaage, J. T., Reynolds, C. W., Reynolds, D., Fossum, S., and Rolstad, B.,** The proliferation and life span of large granular lymphocytes (LGL). Effects of cytokines, *Eur. J. Immunol.,* in press.
65. **Fossum, S.,** Lymph-borne dendritic leukocytes do not recirculate, but enter the lymph node paracortex to become interdigitating cells, *Scand. J. Immunol.,* 27, 97, 1988.
66. **Rogers, A. W.,** *Techniques of Autoradiography,* Elsevier Scientific, Amsterdam, 1973.
67. **Dilley, R. and McGeachie, J.,** Block staining with *p*-phenylenediamine for light microscope autoradiography, *J. Histochem. Cytochem.,* 31, 1015, 1983.
68. **Keyser, A. and Wijfells, C.,** The preparation and use of polyvinylidenechloride protective films in autoradiography, *Acta Histochem. Suppl.,* 8, 359, 1968.
69. **Benestad, H. B.,** A modified technique for combination of cytochemistry and radioautography of haemic cells, *Scand. J. Haematol.,* 18, 54, 1977.
70. **Le Douarin, N. M., Dieterlen-Lièvre, F., and Oliver, P. D.,** Ontogeny of primary lymphoid organs and lymphoid stem cells, *Am. J. Anat.,* 299, 170, 1984.
71. **Mason, D. Y.,** personal communication.

Chapter 4

PHOTOIMMUNOLOGY: ULTRAVIOLET RADIATION EFFECTS ON THE IMMUNE SYSTEM

Edward C. De Fabo and Frances P. Noonan

TABLE OF CONTENTS

I. ULTRAVIOLET RADIATION

Ultraviolet radiation (UVR) comprises about 8% of the total amount of electromagnetic radiation emitted by the sun.[1] Yet this relatively small amount of energy plays an important role in the development and survival of living biological cells. In some cases UVR can be damaging, even to the extent of killing cells. In other cases, UVR can be beneficial, for example, in the induction of vitamin D in human skin.[2] Recently, we have found that UVR can be utilized in a highly specific way to regulate certain protective immune responses.[3]

Solar UVR is generally defined as electromagnetic radiation of wavelengths between 190 and 400 nm. It is now common to refer to regions within this waveband as UVC (wavelengths <280 nm), UVB (280 to 320 nm) and UVA (320 to 400 nm). It needs to be emphasized that these are arbitrary definitions, and one often encounters different designations. This is unfortunate since the terms UVA, UVB, and UVC are often used by dermatologists and biologists to specify a given type of lamp or an effect. This can be problematic since some investigators define UVB differently, e.g., as wavelengths between 280 and 315 nm. Thus, a lamp which emits across the range 315 to 400 nm would be called a "UVA" lamp by the latter definition, but, according to the first definition, the lamp also emits UVB. Therefore, it is highly recommended that the emission spectrum or spectral power distribution (SPD) (see Figure 1) of the lamp being considered for use in an experiment be studied carefully beforehand and published along with usual experimental data when the experiments are completed. SPDs are generally available from the lamp manufacturer, but are often not reliable, however, because linear coordinates are used. Thus, small amounts of radiation of short wavelengths, wavelengths which often have high biological effectiveness (more biological "bang" per photon) tend to be obscured. Logarithmic plots of the SPD are much more desirable (Figure 1A). The SPD of many lamps including the most commonly used can usually be found in the literature.[4,5] One must also be cautious about so-called "solar simulators". No lamp can match the entire spectral output of the sun. Some lamps, however, can shape the shorter wavelength output to simulate closely the solar UVB/A regions. More will be said about UVR-emitting lamps below (Section III.A).

II. EFFECTS OF UV RADIATION ON BIOMOLECULES

There are a number of important biomolecules which absorb UVB. While it is not clear at this time that UVB effects on all of these molecules play a direct role in photoimmunology, all active mechanisms by which UV interacts with the immune system have not yet been elucidated. When considering UVB effects in mammalian systems, it should be remembered that UVB penetrates very little, if at all, beyond the skin,[6] partly because of scattering of the radiation and partly because there are many efficient absorbers of UVB in the skin. UVA, in contrast, penetrates considerably further.[6]

A. UV RADIATION AND DNA: DAMAGE, REPAIR, AND BIOLOGICAL CONSEQUENCES

1. Damage

Damage to DNA may be direct, as, for example, the formation of pyrimidine dimers (thymine-thymine dimerization) by absorption of UVC photons as discussed below, or indirect, via a route involving "excited-state" oxygen. Purine and pyrimidine bases of DNA absorb radiation in the 240 to 300-nm waveband. At the longer UVB wavelengths, 300 to 320 nm, the probability of purine bases (A or G) absorbing a photon in this waveband is at least an order of magnitude less than at 260 nm.[7] Consequently, to some investigators, it is the formation of active oxygen species formed by the splitting of water molecules which is thought to do damage to DNA when it is irradiated with wavelengths between 300 and

320 nm (UVB).[8,9] It is important to keep in mind these differences when comparing the effects of UVC and UVB on DNA.[14-16]

In UV-irradiated DNA, the energy of the triplet levels of the four nucleic acid bases proceeds from highest to lowest (C→G→A→T).[10] Therefore, if energy transfer on UV irradiation occurs at the triplet state in DNA, then the excitation energy should be trapped by thymine which has the lowest triplet state. In fact, this appears to be the case as thymine dimer formation appears to be the most common defect in UVC/B irradiated DNA.[11,12]

In addition to thymine-thymine dimerization, thymine-cytosine and cytosine-cytosine dimers are also formed.[13] The longest wavelength at which dimer formation has been shown to occur is 365 nm.[14]

In general, for UVB absorption, at least up to 313 nm, the mechanism of dimer formation is thought to be due to direct absorption of photons.[17] However, for wavelengths longer than 313 nm, the mechanism of dimer formation may not be clear.

In addition to pyrimidine dimers other damage can occur in UV-irradiated DNA including pyrimidine (6-4) adducts, pyrimidine hydrates, thymine glycols, DNA-protein cross-links, and DNA strand breaks.

Pyrimidine adducts involve the formation of four-membered rings between two consecutive bases on the same strand of DNA. Adduct formation is similar to, but not as efficient as, pyrimidine dimer formation.[18] These adducts can be split by light in the 310 to 340-nm waveband.[19,20] The lethality of this type of damage is not clear,[21] although recent studies suggest it may be a major cytotoxic and/or mutagenic lesion.[22-24]

The hydrate of cytosine is a result of the addition of a water molecule across its 5-6 double bond. Although the water molecule can be quickly lost, the hydrate may be deaminated during this process, forming uracil, and this may lead to a mutation in newly replicated DNA.[10] There appears to be no known repair mechanism for this type of lesion.[26]

Ring saturated lesions of the 5,6-dihydroxydihydrothymine type have been detected in the DNA of UV-irradiated human cells.[25] These compounds make up about 20% of the total ring saturated thymine products and may decay spontaneously to leave apyrimidinic sites.[27] The lethal effects of these compounds are not clear.

Several types of proteins from eukaryotic and prokaryotic cells have been cross-linked to DNA by UV irradiation.[28] Significant DNA-protein cross-linking in mammalian cells by wavelengths longer than 345 nm has been reported.[26,29] Unless repaired, it seems that these cross-links could present a problem for replicating DNA. It is unclear whether or not repair of DNA-protein cross-links can take place.[30,31]

DNA strand breakage makes up an important class of cell damage which can be induced by UV as well as by ionizing radiation. Generally, strand breaks are minority lesions even at the longer wavelengths such as 365 nm. However, they do become increasingly important as the wavelengths increase in the solar UV range.[32] Strand breaks appear to be the only form of photochemical damage in DNA after irradiation with visible light (>405 nm).[26] For example, in human fibroblasts, breaks have been detected with wavelengths as long as 546 nm.[33] Double strand breakage occurs at a rate about 80 times less than that of single-strand breakage after 313-nm irradiation in bacteria[26] but does not seem to be induced following 254-nm radiation. Repair of UV-induced strand breakage in DNA has been reported.[32,34] For strand breaks induced by short wavelength UVR, cell survival is probably not impaired because of the fast and efficient repair mechanisms available. However, for wavelengths longer than 313 nm, oxygen-dependent strand breakage resulting in cell death may occur.[35,36] DNA-DNA cross-links could be a serious hazard if not repaired, but little is known about their formation.[9,26]

2. Repair

It is now well known that several mechanisms exist for the repair of DNA damage due to pyrimidine dimer formation. The first repair mechanism to be discovered was photoreactivation. It is specific for pyrimidine dimers and monomerizes them *in situ* by a light-activated enzyme (photolyase; photoreactivating enzyme). Action spectra indicate light in the visible and near UV range is responsible for the activation of this enzyme. (For review see References 37 and 50.) In the process of photoreactivation the enzyme binds to DNA which contains a dimer(s). This complex is then able to absorb visible light and UVA causing the dimer to split or monomerize, thereby removing the damage. Because of the high specificity for pyrimidine dimers, if a particular biological response such as survival or mutation can be "photoreactivated", it is taken as evidence that dimer formation was the inital cause of the effect.[38]

The role of the pyrimidine dimer in higher organisms is difficult to establish.[39,40] Photoreactivating enzymes are present in animal cells,[41] including human cells.[42] However, photoreactivating enzyme levels in animal cells tend to be low so that photoreactivation of the pyrimidine dimers is very difficult to demonstrate. Photoreactivation of UV damage by wavelengths less than 300 nm in frog cells have been demonstrated,[43] and photorepair of pyrimidine dimer formation in human skin has been reported.[44] In 1977, Hart and colleagues[45] reported that UV-induced tumors growing in *Poecilia*, a species of fish which grows in clones, could be prevented if the UV-irradiated cells were re-irradiated with light in the photoreactivating waveband. These data support the argument that it is the pyrimidine dimer which is responsible for the tumor development, although the exact mechanism is still unknown.[10] It should also be kept in mind that mutations in DNA are known to occur at sites other than pyrimidine molecules.[46]

A second type of repair mechanism, known as excision or dark repair, removes pyrimidine dimers from DNA.[47,48] Most normal cells possess a very efficient mechanism for this type of repair. In excision repair, the damage in DNA is removed from one strand by a series of enzymes, each specific for one of the several steps involved. The opposite undamaged strand is then used as a template for replicating a new strand of DNA without the damage. For a review of the actual steps involved in this process see Reference 49. In *Escherichia coli* this process is under the control of a complex system of genes designated uvr A, B, C.[48] Individuals with the inherited, sunsensitive cancer-prone disease called xeroderma pigmentosum are generally defective in excision repair. The high prevalence of skin cancer in these individuals is considered due to this defective repair mechanism.[49]

Although DNA replication is blocked at the sites of pyrimidine dimer formation, it can resume further away leaving a "gap" in front of the dimer. If the other DNA strand has been correctly replicated, an exchange of DNA fragments can occur which fills in the gap opposite the lesion. A new gap is created in the second daughter strand and can be filled by DNA polymerase using the correct information on the original undamaged parent strand. This process is known as postreplication (recombination) repair.[10] In *E. coli*, this process is under control of the recA gene. The final fate of the damage may be removal by excision repair or it may be "diluted out" by being transmitted to one of the daughter cells.

The three types of repair processes described above are constitutively expressed, albeit at low levels. Another type of repair known as "SOS" (induced) repair[10] can be induced by the UVR itself. In *E. coli* the recA and lexA genes control "SOS" repair. This type of repair tends to be "error prone" (increased mutations), and its presence in mammalian systems is unclear.

B. EFFECT OF UV RADIATION ON PROTEINS

The effects of UV absorption on proteins are complex as well as manifold. Only a brief mention of some of the absorption characteristics of these effects will be made here, particularly those effects which may have an impact on photoimmunologic responses. However, the reader is encouraged to consult the literature for further details.[50,51]

The most serious biological consequences of UV irradiation on proteins are due to absorption by a relatively few amino acids - tryptophan, tyrosine, phenylalanine, cystine, and cysteine - and the peptide bond.

The aromatic amino acids tyrosine and phenylalanine show strong absorption at wavelengths longer than 240 nm. Tyrosine has a peak of absorption about 275 nm, and phenylalanine shows a wavelength maximum around 257 nm.[51] Tryptophan, another good UV absorber shows peak absorption around 280 nm.[37,50] Cysteine and cystine will show appreciable absorption at wavelengths longer than 250 nm at neutral or alkaline pH and become very labile following absorption.[51] Cystine with its chain-linking disulfide bond, critical in tertiary structure formation, is probably the most important target in the inactivation of proteins.[50,51]

The peptide bond can be broken by UVR of wavelengths less than 240 nm (peak activity at 180 to 190 nm).[51] Therefore, these wavelengths should be screened (see below) from sources emitting across this region to avoid this problem.

Little interaction occurs between the amino acids in a protein structure in regard to UV absorption so that the total absorbancy of a protein in solution is similar to the sum of the absorbancies of the individual solubilized amino acids.[37] In general, proteins show a peak UV absorption around 280 nm.

Some changes which occur following UV absorption by proteins include increased sensitivity to heat, enzyme inactivation, and antigenic changes due to altered amino acid residues. The latter two effects should be of particular interest to immunologists.

C. EFFECT OF UV RADIATION ON SKIN LIPIDS

The most significant biological effect in this category appears to be the formation of pre-vitamin D in the skin from 7-dehydrocholesterol. This is caused by UVB.[2]

D. EFFECT OF UV ON OTHER SKIN MOLECULES

UVB causes the isomerization in the skin of urocanic acid (UCA) (deaminated histidine) which is a major UV absorbing component of the stratum corneum.[52] UCA is formed by the action of the enzyme histidine-ammonia lyase in the stratum corneum and is naturally present as the *trans* isomer. On UV irradiation of *trans* UCA either *in vivo* or *in vitro* the *cis* isomer is formed and, with enough UV, a photostationary state is reached.[53] It was originally thought that UCA had a role as a natural sunscreen, but it now appears that the UV-induced formation of *cis* UCA has a major role in initiating UV-induced immunosuppression (Section IV.B.3.).

III. EXPERIMENTAL METHODS OF UV IRRADIATION

In this section, experimental UV sources are described, and methods of UV irradiating experimental animals and of UV irradiating cells *in vitro* are given. A discussion is also given of detectors used to measure UVR, and the concept of biologically effective dose is introduced.

A. SOURCES OF UV RADIATION

Various types of lamps exist which emit across the UV region. They are relatively easy to obtain and do not have to be expensive to be used in UV photobiological experiments. Which type of lamp to use depends on the question being asked by the experimenter. For example, if one is concerned about the effects of UVB (280 to 320 nm), one can use the Westinghouse FS40 sunlamp,[4] or an equivalent such as a Philips TL40W/12 or a Sylvania F20 T12 sunlamp. These are fluorescent lamps, which means that the "light" actually emitted (fluorescence or phosphorescence) is a product of 254 nm radiation (given off by the electronic excitation of mercury vapor at low pressure within the lamp) striking a specially formulated phosphor which coats the inside of the glass housing. The chemical nature of the phosphor (usually a trade secret) is what determines the actual spectral output of the lamp. In the case of UVB-emitting lamps this glass must be able to transmit shortwave UV light since ordinary "window" glass can be an effective UVB filter. Although this type of UV source is convenient, there are disadvantages. As can be seen from the SPD of the UVB sunlamp (Figure 1), radiation on both sides of the UVB band, i.e., in the UVA and UVC, is also being emitted. Thus, if one irradiates a mouse with an unfiltered sunlamp, the animal is exposed simultaneously to other wavelengths of radiation in addition to UVB. To determine if an observed effect, e.g., UV-induced suppression of the contact hypersensitivity (CHS) immune response is truly caused by UVB radiation, one can remove this waveband by filtering the lamp output with a plastic sheet of Mylar and then irradiating the animals. This filter prevents nearly all of the radiation with wavelengths less than 315 nm from striking the animals.[54] Loss of the UV effect (immunosuppression, in this example) by this technique is taken as strong evidence for the involvement of UVB in eliciting the response.

For studies involving UVA radiation, a useful lamp is the "PUVA" lamp. This lamp emits radiation principally in the UVA region, although small amounts of radiation in the UVB region can be emitted as well.[5] The PUVA lamp is commonly used for the treatment of psoriasis. Individuals are given a photosensitizing drug (e.g., 8-methoxypsoralen) and are exposed to radiation emitted by the lamp (hence, the name PUVA, psoralen + UVA). As a source of radiation emitting primarily within the UVA waveband, these lamps are quite reliable. However, just as with the so-called "UVB" lamps, they contain radiation outside the defined limits of the UVA wavelength range.[5]

Only an accurate SPD can actually tell which wavelengths are being emitted by any particular lamp. With fluorescent lamps, changes in the phosphor, lamp envelope, and temperature can affect the type and amount of radiation being emitted. Changes in wavelength are particularly important to watch for since biological "effectiveness" can change sharply over a very narrow wavelength range (see Section III.C). Access to a spectroradiometer, a device which measures spectral output as a function of wavelength, should be seriously considered when contemplating long-term experiments in photobiology.

Other lamps which emit principally in the UVA range are also available. One of the most common is the so-called "black-light" lamp which emits principally across the 300 to 400-nm range with a peak close to 360 nm.

For studying UVC effects, a low-pressure mercury lamp called the "germicidal" lamp is often used. Since the pressure of the mercury vapor is low, the principal spectral line of emission is at 253.7 nm. By being encased within a quartz or vycor housing this lamp will emit nearly all of the radiation at this wavelength. Fortuitously, the absorption maximum for DNA lies at 260 nm. This lamp, therefore, is an excellent source for inducing mutagenic or lethal changes in DNA. Bacterial, viral, or cell inactivation dose-response curves can easily be determined with this type of lamp. It is inexpensive, easy to install and use. However, its utility in photoimmunologic experiments is probably limited. As with all UVR-producing lamps, exposure to eyes and skin needs to be avoided.

Many other more complex sources such as medium and high pressure mercury and xenon arc lamps are available, but detailed descriptions are beyond the scope of this chapter. One such system, however, which we have used with much success is the high-pressure 2500-W Xenon arc. This lamp has been coupled to a housing unit which can utilize specially designed interference filters to produce selected narrow-band radiation in the UVC, B, A, or visible range. The unique advantages of this system[3] are that it allows for the production of relatively narrow band-pass radiation, 2.5-nm HBW (half bandwidth), and a large area of exposure, about 50 to 60 cm². Thus, it is possible to irradiate the entire dorsal surface of three mice simultaneously with narrow bands of UV across the UV spectrum in 5-nm steps. One can thus control the intensity and wavelengths of the radiation to prevent inflammation and sunburning effects which could interfere with the response being measured. The drawback for such a system is that it is expensive. It requires a large (750-lb) power supply, special housing and cooling apparatus, and custom-made interference filters with a limited lifetime. The relatively low irradiances in the UVC region can also present a problem. Nonetheless, it has distinct advantages over other types of monochromatic systems in that it can provide a much larger field of exposure with very narrow band-pass radiation. These characteristics make it possible to conduct experiments with high wavelength resolution in animals such as mice. It was this capability which enabled us to determine the existence of a specialized immune-regulating skin photoreceptor in mammalian skin.[3]

B. UV DETECTORS

Determining the type of lamp or spectral output is only the first part of setting up the source of radiation for photoimmunologic or photobiological experiments. Second, and perhaps more important, is selection of instrumentation to measure correctly the energy output of the source. This means measuring *both* the wavelength *and* the "intensity" output of the lamp. In photobiology, one is interested in measuring the energy, of the beam *striking* an object of interest, e.g., skin, rather than the amount of energy *absorbed* by the object. This can be more accurately thought of as the power of the beam falling on the object *per unit area* and is called the radiant flux density, or "irradiance". The unit of irradiance commonly used is the watt per square centimeter, and many radiometers give their output in these units.

Conceptually, the watt is the unit for power, and power can be thought of as the rate at which energy is dispersed (photon flux). It is extremely helpful to commit to memory that 1 W is equivalent to 1 J/s. Thus, to determine the total dose, usually given in J/m^2, striking an irradiated surface, multiply the irradiance by exposure time and convert square centimeters to square meters. Thus, it is seen that the term dose as used in photobiology is quite different from that used in radiobiology since the concept of incident dose rather than absorbed dose is used.

In addition to not accurately calculating the dose, not choosing the correct detector is an even more common problem. Just because a detector might be listed by a manufacturer as being a "UVB" or "UVA" type of detector does not mean it will automatically correctly measure the amount of UVA or UVB being emitted by the lamp you have chosen (see Figures 1A and B). Unless one uses a thermopile, a detector whose spectral response is *flat* over the entire wavelength range of interest, caution is required and corrections must be applied to obtain a valid measurement. (Note: Thermopiles are not commonly used because they are not very sensitive to low UV fluxes and because their flat spectral response can extend from 200 to greater than 50,000 nm. Thus, they are sensitive to infrared, including that from one's own body heat. For these and other reasons they are not usually used by photobiologists even though they are theoretically ideal.)

Because other detector types, e.g., silicon photodiodes, vacuum photodiodes, or photomultipliers, are not spectrally flat, accurate absolute measurements of irradiance and dose are not usually possible without corrections. Unless properly calibrated, when using these types of detectors the number "read" on the meter and the amount of energy actually striking the object of interest may be two different things. The following will illustrate this point. Figure 1A shows the SPD of a single UVB emitting "sunlamp". Figure 1B shows the response function (how sensitive the detector is to each wavelength) of a so-called "UVB" detector. Note that for both lamp and detector there is a peak at one wavelength (λ max) and a drop toward zero on both sides of this peak. Note also that the peak is different for lamp and detector. It is obvious, therefore, that, for this detector, not all wavelengths have the same "sensitivity" or likelihood of being detected. Therefore, only a calibrated "ideally matched" detector or a spectrally "flat" detector would be capable of reacting correctly to

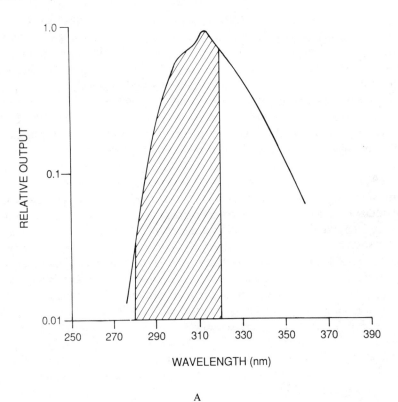

A

FIGURE 1.   (A) Spectral power distribution for the FS40 sunlamp. Hatched area represents the ultraviolet-B region (UVB; 280 to 320 nm). Log plot clearly shows radiation is emitted on both sides of the UVB range. (B) Spectral sensitivity of a UVB detector. Hatched area represents the UVB region. Log plot clearly shows wavelength sensitivity on both sides of the UVB range. A lack of "1:1 correspondence" is easily noted between the wavelength sensitivity of the detector and the spectral output of the lamp. To correct for this and to avoid errors in reporting UVB doses, the detector needs to be calibrated against the specific sunlamp used in the experiment. See text for details.

all the wavelengths emitted by this particular lamp. To measure accurately the UVB output of the lamp using this particular detector, a correction is needed. This correction is called curve multiplying or curve convolution. A detailed description of how it is carried out is beyond the scope of this chapter. However, many texts on radiometry can provide the necessary procedure for carrying out such a calibration, or one may write to any number of manufacturers of detectors and radiometers for similar information. (*Note:* For a comprehensive listing of international manufacturers of optical equipment and related accessories, Reference 55 is highly recommended. In addition, important optical definitions, conversion units, and other valuable information can be found within this three-volume set.)

An alternative method for accurately calibrating a detector is to obtain an SPD of the lamp(s) to be used, preferably measured by yourself with a spectroradiometer or else obtained from the lamp manufacturer, and send it along with the detector to the detector manufacturer. Some companies, for example, International Light Corporation, will, for a fee, make the necessary calibration and return the detector with a "sensitivity" factor(s). Absolute measurements can be made over the wavelength range of interest simply by "dialing in" the appropriate factor. Keep in mind, however, such a calibration is accurate only for the lamp the SPD of which was used in the calibration and will not give correct absolute measurements for lamps with other types of spectral output no matter how similar the spectral output is.

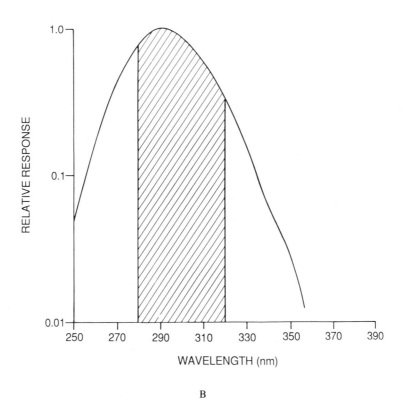

RELATIVE RESPONSE

WAVELENGTH (nm)

B

This is very important if one hopes to report correct doses used in a particular experiment. Reproducibility by other investigators without this type of information is very difficult.

Clearly, referring to a lamp or detector simply by its generic description such as "a UVB lamp" or a "UVB detector" can lead to serious error.

Finally, another important parameter which needs to be considered when measuring light is the "spatial response" of the detector. That is, how well does a detector "see" light coming in from all directions and therefore measure it accurately?

The irradiance from a light source depends on the angle between the light and a point on the light-sensitive surface of the detector. The irradiance will be greatest when the "point" of irradiance on the plane of the detector surface is directly perpendicular to its "point" of emission from the light source. As the light angle increases with respect to the perpendicular, the irradiance decreases. This is because the same amount of light covers a larger surface when it strikes at an oblique angle. Thus, the amount of light energy per unit area (irradiance) decreases. Furthermore, this decrease in irradiance is proportional to the cosine of the angle of incidence. Therefore, in order to compensate for this drop in irradiance for light coming in "from the side", only those detectors calibrated with a "cosine correcting" diffuser should be used.

Spatial response is particularly important when measuring the irradiance "close to" an extended source emitter such as a bank of sunlamps. This error is due primarily to differences between an ideal cosine response and the cosine response of the detector actually used. Consultation with a technical representative from the company from which the detector was purchased is highly recommended to determine critical positioning of the detector.

C. BIOLOGICALLY EFFECTIVE DOSES

Another potential source of error is the incorrect use of biologically effective dose (BED). As stated above, a detector is able to measure accurate absolute doses only for the specific lamp against which it has been calibrated. Consider now that that such a detector will be used to measure the output of another "UVB" type sunlamp with similar, but not identical wavelength output. As indicated, inaccurate measurement of the absolute irradiance will occur. However, if the detector is recalibrated, this time against the second lamp, using the curve-multiplying technique mentioned above, accurate comparisons in absolute irradiance between both lamps can be made. By doing this, one is using the response of the detector as a "weighting" function in order to accurately measure the output of each lamp.

This same concept can be applied if one wishes to make comparisons between laboratory experiments conducted with sunlamps and what is actually happening in the "real world", e.g., with sunlight. Sunlamps or solar simulators cannot produce a spectrum which is identical to that of the sun. Therefore, the dose which produces, e.g., skin cancer in 50% of a group of sunlamp-irradiated mice will be different from the dose needed to produce a similar result in sunlight-exposed mice. So how can such comparisons be made accurately? It is done once again by curve multiplying. In the case of biological effects the "multiplier" or "weighting function" is generally known as a "biological action spectrum" or wavelength effectiveness spectrum. In theory, the shape of the action spectrum should be exactly congruent with the *in vivo* absorption spectrum of the chromophore or photoreceptor which absorbs the UVR.[56] The photoreceptor may then act as a signal transducer and convert the radiant energy signal into a biochemical one which can then enter a biological path.

Action spectroscopy is a powerful technique which, given the assumptions and difficulties in determining action spectra,[57] has been surprisingly accurate in its ability to describe the absorption characteristics of unknown chromophores. In 1930 Gates[59] determined an action spectrum for the killing of bacteria by UVR which was one of the earliest pieces of evidence to implicate DNA as the lethally sensitive target.

A detailed action spectrum can provide very important information concerning the existence of a discrete "target" or photoreceptor for a particular biological response and the absorption spectrum for that photoreceptor. Unfortunately, not all important biological photoresponses have had an action spectrum completed, nor may it even be possible in some cases to determine one. Thus, in these cases, without such a weighting function, comparison in absolute dose is difficult to make between artificial and sunlight exposure or between other sources. Other complications may also exist. For example, in the skin cancer example cited above, more than one photoreceptor may be involved. Although alterations to DNA by UV irradiation are almost certainly involved in skin cancer induction, the actual outgrowth of the skin tumor may involve activating another target molecule which compromises the immune surveillance mechanism (Section IV below).

Despite the difficulties described, some important biological action spectra do exist and have provided important information.[56-58] In our own action spectrum on the effect of UVR on immune suppression (50% level) in mice,[3] the *in vivo* action spectrum, normalized to 270 nm, the most effective wavelength, shows a clearly defined wavelength dependence. This indicated for the first time the existence of a discrete photoreceptor molecule capable of mediating interaction between UVB and the immune system and gave critical information as to its chemical nature.[3] Subsequent experiments have lent additional support to this model (Section IV.B.3).

With information derived from an action spectrum, one can calculate the important BED. This is done by multiplying the action spectrum, with, e.g., a solar irradiance spectrum for some specific time and place, (e.g., average irradiance from 1:00 to 2:00 p.m. at latitude 40° north on June 21st) or some other source. The product of this convolution defines the biological effective irradiance (BEI). In our example of immune suppression, the BEI would be for 50% immune suppression of the CHS response in BALB/c mice (BEI_{imms}). Using the same "action spectrum", one can also determine the BEI_{imms} for any fluorescent sunlamp or solar simulator used in a UVB-induced immune suppression experiment. Once the BEI_{imms} is determined, the BED is obtained by multiplying the BEI_{imms} by exposure time. By using this technique valid comparisons in BED can be made by different investigators, using different sources and/or different detectors. Conversely, one can determine the exposure time needed if both the BED and BEI are known simply by dividing the former by the latter. Keep in mind that in the immune suppression experiments, the action spectrum was normalized to 270 nm, the most effective wavelength for inducing immune suppression. Normalizing to some other wavelength would change the amounts for both the BED and BEI. Thus, specifying the normalization wavelength is important.

D. UV IRRADIATION OF MICE

Unless congenitally hairless mice are used, mice are shaved before UV exposure. Animal clippers with a 40 blade are suitable. The shaved area should have a pinkish appearance. Animals which are in an active hair-growth cycle with fine dense hair which cannot be removed by shaving should not be used as the expected UV dose will not be delivered to the skin. Depilatory agents should also not be used since they may interact with skin components and/or leave photoactive residues on the skin.

Animals should be effectively separated during irradiation to prevent shielding by cage mates. The simplest method is to place plastic dividers in a mouse cage to make several individual compartments. The UV dose rate should be measured within the actual UV irradiation setup. A fan should be included just outside the irradiation area to disperse any heat or ozone which may be generated by the UV source (ozone absorbs UVB and is toxic to cells). Since any UVB source will cause sunburn and corneal damage ("snow blindness"), the UV device must be effectively screened, usually with black cloth so that personnel are not exposed to UV.

If systemic suppresion of delayed-type hypersensitivity in mice is to be investigated, ears may be shielded during irradiation with small pieces of black electrical insulating tape. Mice can also be anesthetized and restrained for a limited time to allow irradiation of only a certain area. Particular care should be taken to ensure that light/dark cycles in animal rooms are regular, and irradiations should be done at approximately the same time each day to minimize potential perturbations due to circadian rhythms. Plastic cutoff filters can be used to remove shorter wavelengths from a broadband UV source as indicated below.

E. UV IRRADIATION OF CELLS *IN VITRO*

For *in vitro* studies, cells must be diluted to effect a "monolayer" to prevent shielding or screening of underlying cells. Such screening will produce spurious dose-response curves. The UV absorbing plastic lid of the petri dish must be removed or replaced with a non-UV absorbing material, e.g., quartz. The cell culture medium must be non-UV absorbing, e.g., phosphate buffered saline, and, as a control, the medium should be UV irradiated first and then cells added as a control for the formation of toxic photoproducts. If possible, cells should be stirred with a small magnetic flea during irradiation, or the petri dish should be slowly rotated by means of a turntable. It should be remembered that cells which are flattened by attachment to a culture surface may well have different susceptibility to UV than cells in suspension. Dose-response studies for survival should be carried out to determine sensitivity to UV radiation.

Since fluorescent sunlamps are the commonest UV sources used in photoimmunology, plastic cutoff filters, e.g., Mylar or polystyrene,[54] can be used to eliminate shorter wavelengths. This material can be placed around the lamp and "solarized" by exposure to the UV source for 72 h before use.[54] The transmission spectrum should be determined before use.

Irradiance measurements must be made before and after each experiment, and when the irradiance begins to decrease, the plastic filter must be replaced. Access to a spectroradiometer is recommended, if possible, to detect change in wavelength output. Careful attention should be paid to ambient lighting in the room because of the possibility of photoreactivation as discussed above.

F. SUMMARY

Photobiological techniques can be powerful tools in getting at underlying mechanism in UVR or visible light effects on mammalian or other immune systems. However, some knowledge of the various effects photons can have on important biomolecules as well as knowledge in understanding and carrying out accurate optical experiments is critical to achieving valid experimental results. Furthermore, using the concept of BED, legitimate comparisons in absolute dose can be made by different investigators using different sources and/or different detectors for a given biological response, the action spectrum of which is known. This could help eliminate much of the disparity and confusion about dose levels and would either accurately define or eliminate such vague terms as ''high'' or ''low'' dose effects.

IV. REVIEW OF IMMUNE SUPPRESSION BY UVB RADIATION

A. HISTORICAL BACKGROUND

Although early studies indicated that UVB plays a critical role in the formation of skin cancers,[60] the effects of UV irradiation on the immune system have, in contrast, only recently been described. Hanisko and Suskind in 1963[61] reported that the contact hypersensitivity (CHS) response of UV-irradiated guinea pigs was significantly lower than that of their control counterparts. Graffi et al. in 1964[62,63] observed that UV-induced tumors were extremely antigenic. These observations were not connected until further studies were commenced on transplantation of tumors which had been induced in mice by a course of UV irradiation.[64-70] A large number of these UV-induced tumors did not grow on transplantation into syngeneic mice, but were immunologically rejected; the tumors could, however, be successfully transplanted into immunosuppressed syngeneic recipients. This observation became particularly interesting when Fisher and Kripke[68] found that the UV-induced tumors could also be transplanted into syngeneic mice which had been given a subcarcinogenic dose of UVR, suggesting that UVR compromised the immune system of the host. More startling, they demonstrated that the UV-induced susceptibility to transplanted UV tumors could be transferred to nonirradiated animals with lymphoid T cells from UV-irradiated mice, indicating that UV irradiation generated in the lymphoid organs a population of suppressor T cells.[69] These observations were rapidly confirmed.[71-73]

Subsequent studies by Fisher and Kripke[74] showed that the suppressor cells also played a role in the development of primary UV-induced tumors. The suppresor T cells have antigenic specificity for UV-induced tumors as a group, in contrast to the rejection antigens which are unique to each UV tumor.[75] Evidence has been produced in hyperimmunized mice for the existence of cross-reactive rejection antigens on UV-tumors,[73] but the consensus now is that regressor UV tumors carry unique immunodominant antigens.[76] For one tumor this immunodominant antigen was shown to be an altered Class I antigen,[77,78] but this may not be a general case.[79]

B. MECHANISM OF UV-INDUCED SUPPRESSION
1. Photobiologic Characteristics

De Fabo and Kripke[54,80] undertook dose and wavelength studies of the UV suppresion of tumor rejection. They found that the wavelengths responsible for this form of suppression are in the UVB and UVC range, as are the wavelengths responsible for UV carcinogenesis, but that the dose-response characteristics for UV suppression were very different from those for UV carcinogenesis, suggesting a different mechanism. Suppression of tumor rejection was proportional to $\log_{10}$ UV dose and was dependent only on total dose. In contrast to UV carcinogenesis, where the tumor yield can be critically dependent both on dose rate and on number of treatments used to administer any given dose, UV suppression was independent of dose rate over a tenfold range, and of dose fractionation from 1 to 12 treatments, i.e., reciprocity held over the range tested. A time delay of 1 to 3 d after UV irradiation was necessary for induction of UV suppression. These studies suggested the presence of a unique and specific mechanism which initiates UV-induced suppression.

To investigate further the mechanism of this effect, another form of UV suppression requiring lower doses of UV and using a defined antigen was employed. It had originally been shown that UV-induced immunosuppression was not a generalized phenomenon, many immune responses, most notably antibody responses, being normal in UV-irradiated animals.[70,71] Jessup et al.[81] had, however, shown that mice which had been UV irradiated had depressed DTH responses to a contact sensitizer which had been injected subcutaneously. Subsequently, a systemic suppression of contact hypersensitivity was demonstrated.[82,83] In these experiments, mice were given a single dose of UV on the back and were contact sensitized on the abdomen. A CHS reaction subsequently was elicited on the ears, which had been protected from UV. This form of suppression showed similar characteristics to the suppression of tumor rejection, being proportional to $\log_{10}$ UV dose and independent of dose rate and of dose fractionation, although the doses needed to cause suppression of CHS were about tenfold lower. A time delay of between 1 and 3 d after UV irradiation was also necessary for generation of suppression, and when a contact sensitizer was applied to a UV-irradiated animal, transferable suppressor T cells with antigenic specificity for the contact sensitizer were generated. These similarities to UV suppression of tumor rejection suggested a common mechanism of initiation of these two forms of suppression.[4]

2. Immunologic Characteristics

In terms of immunolgic mechanism, a major contribution was made by the work of Greene and associates,[83,84] who showed an antigen-presenting cell (APC) defect in the spleens of UV-irradiated mice. Their initial experiments found that UV-irradiated mice could be immunized to give a DTH response if antigen was given conjugated to APCs from normal animals, suggesting the underlying problem in UV-irradiated mice was in antigen presentation and/or processing. Subsequent studies confirmed that APCs from UV mice did not present antigen as well as APC from normal mice in a number of *in vitro* systems.[85,86] It has since been shown that this "defect" or alteration in antigen presentation is not simply a change in the numbers of APC in the spleens of UV animals, but that a specific defect occurs in the antigen-presenting function of dendritic cells of the spleen.[87] The mechanism is not clear, but it is not caused by an alteration to MHC class II glycoprotein expression on these cells and is not reversible *in vitro* by indomethacin.

The finding of an APC defect in UV animals provides an hypothesis for the mechanism by which suppressor T cells are formed when antigen is administered to a UV-irradiated animal. In the hypothesis,[83,84,88] antigens which induce cell-mediated immunity are presented in UV-irradiated animals in such a way that suppressor T cells rather than positive effector cells are generated. This hypothesis implies that antigen presentation for the generation of antibody responses occurs by a different a pathway(s) from that used for cell-mediated immunity and is supported by the recent literature indicating that B cells, particularly immune B cells, are efficient APCs.[89] It is also necessary to postulate that suppressor T cells specific for UV-tumor rejection are formed when UV antigens generated in the skin interact with the UV-altered APCs to produce suppressor cells specific for UV tumors.[3,4,88] Recent evidence has been obtained for the generation of such UV antigens.[79,90] In the initial studies showing the generation of antigen-specific suppressor T cells, suppressors of the induction phase of the immune response were demonstrated, but, subsequently, suppressors of the elicitation phase of the DTH response were also described.[91,92]

3. Possible Mechanisms of UV Suppression

The finding that UV-induced suppression is initiated by UVB and UVC, i.e., wavelengths between 260 and 320 nm, posed a critical problem in consideration of mechanism. These wavelengths do not penetrate beyond the skin,[6] most certainly not to the spleen, and very little if at all even to the capillaries of the dermal/epidermal junction. How, then, does the generation of the APC defect and of the suppressor T cells occur? De Fabo and Noonan[3] put forward the hypothesis from a wavelength dependence study (action spectrum) for the UV-induced systemic suppression of CHS that it is initiated by an interaction between UV radiation and a specific photoreceptor molecule in the upper layers of the skin. From arguments detailed in that paper, we postulated that the nature of this photoreceptor is UCA (deaminated histidine), a major UV-absorbing component of the stratum corneum, which isomerizes from the naturally occurring *trans* form to the *cis* isomer on UV irradiation. We postulated that *cis*-UCA interacts with the immune system, initiating immunosuppression by the generation of an APC defect, which, on antigenic challenge, results in the formation of antigen specific suppressor T cells.[3] Evidence for three steps of this hypothesis has since been presented. Administration of *cis*-UCA initiates a splenic APC defect *in vivo*,[87] and administration of a UV-irradiated mixture of *cis*- and *trans*-UCA or of *cis*-UCA alone initiates suppression of the DTH response to herpesvirus[93,94] or to contact sensitizers.[95,96] The generation of antigen specific suppressor T cells has been shown in the herpesvirus system.[93,94] Preliminary experiments with animals genetically deficient in UCA have indicated that they do not show UV-induced suppression.[97]

UV irradiation also causes a number of changes both to epidermal Langerhans cells (LC) and to keratinocytes which have also been postulated to have a role in UV-induced suppression. The studies of Toews et al.[98] showed that contact sensitization of mice directly through the UV-irradiated site caused suppression of the CHS response. Since this effect correlated with loss of epidermal LC at the UV-irradiated site it was initially concluded that this form of "local" suppression was due to local depletion of LC which are antigen-presenting cells. Streilein and Bergstresser have since reported that "local" suppression does not occur in all strains of mice, despite UV-induced depletion of LC, indicating loss of LC does not necessarily correlate with UV-induced "local" suppression of CHS.[99]

Systemic suppression of CHS, in contrast, occurs in all strains of mice so far tested, with the exception of the UCA-deficient animals referred to above, and can be clearly separated from UV effects on LC numbers by the use of narrow bands of UV irradiation.[100] A recently described class of epidemal cells which carry the T cell marker Thy.1 and have rearranged gamma/delta T cell receptors has been described. These cells are depleted *in vivo* by UV irradiation and appear to repopulate slowly, if at all, after UV irradiation.[101] An IJ[+] "UV-resistant" cell has been postulated to be responsible for the generation of UV-induced suppression [102] but has not yet been isolated and characterized, especially in regard to its UV sensitivity.

UV irradiation stimulates interleukin 1 (IL-1) production both *in vivo* and *in vitro*.[103] More relevant to this discussion are the *in vivo* studies which showed a biphasic dose response with an initial rise in IL-1 production with increasing UV dose, followed by a decrease in production presumably as cell killing occurs.[103] Comparable dose-response curves have also been obtained *in vivo* for the formation of inflammatory mediators after UV irradiation.[104] Thus, the systemic appearance of these meditors does not correlate directly with suppression. Immunosuppressive factors have been found in the serum of UV-irradiated mice[105,106] and in the culture supernatants of UV-irradiated PAM 212 cells (a mouse keratinocyte cell line).[107] These may be related to the "Contra-IL-1" activity also found in the same supernatants.[108] Prostaglandins have also been implicated since very high doses of indomethacin can reverse UV-induced suppression.[109] Our own studies with narrow bands of UV irradiation indicated, however, that wavelengths of UV which were the most inflammatory were not the most immunosuppressive.[3] Alterations to lymphocyte recirculation have also been described in UV-irradiated animals,[110] but in these studies broadband UV irradiation was used, so whether these are sequelae of inflammation or are relevant to the generation of UVB-induced suppression is as yet unresolved.

C. SCOPE AND RELEVANCE OF UV SUPPRESSION

UV irradiation has been shown to suppress cell-mediated immunity to UV tumors and to contact sensitizers or to hapten-conjugated cells as discussed above, to viruses,[91] to parasites,[111,112] and to alloantigens.[92] UV irradiation also prevents the induction of experimental allergic encephalomyelitis, an autoimmune disease,[113,121] and UVB is lethal for male but not female mice of the autoimmune strain BxSb.[114] It should be noted that it has not been established that all these effects are mediated by the same mechanism.

UV radiation effects on transplantation have been recently reviewed[115] and are complicated, as are the CHS studies discussed above, by the differing methodologies used. Initial studies of UV-induced suppression could find no effect on skin graft survival or on the survival of allogeneic UV tumors in UV-irradiated mice, suggesting that UV radiation did not alter responses to alloantigens. In a recent study,[92] we confirmed that skin grafts are not prolonged in UV-irradiated recipients, but that heart allografts — a primary vascularized graft — can be prolonged by a single prior dose of UV irradiation. This finding is in accord with (1) the observations that UV irradiation depresses the DTH response to alloantigens with the formation of antigen-specific suppressor T cells and (2) previous studies which had found primarily vascularized grafts to be more readily prolonged than skin grafts.

A number of studies[115] have reported that UV irradiation of the graft itself (platelets, bone marrow, pancreatic islets, or heterotopically transplanted corneas) rather than of the recipient prolonged graft acceptance. Administration of UV-irradiated donor blood prior to transplantation has also been reported to prolong allografts.[115] The mechanism has been postulated to be an effect on APC function in the irradiated tissue. In a rabbit model of orthotopic transplantation of vascularized corneal grafts UV irradiation of both the graft and the recipient cornea prolonged graft acceptance.[116] This was not accompanied by a loss of cells positive for leukocyte common antigen, but may be related to loss of antigen-presenting function in the irradiated tissues.

The effects of UV irradiation on autoimmunity are of particular interest since we have postulated[3] that a possible reason for the existence of a regulatory skin photoreceptor (UCA) controlling interaction between UVB and the immune system may be to regulate against autoimmune attack on sun-damaged skin containing "photoantigens". All mammalian systems tested to date have UCA in the skin; thus, isomerization to *cis*-UCA in the presence of UVB will occur. This regulatory circuit may be an evolutionary consequence of an adaptive response to an environment high in UVB. An unfortunate ancillary consequence of such a mechanism would be the inability to stop the growth of UV-induced skin tumors which also bear UVB-induced "photoantigens".

The levels of solar UVB which reach the surface of the earth are attenuated by stratospheric ozone. Convolution of the action spectrum for systemic suppression of CHS in mice with solar UVB levels indicates that about 30 min exposure to sunlight at noon in summer at latitude 40° degrees north is sufficient for 50% systemic suppression of CHS in the BALB/c mouse.[122] Recent studies in the Antarctic have indicated large (around 40%) depletions in stratospheric ozone in the polar spring. This is believed to be due to unusual chemistry between chlorofluorocarbons and ice crystals formed in the extremely cold atmosphere of the South Pole.[117-119] Although such huge depletions appear to be a result of unusual chemistry occurring in polar regions, much smaller but significant depletions of stratospheric ozone have been reported at more temperate latitudes.[120] Thus, studies of UVB effects on biological systems, particularly recently described effects on the mammalian immune system, are of considerable current importance.

ACKNOWLEDGMENT

We thank Dr. Paul T. Strickland for valuable comments and suggestions.

REFERENCES

1. **Frederick, J. E., Snell, H. E., and Haywood, E. K.,** Solar ultraviolet radiation at the earth's surface, *Photochem. Photobiol.,* in press.
2. **Holick, M. F., Smith, E., and Pincus, S.,** Skin as the site of vitamin D synthesis and target tissue for 1,25-dihydroxyvitamin D3. Use of calcitriol (1,25-dihydroxyvitamin D3) for treatment of psoriasis, *Arch. Dermatol.,* 123, 1677, 1987.
3. **De Fabo, E. C. and Noonan, F. P.,** Mechanism of immune suppression by ultraviolet irradiation in vivo. I. Evidence for the existence of a unique photoreceptor in skin and its role in photoimmunology, *J. Exp. Med.,* 158, 84, 1983.
4. **Noonan, F. P., De Fabo, E. C., and Kripke, M. L.,** Suppression of contact hypersensitivity in mice by ultraviolet irradiation. An experimental model. *Springer Semin. Immunopathol.,* 4, 293, 1981.
5. **Morison, W. L. and Strickland, P. T.,** Environmental UVA radiation and eye protection during PUVA therapy, *J. Am. Acad. Dermatol.,* 9, 522, 1983.
6. **Agin, P., Rose, A. P., III, Lane, C. C., Akin, F. J., and Sayre, R. M.,** Changes in epidermal forward scattering absorption after UVA or UVA-UVB irradiation, *J. Invest. Dermatol.,* 76, 174, 1981.
7. **Setlow, R. B.,** The wavelengths in sunlight effective in producing skin cancer: a theoretical analysis, *Proc. Natl. Acad. Sci. U.S.A,* 71, 3363, 1974.
8. **Cerutti, P. and Netrawali, M.,** in *Proc. 6th Int. Congr. on Radiation Research,* Okada, S., Inamura, M., Terashima, T., and Yamaguchi, H., Eds., Toppan, Tokyo, Japan, 1979, 423.
9. **Peak, M. J., Peak, J. G., and Carnes, B. A.,** Induction of direct and indirect single strand breaks in human cell DNA by far and near ultraviolet radiations: action spectrum and mechanisms, *Photochem. Photobiol.,* 45, 381, 1987.
10. **Helene, C.,** Photo-induced responses in UV-irradiated cells, in *Molecular Models of Photoresponsiveness,* Montagnoli, G. and Erlanger, B. F., Eds., Plenum Press, New York, 1983, 57.
11. **Beukers, R. and Berends, W.,** Isolation and identification of the irradiation product of thymine, *Biochim. Biophys. Acta,* 41, 550, 1960.
12. **Wulff, D. L. and Frankel, G.,** On the nature of thymine photoproduct, *Biochim. Biophys. Acta,* 51, 332, 1961.
13. **Setlow, R. B. and Carrier, W. L.,** Pyrimidine dimers in ultraviolet-irradiated DNAs, *J. Mol. Biol.,* 17, 237, 1966.
14. **Tyrrell, R.,** Induction of pyrimidine dimers in bacterial DNA by 365nm radiation, *Photochem. Photobiol.,* 17, 69, 1973.
15. **Elkind, M. and Han, A.,** DNA single strand lesions due to sunlight and UV light: a comparison of this induction in Chinese hamster and human cells, and their fate in Chinese hamster cells, *Photochem. Photobiol.,* 27, 717, 1978.
16. **Webb, R.,** Lethal and mutagenic effects of near-ultraviolet radiation, in *Photochemical Photobiological Reviews,* Vol. 2, Smith, K., Ed., Plenum Press, New York, 1977, 169.
17. **Ellison, M. J. and Childs, J. D.,** Pyrimidine dimers induced in *E. coli* DNA by ultraviolet radiation present in sunlight, *Photochem. Photobiol.,* 34, 465, 1981.
18. **Patrick, M. H. and Rahn, R. O.,** in *Photochemistry and Photobiology of Nucleic Acids,* Vol. 2, Wang, S. Y., Ed., Academic Press, New York, 1976, 147.
19. **Patrick, M. H.,** Near-UV photolysis of pyrimidine heteroadducts in *E. coli* DNA, *Photochem. Photobiol.,* 11, 477, 1970.
20. **Ikenaga, M., Patrick, M. H., and Jagger, J.,** Action of photoreactivating light on pyrimidine heteroadduct in bacteria, *Photochem. Photobiol.,* 11, 487, 1970.
21. **Patrick, M. H.,** Studies of thymine-derived UV photoproducts in DNA. I. Formation and biological role of pyrimidine adducts in DNA, *Photochem. Photobiol.,* 25, 357, 1977.
22. **Taylor, J.-S., Garrett, D. S., and Cohrs, M. P.,** Solution-state structure of the Dewar pyrimidinone photoproduct of thymidylyl-(3'-5')-thymidine, *Biochemistry,* 27, 7206, 1988.
23. **Husain, I., Carrier, W. L., Regan, J. D., and Sabcar, A.,** Photoreactivation of killing in *E. coli* K-12 *phr-* cells is not caused by pyrimidine dimer reversal, *Photochem. Photobiol.,* 48, 233, 1988.
24. **Glickman, B. W., Schaaper, R. M., Haseltine, W. A., Dunn, R. L., and Brash, D. E.,** The C-C (6-4) UV photoproduct is mutagenic in *E. coli, Proc. Natl. Acad. Sci. U.S.A.,* 83, 6945, 1986.

25. **Hariharan, P. and Cerutti, P. A.,** Formation of photoproducts of the 5,6-dihydroxydihydrothymine type by ultraviolet light in HeLa cells, *Biochemistry,* 16, 2791, 1977.

26. **Tyrrell, R.,** Damage and repair from non-ionizing radiations, in *Repairable Lesions in Microorganisms,* Hurst, A. and Nasin, A., Eds., Academic Press, New York, 1984.

27. **Dunlap, B. and Cerutti, P. A.,** Apyrimidinic sites in gamma-irradiated DNA, *FEBS. Lett.,* 51, 188, 1975.

28. **Shetlar, M. D.,** Cross-linking of proteins to nucleic acids by ultraviolet light, *Photochem. Photobiol. Rev.,* Vol. 5, Smith, K., Ed. 1980, 105.

29. **Bradley, M. A. and Kohn, K. W.,** X-Ray induced DNA double strand break production and repair in mammalian cells as measured by neutral filter elution, *Nucleic Acids Res.,* 7, 793, 1979.

30. **Peak, M. J., Peak, J. A., and Jones, C. A.,** Different (direct and indirect) mechanisms for the induction of DNA-protein cross-links in human cells by far and near ultraviolet irradiation (290 and 405nm), *Photochem. Photobiol.,* 42, 141, 1985.

31. **Peak, J. G., Peak, M. J., Sikorski, R. S., and Jones, C. A.,** Induction of DNA-protein cross-links in human cells by ultraviolet and visible radiations: action spectrum, *Photochem. Photobiol.,* 41, 295, 1985.

32. **Miguell, A. G. and Tyrrell, R. M.,** Induction of oxygen-dependent lethal damage by monochromatic UVB (313 nm) radiation: strand breakage, repair and cell death, *Carcinogenesis,* 4, 375, 1983.

33. **Rosenstein, B. S. and Dunmore, H. M.,** 10th Annu. Meet. American Society for Photobiology, Vancouver, Canada, June 27 to July 1, 1982.

34. **Ley, R. D., Sedita, B. A., and Boye, E.,** DNA Polymerase I mediated repair of 365nm induced single strand breaks in the DNA of *E. coli, Photochem. Photobiol.,* 27, 323, 1978.

35. **Peak, M. J. and Peak, J. G.,** Single strand breaks induced in *Bacillus subtilis* DNA by ultraviolet light: action spectrum and properties, *Photochem. Photobiol.,* 35, 675, 1982.

36. **Webb, R. B., Brown, M. S., and Ley, R. D.,** Non-reciprocal synergistic lethal interaction between 365nm and 405nm radiation in wild type and *uvr*A strains of *E. coli., Photochem. Photobiol.,* 35, 697, 1982.

37. **Jagger, J.,** *Induction to Research in Ultraviolet Photobiology,* Prentice-Hall, Englewood Cliffs, NJ, 1967.

38. **Setlow, R. B. and Setlow, J. K.,** Evidence that ultraviolet-induced thymine dimers in DNA cause biological damage, *Proc. Natl. Acad. Sci. U.S.A.,* 48, 1250, 1962.

39. **Harm, W.,** *Biological Effects of Ultraviolet Radiation,* Cambridge University Press, London, 1980.

40. **Brash, D. E., Seetheram, S., Kraemer, K. H., Seidman, M. M., and Bredberg, A.,** Photoproduct frequency is not the major determinant of UV base substitution hot spots or cold spots in human cells, *Proc. Natl. Acad. Sci. U.S.A,* 84, 3782, 1987.

41. **Cook, J. S.,** Photoreactivation in animal cells, *Photophysiology,* 5, 191, 1970.

42. **Sutherland, B. M.,** Enzymic photoreactivation of DNA, in *DNA Repair Mechanisms,* Hanawalt, P. C., Friedberg, E. C., and Fox, C. F., Eds., Academic Press, New York, 1978, 113.

43. **Rosenstein, B. W. and Setlow, R. B.,** Photoreactivation of ICR2A frog cells after exposure to monochromatic ultraviolet radiation in the 252-313nm range, *Photochem. Photobiol.,* 32, 361, 1980.

44. **D'Ambrosio, S. M., Whetstone, J. W., Slazinski, L., and Lowney, E.,** Photorepair of pyrimidine dimers in human skin *in vivo, Photochem. Photobiol.,* 34, 461, 1981.

45. **Hart, R. W., Setlow, R. B., and Woodhead, A. D.,** Evidence that pyrimidine dimers in DNA can give rise to tumors, *Proc. Natl. Acad. Sci. U.S.A.,* 74, 5574, 1977.

46. **Le Clerc, J. E. and Istock, N. L.,** Specificity of UV mutagenesis in the *lac* promotor of M13 *lac* hybrid phage DNA, *Nature (London),* 297, 596, 1982.

47. **Setlow, R. B. and Carrier, W. L.,** The disappearance of thymine dimers from DNA: an error correcting mechanism, *Proc. Natl. Acad. Sci. U.S.A.,* 51, 226, 1964.

48. **Boyce, R. P. and Howard-Flanders, P.,** Release of ultraviolet light-induced thymine dimers from DNA in *E. coli* K-12, *Proc Natl. Acad. Sci. U.S.A.,* 51, 293, 1964.

49. **Kraemer, K. H. and Slor, H.,** Xeroderma pigmentosum, *Clin. Dermatol.,* 3, 33, 1985.

50. **Jagger, J.,** *Solar-UV Actions on Living Cells,* Praeger, New York, 1985.

51. **Smith, K. C. and Hanawalt, P. C.,** *Molecular Photobiology: Inactivation and Recovery,* Academic Press, New York, 1969, chap. 5.

52. **Anglin, J. R., Bever, A. T., Everrett, M. A., and Lamb, J. H.,** Ultraviolet light induced alterations in urocanic acid *in vivo, Biochim. Biophys. Acta,* 53, 408, 1961.

53. **Morrison, H., Avnir, D., Bernasconi, C., and Fagan, G.,** Z/E photoisomerization of urocanic acid, *Photochem. Photobiol.,* 32, 711, 1980.

54. **De Fabo, E. C. and Kripke, M. L.,** Wavelength dependence and dose-rate independence of UV radiation-induced suppression of immunologic unresponsiveness of mice to a UV-induced fibrosarcoma, *Photochem. Photobiol.,* 32, 183, 1980.

55. *The Photonics Buyers Guide,* P.O. Box 1146, Dept. BOF-88, Pittsfield, MA 01202-9985.

56. **Shropshire, W., Jr.,** Action spectroscopy, in *Phytochrome,* Mitrakos, K. and Shropshire, W., Jr., Eds., Academic Press, New York, 1972, 161.

57. **De Fabo, E. C.,** On the nature of the blue light photoreceptor: still an open question, in *The Blue Light Syndrome,* Senger, H., Ed., Springer-Verlag, Berlin, 1980, 187.

58. **Urbach, F.**, *The Biological Effects of Ultraviolet Radiation*, Pergamon Press, New York, 1969.
59. **Gates, F. L.**, A study of the bactericidal action of ultraviolet light. III. The absorption of ultraviolet light by bacteria, *J. Gen. Physiol.*, 14, 31, 1930.
60. **Blum, H. F.**, *Carcinogenesis by Ultraviolet Light*, Princeton University Press, Princeton, NJ, 1959.
61. **Hanisko, J. and Suskind, R. R.** The effect of ultraviolet radiation on experimental cutaneous sensitization in guinea pigs, *J. Invest. Dermatol.*, 40, 183, 1963.
62. **Graffi, A., Pasternak, G., and Horn, K.-H.**, Die Erzeugung von Resistenz gegen isologe Transplantate UV-induzierter Sarkome der Maus, *Acta Biol. Med. Ger.*, 12, 126, 1964.
63. **Pasternak, G., Graffi, A., and Horn, K.-H.**, Der Nachwies individualspezifischer Antigenitat bei UV-induzierten Sarkomen der Maus, *Acta Biol. Med. Ger.*, 13, 276, 1964.
64. **Kripke, M. L.**, Antigenicity of murine skin tumors induced by ultraviolet light, *J. Natl. Cancer Inst.*, 53, 1333, 1974.
65. **Kripke, M. L.**, Latency, histology, and antigenicity of tumors induced by ultraviolet light in three inbred mouse strains, *Cancer Res.*, 37, 1395, 1977.
66. **Spikes, J. D., Kripke, M. L., Connor, R. J., and Eichwald, E. J.**, Time of appearance and histology of tumors induced in the dorsal skin of C3H female mice by ultraviolet radiation from a mercury arc lamp, *J. Natl. Cancer Inst.*, 59, 1637, 1977.
67. **Kripke, M. L. and Fisher, M. S.**, Immunologic parameters of ultraviolet carcinogenesis, *J. Natl. Cancer Inst.*, 57, 211, 1976.
68. **Fisher, M. S. and Kripke, M. L.**, Systemic alteration induced in mice by ultraviolet light irradiation and its relationship to ultraviolet carcinogenesis, *Proc. Natl. Acad. Sci. U.S.A.*, 74, 1688, 1977.
69. **Fisher, M. S. and Kripke, M. L.**, Further studies on the tumor-specific suppressor cells induced by ultraviolet radiation, *J. Immunol.*, 121, 1139, 1978.
70. **Kripke, M. L., Lofgreen, J. S., Beard, J., Jessup, J. M., and Fisher, M. S.**, *In vivo* immune responses of mice during carcinogenesis by ultraviolet irradiation, *J. Natl. Cancer Inst.*, 59, 1227, 1977.
71. **Spellman, C. W., Woodward, J. G., and Daynes, R. A.** Modification of immunologic potential by ultraviolet radiation. I. Immune status of short-term UV-irradiated mice, *Transplantation*, 24, 112, 1977.
72. **Spellman, C. W. and Daynes, R. A.**, Modification of immunologic potential by ultraviolet radiation. II. Generation of suppressor cells in short-term UV-irradiated mice, *Transplantation*, 24, 120, 1977.
73. **Spellman, C. W. and Daynes, R. A.**, Properties of ultraviolet light-induced suppressor lymphocytes within a syngeneic tumor system, *Cell. Immunol.*, 36, 383, 1978.
74. **Fisher, M. S. and Kripke, M. L.**, Suppressor T lymphocytes control the development of primary skin cancers in ultraviolet irradiated mice, *Science*, 216, 1133, 1982.
75. **Kripke, M. L.**, Immunologic mechanisms in UV radiation carcinogenesis, *Adv. Cancer Res.*, 34, 69, 1981.
76. **Urban, J. L., Van Waes, C., and Schreiber, H.**, Pecking order among tumor-specific antigens, *Eur. J. Immunol.*, 14, 181, 1984.
77. **Philips, C., McMillan, M., Flood, P. M., Murphy, D. B., Forman, J., Lancki, D., Womack, J. E., Goodenow, R. S., and Schreiber, H.**, Identification of a unique tumor-specific antigen as a novel Class I major histocompatibility molecule, *Proc. Natl. Acad. Sci. U.S.A.*, 82, 5140, 1985.
78. **Stauss, H. J., Van Waes, C., Fink, M. A., Starr, B., and Schreiber, H.**, Identification of a unique tumor antigen as rejection antigen by molecular cloning and gene transfer, *J. Exp. Med.*, 164, 1516, 1986.
79. **Ananthaswamy, H.**, Relationship between expression of tumor-specific transplantation antigens and neoplastic transformation in an ultraviolet radiation-induced murine skin cancer, *Cancer Res.*, 46, 6322, 1986.
80. **De Fabo, E. C. and Kripke, M. L.**, Dose-response characteristics of immunologic unresponsiveness to UV-induced tumors produced by UV irradiation of mice, *Photochem. Photobiol.*, 30, 385, 1979.
81. **Jessup, J. M., Hanna, N., Palaszynski, E., and Kripke, M. L.**, Mechanisms of depressed reactivity to dinitrochlorobenzene and ultraviolet-induced tumors during ultraviolet carcinogenesis in BALB/c mice, *Cell. Immunol.*, 38, 105, 1978.
82. **Noonan, F. P., De Fabo, E. C., and Kripke, M. L.**, Suppression of contact hypersensitivity by UV radiation and its relationship to UV-induced suppression of tumor immunity, *Photochem. Photobiol.*, 34, 683, 1981.
83. **Noonan, F. P., Kripke, M. L., Pedersen, G. M., and Greene, M. I.**, Suppression of contact hypersensitivity in mice is associated with defective antigen presentation, *Immunology*, 43, 524, 1981.
84. **Greene, M. I., Sy, M.-S., Kripke, M. L., and Benacerraf, B.**, Impairment of antigen-presenting function by ultraviolet radiation, *Proc. Natl. Acad. Sci. U.S.A.*, 76, 6592, 1979.
85. **Letvin, N. L., Greene, M. I., Benacerraf, B., and Germain, R. N.**, Immunologic effects of whole-body ultraviolet irradiation: selective defect in splenic adherent cell function in vitro, *Proc. Natl. Acad. Sci. U.S.A.*, 77, 2881, 1980.
86. **Letvin, N. L., Nepom, J. T., Greene, M. I., Benacerraf, B., and Germain, R. N.**, Loss of Ia-bearing splenic adherent cells after whole body ultraviolet irradiation, *J. Immunol.*, 125, 2550, 1980.

87. **Noonan, F. P., De Fabo, E. C., and Morrison, H.,** Cis-urocanic acid, a product formed by ultraviolet B irradiation of the skin, initiates an antigen presentation defect in splenic dendritic cells in vivo, *J. Invest. Dermatol.,* 90, 92, 1988.

88. **Noonan, F. P. and De Fabo, E. C.,** Immune suppression by ultraviolet radiation and its role in ultraviolet radiation induced carcinogenesis in mice, *Aust. J. Dermatol.,* 26, 4, 1985.

89. **Chesnut, R. W., and Grey, H. M.,** Antigen presenting cells and mechanisms of antigen presentation, *CRC Crit. Rev. Immunol.,* 5, 263, 1985.

90. **Hostetler, L. W., Ananthaswamy, H. N., and Kripke, M. L.,** Generation of tumor-specific transplantation antigens by UV radiation can occur independently of neoplastic transformation, *J. Immunol.,* 15, 2721, 1986.

91. **Howie, S. E. M., Norval, M., Maingay, J., and Ross, J. A.,** Two phenotypically distinct T cells (Ly 1^+2^- and Ly1^-2^+) are involved in ultraviolet B light induced suppression of the efferent DTH response to HSV-1 in vivo, *Immunology,* 58, 653, 1986.

92. **Mottram, P. M., Mirisklavos, A., Clunie, G. J. A., and Noonan, F. P.,** A single dose of UV radiation suppresses delayed type hypersensitivity responses to alloantigens and prolongs heart allograft survival in mice, *Immunol. Cell Biol.,* 66, 377, 1988.

93. **Ross, J., Howie, S. E. M., Norval, M., and Maingay, J., and Simpson, T.,** UV irradiation of urocanic acid suppresses the delayed type hypersensitivity response to H. Simplex virus in mice, *J. Invest. Dermatol.,* 87, 630, 1986.

94. **Ross, J. A., Howie, S. E. M., Norval, M., and Maingay, J.,** Two phenotypically distinct T cells are involved in ultraviolet-irradiated urocanic acid-induced suppression of the efferent delayed-type hypersensitivity response to Herpes Simplex Virus Type I in vivo, *J. Invest. Dermatol.,* 89, 230, 1987.

95. **Harriott-Smith, T. G. and Halliday, W. J.,** Suppression of contact hypersensitivity by short-term ultraviolet irradiation. II. The role of urocanic acid, *Clin. Exp. Immunol.,* 72, 174, 1988.

96. **Hug, D. H., Morrison, H., Tessman, I., De Fabo, E. C., and Noonan, F. P.,** Photobiology of urocanic acid, in *Photobiology, 1984,* Longworth, J. W., Jagger, J., and Shropshire, W., Jr., Eds., Praeger, New York, 1985, 141.

97. **De Fabo, E. C., Noonan, F. P., Fisher, M. S., Burns, J., and Kacser, H.,** Further evidence that the photoreceptor mediating UV-induced systemic immune suppression is urocanic acid, *J. Invest. Dermatol.,* 80, 319, 1983.

98. **Toews, G. B., Bergstresser, P. R., and Streilein, J. W.,** Epidermal Langerhans cell density determines whether contact hypersensitivity or unresponsiveness follows skin painting with DNFB, *J. Immunol.,* 124, 445, 1980.

99. **Streilein, J. W. and Bergstresser, P. R.,** Genetic basis of ultraviolet-B effects on contact hypersensitivity, *Immunogenetics,* 27, 252, 1988.

100. **Noonan, F. P., Bucana, C., Sauder, D. N., and De Fabo, E. C.,** Mechanism of systemic immunosuppression by UV irradiation in vivo. II. The UV effects on number and morphology of epidermal Langerhans Cells and the systemic suppression of contact hypersensitivity have different wavelength dependencies, *J. Immunol.,* 132, 2408, 1984.

101. **Aberer, W., Romani, N., Elbe, A., and Stingl, G.,** Effects of physicochemical agents on murine epidermal Langerhans cells and Thy $-1+$ dendritic epidermal cells, *J. Immunol.,* 136, 1210, 1986.

102. **Granstein, R. and Greene, M. I.,** Epidermal antigen presenting cells in activation of suppression, *J. Immunol.,* 132, 563, 1984.

103. **Ansel, J. C., Luger, T. A., and Green, I.,** The effect of in vitro and in vivo UV irradiation on the production of ETAF Activity by human murine keratinocytes, *J. Invest. Dermatol.,* 81, 519, 1983.

104. **Gahrig, L. C. and Daynes, R. A.,** Desensitization of animals to the inflammatory effects of ultraviolet radiation is mediated through mechanisms which are distinct from those responsible for endotoxin tolerance, *J. Immunol.,* 136, 2868, 1986.

105. **Swartz, R. P.,** Role of UVB-induced serum factor(s) in suppression of contact hypersensitivity in mice, *J. Invest. Dermatol.,* 83, 305, 1984.

106. **Harriott-Smith, T. G. and Halliday, W. J.,** Circulating suppressor factors in mice subjected to ultraviolet irradiation and contact sensitisation, *Immunology,* 57, 207, 1986.

107. **Schwartz, T., Urbanska, A., Gschanit, F., and Luger, T. A.,** Inhibition of the induction of contact hypersensitivity by a UV-mediated epidermal cytokine, *J. Invest. Dermatol.,* 87, 289, 1986.

108. **Schwartz, T., Urbanska, A., Gschnait, F., and Luger, T.,** UV-irradiated epidermal cells produce a specific inhibitor of interleukin 1 activity, *J. Immunol.,* 138, 1457, 1987.

109. **Chung, H.-T., Burnham, D. K., Robertson, B., Roberts, L. K., and Daynes, R. A.,** Involvement of prostaglandins in the immune alterations caused by the exposure of mice to ultraviolet radiation, *J. Immunol.,* 137, 2478, 1986.

110. **Spangrude, G. J., Bernhard, E. J., Ajioka, R. S., and Daynes, R. A.,** Alterations in lymphocyte homing patterns within mice exposed to ultraviolet radiation, *J. Immunol.,* 130, 2974, 1983.

111. **Giannini, M. S. H.,** Suppression of pathogenesis in cutaneous Leishmaniasis by UV irradiation, *Infect. Immun.,* 51, 838, 1986.

112. **Giannini, M. S. H. and De Fabo, E. C.,** Abrogation of skin lesions in cutaneous Leishmaniasis by ultraviolet B radiation, in *Leishmaniasis: The First Centenary,* Hart, D. T., Ed., NATO ASI Ser. A: Life Sciences, Plenum Publishing, London, 1988, 65.

113. **Hauser, S. L., Weiner, H. L., Che, M., Shapiro, M. E., Gilles, F., and Letvin, N.,** Prevention of experimental allergic encephalitis (EAE) in the SJL/J mouse by whole body ultraviolet irradiation, *J. Immunol.,* 132, 1276, 1984.

114. **Ansel, J. C., Mountz, J., Steinberg, A. D., De Fabo, E. C., and Green, I.,** Effects of UV radiation on autoimmune strains of mice: increased mortality and accelerated autoimmunity in BxSB male mice, *J. Invest. Dermatol.,* 85, 181, 1985.

115. **Deeg, H. J.,** Ultraviolet irradiation in transplantation biology, *Transplantation,* 45, 845, 1988.

116. **Williams, K. A., Ash, J. K., Mann, T. S., Noonan, F. P., and Coster, D. J.,** Cells infiltrating inflamed and vascularized corneas, *Transplant. Proc.,* XIX(2), 2889, 1987.

117. **Farman, J. C., Gardiner, B. G., and Shanklin, J. D.,** Large losses of total ozone in Antarctica reveal seasonal CLO_x/NO_x interaction, *Nature (London),* 315, 207, 1985.

118. **Stolarski, R. S., Krueger, A. J., Schoeberl, M. R., McPeters, R. D., Newman, P. A., and Alpert, J. C.,** Nimbus satellite measurements of the springtime Antarctic ozone decrease, *Nature (London),* 322, 808, 1986.

119. **Solomon, P. M., Connor, B., de Zafra, R. L., Parrish, A., Barrett, J., And Jaramillo, M.,** High concentrations of chlorine monoxide at low altitudes in the Antarctic spring stratosphere: secular variation, *Nature (London),* 328, 411, 1987.

120. **Watson, R. T., Geller, M. A., Stolarski, R. S., and Hampson, R. F.,** Present state of knowledge of the upper atmosphere: an assessment report, *NASA Ref. Publ. No.* 1162, Washington, D.C. May, 1986. (Available from the National Technical Information Service, Springfield, Virginia, 22161.)

121. **Noonan, F. P. and McCarron, R. M.,** UV irradiation suppresses EAE, an autoimmune demyelinating disease in mice, *Photochem. Photobiol.,* 495, 925, 1989.

122. **DeFabo, E. C., Noonan, F. P., and Frederick, J. E.,** Biologically effective doses of sunlight for immune suppression and their relationship to changes in stratospheric ozone, submitted.

Antigen Detection in Tissues and Cells

Chapter 5

ULTRASTRUCTURAL DEMONSTRATION OF CELL SURFACE AND INTRACELLULAR CLASS II MAJOR HISTOCOMPATIBILITY ANTIGENS BY MONOCLONAL ANTIBODIES USING A PREEMBEDDING INDIRECT IMMUNOPEROXIDASE TECHNIQUE

Llewellyn D. J. Spargo and Graham Mayrhofer

TABLE OF CONTENTS

I. INTRODUCTION

The Class II antigens encoded within the major histocompatibility complex (MHC) are glycoproteins expressed on the surfaces of a restricted range of cells[1-3] including some macrophages, Langerhans cells, dendritic cells, interdigitating cells, and B lymphocytes. More recent studies have indicated a wider distribution, including vascular endothelium[4] and a variety of epithelial cells.[5-8] MHC Class II antigens are also referred to as "Ia antigens" and are products of genes located in the I-A and I-E subregions in the mouse MHC and in the RTIB and HLA-D subregions of the rat and human MHC, respectively. The molecules which are synonymous with these antigens play a vital role in the presentation of exogenous antigens to T lymphocytes of the helper/inducer type. The genes encoding them in mice are now recognized as the immune response (Ir) genes,[9] and molecular explanations for the effects of these genes are now forthcoming in terms of the ability of the corresponding Class II molecules to bind fragments of antigen containing the relevant epitopes recognized by T cells.[10] MHC molecules are thought of increasingly as receptors with roles in capturing and transporting processed antigen (ligand) through specific intracellular compartments, presenting both endogenous and exogenous antigens in processed form on the surfaces of cells.[11] They may have a fundamental importance in selecting and salvaging epitope-bearing fragments from intracellular proteolysis,[12] and they exert a profound influence on the selection of maturing T lymphocytes.[13]

Despite the acknowledged importance of Class II MHC molecules in associative recognition of antigens by T helper/inducer lymhocytes, relatively little work has been published on the cell biology surrounding the interaction between processed exogenous antigen and its intersection with Class II molecules. Furthermore, there are very few accounts of the ultrastructural localization of Class II molecules in any cell type, although this has been the subject of some studies on intestinal epithelium.[14,15] Functional studies have suggested that endocytosed transferrin, which enters the compartment for uncoupling of receptor and ligand (CURL) and, subsequently, the lysosomes, intersects with a compartment through which newly synthesized Class II molecules are transported to the cell surface.[16] Pernis[17] has described cycling of cell-surface Class I and Class II molecules into an acid intracellular compartment in activated T cells and B cells, respectively. However, there have been no ultrastructural studies to demonstrate the intracellular localization of Class II molecules and, in particular, to identify the intracellular sites at which Class II molecules and processed antigens might interact.

We have examined the subcellular localization of Class II molecules in epithelial cells (enterocytes) of the rat and the human small intestine. The study was undertaken to gain information about the possible functions of Class II molecules in these cells. We encountered great difficulty in preserving the antigenicity of the molecules for ultrastructural studies and have developed techniques to overcome the problems, thus allowing for the first time (in any cell type) the localization of internal Class II MHC molecules.

II. TECHNICAL PROCEDURES

Methods are described which allow demonstration of Class II MHC antigens at both light-microscopic and ultrastructural levels. The former are included because demonstration of Class II antigens is a necessary preliminary to any ultrastructural study. Immunohistochemistry on frozen sections fixed lightly with a fixative such as ethanol affords the best chance of detecting antigen, which may be lost during the many manipulations required to prepare tissue for ultrastructural studies. Such material may also give important clues to the expected subcellular distribution of antigen,[8] and we have found it to be a useful standard against which to compare the results of more denaturing methods of fixation (see Section

TABLE 1

Summary of the Methods Used in Preembedding Localization of Class II MHC Antigens in Jejunal Mucosa by the Indirect Immunoperoxidase Technique

A.   Preparation of tissue prior to immunohistochemistry
      1.   Specimens flattened by adherence to nitrocellulose membrane via the serosal surface
      2.   Preliminary fixation
      3.   Preparation of tissue blocks
      4.   Completion of primary fixation
      5.   Wash in buffer
      6.   Cryopreservation
      7.   Embedding in OCT and rapid freezing in liquid Freon
      8.   Frozen sections (20 μm) prepared and picked up flat on nitrocellulose membrane
      9.   Transfer, without drying, to 0.1% bovine serum albumin to block nitrocellulose

B.   Immunohistochemistry
      1.   Sequential incubations in primary and secondary antibodies, washes after each incubation
      2.   Detection of bound horseradish peroxidase conjugate with 3-3′-diaminobenzidine and H_2O_2
      3.   Washes
      4.   Postfixation in 4% paraformaldehyde/2% glutaraldehyde
      5.   Washes
      6.   Postfixation in 1% OsO_4

C.   Processing for electron microscopy
      1.   Washing in distilled water
      2.   Dehydration through graded ethanol series
      3.   Infiltration with resin and embedding
      4.   Preparation of silver-gold sections
      5.   Counterstaining with Reynolds' lead citrate

III). Light microscopy can also be used to assess the effects on antigenicity of each stage in the preparation of tissue for electron microscopy, and we have used tissue fixed as blocks with ethanol as our standard to determine at which point antigenicity is lost during processing. The steps in preparing tissue for ultrastructural localization of Class II MHC molecules are summarized in Table 1.

A. IMMUNOLOGIC REAGENTS

1. Primary Antibodies

In our work, primary antibodies have all been mouse monoclonal IgG1 antibodies directed against monomorphic determinants on Class II MHC antigens of rat or human. The antibodies have been used as undiluted culture supernatants from the appropriate hybridomas. Cocktails of monoclonal antibodies can be used to obtain greater sensitivity. We have used mixtures consisting of equal parts of MRC OX4 and MRC OX6 antibodies for rat tissue[18] and FMC4, FMC14, FMC15 and FMC52 antibodies for human tissue.[19] If available, immunopurified polyclonal antisera could offer advantages of even greater sensitivity. IgG-containing culture supernatants from irrelevant hybridomas have been used as controls for nonspecific binding. Normal homologous serum (rat or human) is added to reagents to a concentration of 10% to reduce the possibility of binding of IgG via Fc receptors in the tissue.

2. Secondary Antibodies

A horseradish peroxidase conjugate of F(ab′)$_2$ sheep anti-mouse IgG antibody is used to detect mouse antibodies. The antibody was raised in a sheep against mouse IgG (all subclasses) that had been purified by absorption to protein A-Sepharose at pH 8.0 and elution at pH 3.5.[20] An immunoglobulin fraction is precipitated using 34% saturated ammonium sulfate. After washing with further ammonium sulfate solution, the precipitate is dissolved in 20 mM NH$_4$HCO$_3$ buffer (pH 7.0) and dialyzed against the same buffer. The immunoglobulin can be freeze-dried (NH$_4$HCO$_3$ sublimes under vacuum).

To prepare $F(ab^1)_2$ fragments, 1 g of freeze-dried immunoglobulin is dissolved in 50 ml of 0.1 M acetate buffer, pH 4.5, containing 20 mg of pepsin (Product No. 7012, 2500-3500 units per mg, Sigma, St. Louis, MO) and incubated at 37°C for 20 h.[21] It is worth noting that many preparations of pepsin contain mainly pepsinogen. For full activity, the pepsin should be activated autocatalytically by preincubation in a small volume of glycine HCl buffer (pH 2.0) at 37°C for several hours prior to use. If this is not done, yields of $F(ab^1)_2$ will be reduced. The pepsin digest of IgG is neutralized with solid Tris, and the immuno-globulin is precipitated with 18% saturated Na_2SO_4 and reconstituted in phosphate-buffered saline (PBS).

A considerable proportion (approximately 40% in our case) of the anti-mouse IgG antibody cross-reacts with rat immunoglobulin. It is, therefore, not sufficient to affinity purify antibodies on a mouse IgG affinity column if they are to be used on rat tissue. The pepsin digest is, therefore, absorbed by passage through a Sepharose-4B column coated with rat serum proteins and a second column coated with rat IgG (all subclasses). The absorbed material is then passed through a final column coated with mouse IgG (all subclasses). After extensive washing with 0.5% Tween 20 in PBS the bound sheep anti-mouse IgG antibody is eluted with 3 M sodium thiocyanate and dialyzed against PBS.

Finally, the $F(ab^1)_2$ fraction of the affinity-purified antibody is separated from aggregated antibody and undigested IgG by gel filtration using a Sephacryl S-200 (Pharmacia, Uppsala, Sweden) column. This order of purification is chosen to ensure that all of the antibody in the final preparation was active and that it is free of aggregate. The conjugate between $F(ab^1)_2$ fragments of sheep anti-mouse IgG and horseradish peroxidase is made according to the method of Farr and Nakane.[21] The aim is to produce a conjugate in which an average of 1 or 2 molecules of horseradish peroxidase are linked to each molecule of $F(ab^1)_2$. This will ensure the maximum diffusion by the conjugate into tissue sections and minimize nonspecific binding. Briefly, 8 mg of horseradish peroxidase (Type VI, 250 to 330 U/mg, Sigma, St. Louis, MO) is dissolved in 2 ml of distilled water and oxidized by the addition of 100 μl of $NaIO_4$ solution (38.5 mg/ml) for 20 min at room temperature. The activated enzyme is separated from the reaction mixture by passage through a desalting column of Sephadex G50 (Pharmacia, Uppsala, Sweden) equilibrated with 0.01 M carbonate buffer, pH 9.5. Six milligrams of activated enzyme in 4 ml of buffer is then added to 10 mg of the immunopurified $F(ab^1)$ antibody (in 1 ml of 0.01 M carbonate buffer, pH 9.5) and reacted for 2 h at room temperature. The reaction is stopped by reduction with 125 μl of freshly prepared sodium borohydride (4 mg/ml), and the mixture is chromatographed on a calibrated Sephacryl S-200 column. The void volume is discarded, and fractions are collected from a relatively broad peak that precedes the small peak of unbound horseradish peroxidase. The latter has a high OD_{403}:OD_{280} ratio relative to the main peak. Representative fractions from the main peak are analyzed by nonreduced SDS-PAGE and compared with $F(ab^1)_2$, IgG, and horseradish peroxidase. The major products of conjugation by this method contain one or two molecules of horseradish peroxidase, and fractions containing these are pooled. The method yields very little free $F(ab^1)_2$ or contamination with free enzyme.

The conjugate is stored frozen at -100°C in small aliquots at a concentration of 200 μg/ml in the presence of bovine serum albumin (10 mg/ml). A working concentration of approximately 50 μg/ml gives strong staining on frozen sections of rat tissue labeled with mouse monoclonal anti-Class II antibodies, while background staining is undetectable.

B. TISSUE SPECIMENS

In our studies we have examined rat and human small intestine, but we believe the same principles should apply to successful demonstration of Class II molecules in other tissues. In general, blocks of tissue should be small, and they should be fixed as soon as possible after removal. Tissue should be chilled during preparation of blocks, and if there is a delay before fixation, it should be kept in an ice-cold container.

1. Animal Tissue

Rat intestine is used as an example for the following steps:

1.   Anesthetize the animal (we use ether) and expose the required organ in the living animal.
2.   Remove the required tissue, in this case a 10-mm segment of small intestine. Avoid inclusion of fat.
3.   To prepare frozen sections from fresh tissue, open the specimen serosa-side down onto filter paper moistened with phosphate buffered saline (Dulbecco's solution A, pH 7.4), using fine scissors to cut along the mesenteric border. This can be done in a depression in dental wax supported on ice. The edges of the resulting rectangle are trimmed with a razor blade to remove tissue damaged by scissors. Blocks approximately 4 × 4 mm are cut with the razor blade, and they are allowed to roll with the mucosa everted. The resulting cylinders are then embedded on their ends in OCT compound (Miles Scientific, Naperville, IL) and frozen (see Section II.D.3).

      With all tissues, it is important to give attention to the orientation of tissue blocks in order to obtain frozen sections in which the tissue architecture is displayed in a plane that best suits subsequent analysis. In the case of rodent gut, where villi are spade-like rather than finger-like, sectioning in the plane of the axis of the intact intestine gives the best profiles of the villous architecture. Advantage can be taken of the natural tendency of a block of small intestine to roll into a cylinder, the axis of which is at right angles to that of the gut lumen. If embedded on end, frozen sections from such blocks display profiles of villi in regular array. Solid organs present greater problems in obtaining reproducible orientation, but if anatomic landmarks can be used, then subsequent analysis and identification of structures can be assisted greatly.
4.   To prepare fixed blocks, the same procedure as described above is used to produce a trimmed rectangle of tissue, in this case on nitrocellulose membrane (type HA, 0.45-μm pore size, Millipore Corp., Bedford, MA). At this stage, the mucosa is covered with a large drop of ice-cold fixative and left for 1 to 2 min. This preliminary fixation prevents subsequent "rolling up" of the tissue. The wax depression is then flooded with ice-cold fixative for a further 1 to 2 min. Blocks 1—2 × 3—4 mm are cut with a razor blade, with the aid of a dissecting microscope. The long axis is in the original axis of the intestine, thus allowing correct orientation for cutting sections. The blocks are then transferred to bottles containing ice-cold fixative to complete fixation (see Section II.C.2).

2. Human Tissue

We have used both resection specimens of human small intestine and duodenal biopsies obtained by Crosby (or similar) capsules. Resection specimens were transported to the laboratory in ice-cold containers and prepared for fixation within 30 min of removal. Biopsy specimens were fixed in the procedures room within 5 min of the sample being removed.

In the case of resection material, specimen thickness is reduced by first dissecting carefully the mucosa from the tunica muscularis of the opened segment of intestine, using fine scissors. Blocks of tissue are then prepared using a razor blade, exactly as described above, to provide fresh tissue for frozen sections and fixed tissue for ultrastructural studies.

In the case of biopsy specimens, the mucosal fragment is picked up by applying a strip of nitrocellulose membrane to the serosal surface. This allows easy handling of the specimen, it is kept flat during processing, and it can be cut easily by microtome knives. The specimen can be divided with a razor blade to prepare both a fresh-frozen block and a fixed block.

C. FIXATION

The following method has been found to produce the best preservation of ultrastructure that is compatible with preservation of the antigenicity of Class II MHC molecules (see below).

1. Fixative

This is 1% paraformaldehyde and 0.0125% glutaraldehyde, in 0.05 M sodium cacodylate buffer, pH 7.2. The buffer (referred to hereafter as cacodylate buffer) contains 0.1 M sucrose, 0.05% (w/v) $CaCl_2$ (anhydrous), and 1% (w/v) polyvinylpyrrolidone, and has a final osmolarity of 215 mOsm.

Preparation of paraformaldehyde — Paraformaldehyde must be prepared freshly. The resulting formaldehyde gradually polymerizes on storage. One gram of paraformaldehyde powder is heated in a conical flask on a hot plate with 40 ml of distilled water. Temperature should not exceed 75°C. The powder dissolves as 1 M NaOH is added dropwise; approximately 4 drops is sufficient and excess must be avoided. The volume is made up to 50 ml with additional distilled water and the solution passed through filter paper before use.

Complete fixative — Equal volumes of the paraformaldehyde solution (2%) and two times concentrated cacodylate buffer are mixed. Electron microscopy-grade glutaraldehyde (24.8%, TAAB Laboratories Equipment, Reading, U.K.) is added to a final concentration of 0.0125% (w/v). The fixative is chilled on ice before use. Although the concentration of glutaraldehyde is low, we find that it improves tissue structure markedly.

2. Tissue Fixation

Fixation times will vary with the size of specimen. For the blocks described above, fixation in 10 ml of fixative for a total of 1 h at 0°C is sufficient. Longer fixation leads to progressive loss of antigenicity. Shorter times (e.g., 15 min) can be used for cell monolayers. After fixation, blocks are washed in two changes of 10 ml of cacodylate buffer for 1 h.

D. CRYOPRESERVATION, EMBEDDING, AND CRYOSECTIONING

Cryopreservation and rapid freezing are essential to preserve good structure during the preparation of thick frozen sections for immunohistochemical staining.

1. Cryopreservation

After washing, fixed blocks are infiltrated with 10% dimethylsulfoxide (DMSO) in cacodylate buffer for 1 h at 0°C. The block is then blotted briefly on filter paper to remove excess buffer and immersed in OCT compound (Miles Scientific, Naperville, IL) containing 2.5% (w/v) DMSO.

2. Embedding

OCT compound is contained in an aluminum foil cup which is fashioned with a flat bottom by wrapping and molding around a pencil or other cylindrical object. The specimen is pushed into the OCT using fine forceps, assuming the correct orientation when standing on edge in contact with the base of the cup.

3. Freezing

The specimen is frozen rapidly by immersing the cup into liquified Freon-22 gas (monochlorodifluoromethane) cooled by liquid nitrogen. This provides heat conduction that is superior to the more commonly used isopentane. Liquified Freon can be prepared by condensation of the gas as it is passed through liquid nitrogen-cooled tubing. Freon gas is toxic, and condensation should be performed in a fume hood. It is also damaging environmentally, so we store it in solidified form for reuse in the vapor phase of a liquid nitrogen container.

Frozen blocks are stored at −100°C in their foil wrapping in air-tight containers. Storage at −20°C leads to sublimation of moisture from the block.

4. Cryosectioning

A cryostat with an operating range down to $-30°C$ is necessary in order to cut sections from cryopreserved tissue. A Bright OTC instrument (Bright Instrument Company, Huntington, U.K.) is suitable. For light microscopy, 5-μm sections are collected onto gelatin-subbed slides and dried for 1 h at room temperature before use. Sections can be stored for several days at 4°C in air-tight boxes containing silica gel as desiccant.

Thick (20-μm) sections are cut for preembedding immunohistochemical staining, in preparation for ultrastructural studies. Frozen sections are used in preference to blocks or vibratome sections because they reduce diffusion distances for antibody penetration. The thick sections are collected by thawing onto small pieces of nitrocellulose membrane (see Section II. B.2) and transferred immediately without drying to wells in a microtiter plate containing chilled 0.1% (w/v) bovine serum albumin (BSA) in cacodylate buffer. After removal of the OCT compound, the sections are transferred to fresh 0.1% (w/v) BSA buffer and incubated for at least 30 min at 4°C to block subsequent binding of antibodies by the nitrocellulose.

E. IMMUNOHISTOCHEMISTRY

We have chosen the immunoperoxidase technique in preference to immunogold because we have been obliged to adopt a preembedding staining technique. The smaller size of the immunoperoxidase conjugate should offer advantages of penetration over immunogold particles.

1. Light Microscopy

Air-dried sections are fixed for 10 min in ice-cold 95% ethanol and then rehydrated by transfer to cold PBS. After excess moisture is dried from around the section with a tissue, remaining PBS is flicked off and replaced with 20 to 50 μl of primary antibody. After incubation at 4°C for 1 h in a humid box, the section is washed through three changes of PBS (using Coplin jars or a slide carrier and baths, depending on numbers of slides), excess moisture is removed, and incubation commenced with 20 to 50 μl of $F(ab^1)_2$ sheep anti-mouse IgG-horseradish peroxidase conjugate (see Section II.A.2).

After a further hour of incubation at 4°C and washes, the slide is covered with a solution containing 3, 3'-diaminobenzidine (0.5 mg/ml, Sigma, St. Louis, MO) and hydrogen peroxide (0.2% v/v) in 0.05 M Tris-HCl buffer (pH 7.6) for approximately 10 min. The solution is dispensed through a Millipore filter to remove any precipitate and all equipment and spent solution is inactivated by oxidation and polymerization with a solution of sodium hypochlorite. Diaminobenzidine (DAB) is a carcinogen!

After rinsing in PBS, the sections are counterstained by immersion in Gill's hematoxylin for 1 min, differentiated in acidified water (0.16% v/v concentrated HCl) for 3 s, dehydrated through 70% and absolute alcohols, and cleared in Histoclear (National Diagnostics, Somerville, NJ). They are then mounted with coverslips using any suitable mounting medium.

A common problem which detracts from the appearance of immunoperoxidase-stained frozen sections occurs when there is excessive evaporation from slides during incubations with small volumes of antibodies or in the brief interval between removing excess fluid and adding the next antibody. This is manifest as loss of distinct cell outlines and, in particular, loss of nuclear contents and nuclear morphology. The problem is eliminated by working quickly, performing all stages in a cold room, and reducing the number of slides in each humid chamber (and, therefore, the length of time each box is open while adding antibodies). It is assumed that the artifact is due to concentration of salts by evaporation.

2. Electron Microscopy

The sections are stained in the wells of microtiter plates. Those wells which are to be used for incubation with antibodies are blocked before use by incubation with 0.1% (w/v)

BSA. The sections are transported serially through the wells, supported on the pieces of nitrocellulose membrane, as described above (Section II.D.4).

The sequence involves incubation with primary antibody (2 h), 3 × 20-min washes in cacodylate buffer, incubation with F(ab^1)$_2$ sheep anti-mouse IgG horseradish peroxidase conjugate (2 to 3 h), and 3 × 20-min washes in cacodylate buffer. After equilibration in 0.05 M Tris HCl for 10 min, bound conjugate is then detected using DAB and H$_2$O$_2$ (as described in Section II.E.1) in a further well, followed by a wash in cacodylate buffer. All procedures are carried out at 4°C, except for the enzyme histochemistry, where the buffer is warmed to about 25°C.

3. Controls

As in all immunohistochemistry, adequate controls must be included to eliminate the possibility of nonspecific staining. Substitution of an irrelevant monoclonal antibody of the same isotype as the primary antibody for test will control for binding through Fc receptors. Omission of the primary antibody will control for nonspecific staining by the secondary antibody. Finally, incubation of a section with the DAB/H$_2$O$_2$ solution alone will reveal the presence of endogenous peroxidase activity. Methods exist to inactivate the latter,[22,23] but caution must be exercised to ensure that they do not affect the antigens in which one is interested.

F. PREPARATION FOR ULTRASTRUCTURE

1. Postfixation

After immunohistochemical staining, sections are postfixed in 2% paraformaldehyde/ 1% glutaraldehyde in cacodylate buffer for 20 min. They are further fixed and stained by incubation in 1.0% OsO$_4$ in cacodylate buffer for 1 h at room temperature. The latter procedure enhances the electron density of the immunoperoxidase reaction product.

2. Embedding

The osmicated sections are rinsed in cacodylate buffer and dehydrated by passage through graded alcohols (30, 50, 70, 75, 80, 85, 90, 95%, 15 min per change) into three changes, 20 min each, of absolute alcohol (dried by storage with solid CuSo$_4$). After infiltration with propylene oxide (two changes, 10 min each), the sections are transferred to a propylene oxide-TAAB resin (TAAB Laboratories Equipment, Reading, U.K.) mixture (2:1) overnight, followed by 8 h in a 1:2 mixture, before infiltration with pure resin for 20 h and transfer to resin contained in the cap of a BEEM capsule (TAAB Laboratories Equipment). At this stage, the nitrocellulose membrane is expanded by the solvent action of the propylene oxide and is no longer rigid. The section is flattened on the bottom of the BEEM capsule lid by gentle pressure with a block of resin previously polymerized in a BEEM capsule. After curing at 70°C for 24 h, the section is flat and contained in a layer of resin polymerized to the surface of the compressing block.

3. Sectioning

After appropriate trimming, the plastic-embedded thick section should be parallel to the knife edge when the block is mounted in the chuck of the ultramicrotome. Semithin sections (1 μm) are cut until the tissue is reached and the general morphology is then examined by light microscopy after staining with hot methylene blue-azure II in borax solution.[24] Silver-gold sections from suitable areas are then collected onto 200-mesh copper grids. The best results are obtained with a good diamond knife.

4. Counterstaining

Some counterstaining is necessary to visualize tissue ultrastructure. However, usual levels of lead citrate and/or uranyl acetate staining make detection of immunoperoxidase reaction product difficult or impossible. Sufficient structure is visible if counterstaining is limited to a 2-min incubation of the thin sections with Reynold's lead citrate,[25] followed by a rinse with deionized water and drying.

5. Electron Microscopy

There are no special requirements or procedures for the examination and photographing of the immunoperoxidase-stained material. Careful examination of control material is necessary to eliminate the possibility of nonspecific staining or the presence of endogenous peroxidase activity (see Section II.E.3).

III. ASSESSMENT OF TECHNIQUE AND RESULTS

We have proceeded from good quality immunohistochemistry at the light-microscopic level, which suggested (but could not prove) that Class II antigens have an intracellular distribution in enterocytes, in addition to a clear representation on the cell surface (Figure 1a). Other studies on renal tubular epithelium alerted us to the difficulties in interpreting the appearance of intracellular antigen where there is extensive interdigitation between adjacent cells.[26] Nevertheless, ethanol-fixation of fresh-frozen sections provided excellent preservation of the antigenicity of Class II molecules and provided a "standard" against which to assess the effects of other fixatives.

Our initial strategy was to attempt postembedding demonstration of antigen by immunoperoxidase or immunogold techniques. This would ensure the best preservation of structure and could potentially eliminate the problems that can arise in preembedding techniques from restricted antibody diffusion into tissue. We embedded tissue with proved Class II MHC antigenicity in several resins — Spurs, London White and Lowicryl K4M — employing a number of polymerization techniques. Fixatives were chosen simply for their ability to preserve antigenicity and included ethanol (excellent antigen preservation), paraformaldehyde-lysine-periodate (PLP),[27] and 1% paraformaldehyde/0.05% glutaraldehyde.[26] We were unable to demonstrate Class II antigens in any resin-embedded material, even after attempts to expose the tissue by etching. Ethanol-fixed tissue blocks could be taken through all steps short of polymerization of the resin and still retained antigenicity after being taken back through the reverse sequence, followed by embedding in OCT and sectioning as conventional frozen sections (Figure 1b). We concluded that the polymerization process either denatured or rendered inaccessible the Class II MHC antigenic determinants recognized by our monoclonal antibodies. However, we have not assessed glycol methacrylate, in which postembedding localization of Class II MHC antigens has been reported recently at the light-microscopic level.[23]

We have adopted a preembedding approach to immunohistochemistry, utilizing staining of thick frozen sections of fixed tissue to facilitate penetration of antibodies. The fixatives chosen initially were PLP and 1% paraformaldehyde/0.05% glutaraldehyde. Tested at the light microscopic level, both revealed a pattern of antigen preservation that was essentially confined to the cell surface (Figure 1c). This raised the possibilities that either Class II MHC molecules are represented only on the cell surface (and the "intracellular" antigen in frozen sections is artifactual), or that fixation with the aldehyde fixative made the cell membrane and/or the cytoplasmic matrix impermeable to antibody.[26]

An additional problem was the quality of structural preservation. With both fixatives, the fixation times (4 to 6 h) required to preserve structure against gross damage by freezing led to extinction of antigenicity. We, therefore, adopted the cryopreservation technique described above (Section II.D.1), together with the use of Freon-22 as a more efficient heat exchanger than isopentane.

Experimentation with fixation conditions led to the finding that short fixation (1 h) in

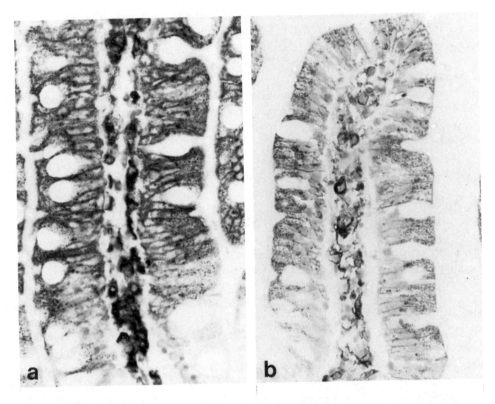

FIGURE 1.   Frozen sections of rat jejunum stained to demonstrate Class II MHC antigens by the indirect immunoperoxidase technique and counterstained with hematoxylin. Mature enterocytes express Class II MHC antigens,
as do macrophages and dendritic cells in the lamina propria. (All × 40 objective.) (a) Base and mid-region of a
villus in fresh-frozen tissue. The section was fixed briefly in ethanol prior to staining. Class II MHC antigens are
expressed as enterocytes and ascend the villus. They are present on the basolateral cell membranes and apparently
within the cytoplasm in a granular distribution. (b) A villus from a block of tissue fixed in cold ethanol and
processed through all stages for embedding in Lowicryl K4M, stopping short of polymerization of the resin. The
block was then taken through the reverse series of steps and frozen to prepare conventional frozen sections.
Antigenicity of Class II MHC molecules has been preserved (but was lost in a replicate block exposed to polymerization of Lowicryl resin) and has a distribution essentially similar to that in freshly frozen tissues. (c) Frozen
section of a villus from a block fixed with a 1% paraformaldehyde/0.05% glutaraldehyde fixative for 2 h prior to
cryopreservation in OCT[26] and freezing. Reaction product indicates a distribution of Class II MHC antigens limited
almost entirely to the basolateral membranes of enterocytes. Overall intensity of staining is reduced in comparison
with that of tissue from fresh-frozen or ethanol-fixed blocks. (d) Base and mid-region of a villus from a block
fixed with 1% paraformaldehyde/0.0125% glutaraldehyde for 1 h prior to cryopreservation and freezing. There
was some loss of staining intensity (photographed using a blue filter to enhance reaction product), but the distribution
of Class II MHC antigens in enterocytes is similar to that seen in fresh-frozen and ethanol-fixed tissues. In particular,
there is preservation of the antigen in the cytoplasm of the cells.

1% paraformaldehyde/0.0125% glutaraldehyde, followed by adequate cryopreservation, produced remarkably good preservation of structure at the light microscopic level. Furthermore, the pattern of antigen localization (Figure 1d) resembled closely that seen in frozen
sections of fresh tissue (Figure 1a) and frozen sections of ethanol-fixed tissue blocks (Figure
1b). Our explanation for this is that antibody can diffuse into the cytoplasm of cells fixed
by this method.

When extended to the ultrastructural level, it is clear that Class II MHC antigens can
be localized to discrete subcellular organelles, as well as to the basolateral cell membranes
of enterocytes. Figures 2a and b illustrate staining of the basolateral cell membranes in
human epithelium. The staining is patchy and tends to be strongest at the basal areas of the

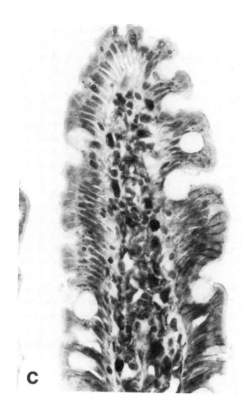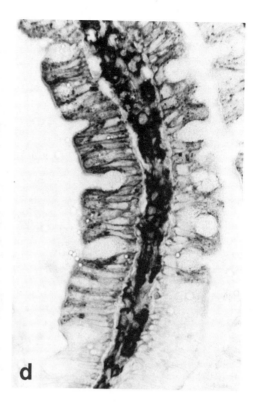

FIGURE 1c and d.

cells where the adjacent membranes are separated widely (Figure 2a). We are undecided as to whether this reflects a real polarization of Class II molecules to certain areas of the cell membrane or whether some barriers to diffusion of antibodies still exist.

We have obtained strong staining of intracellular organelles which are broadly of the endocytic pathway. Figure 3a and b illustrate stained multivesicular bodies in rat and human enterocytes, respectively, while larger bodies identified as secondary lysosomes are also stained (not shown).

We have not detected Class II MHC antigens associated with either rough endoplasmic reticulum or with Golgi, and not all multivesicular bodies within a given cell are necessarily stained. The usual caveats in immunohistochemistry apply here. Positive staining (with satisfactory controls) of a structure is significant, but absence of staining does not rule out definitely the association of antigen with that structure. We are unsure whether the diffusion barriers to antibodies are heterogeneous even within the cytoplasm of a single cell. However, the technique as described has allowed identification of intracellular antigens that could not be detected by any other method, while preserving acceptable tissue morphology. In this way, we have achieved the first ultrastructural localization of Class II MHC molecules to the endocytic pathway of a cell.

ACKNOWLEDGMENTS

The authors thank Mr. Chris Leigh for his advice and assistance with ultramicrotomy and Mrs. Glenys King for processing the manuscript. The work was supported by grants from the National Health and Medical Research Council of Australia.

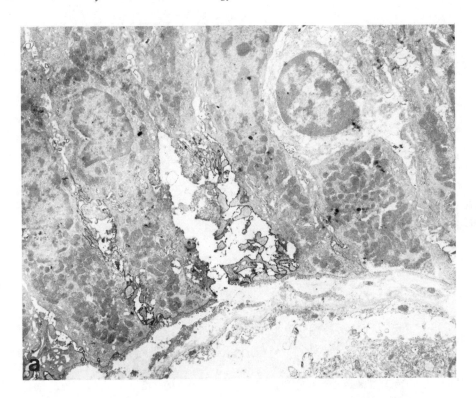

FIGURE 2.   Ultrastructural localization of Class II MHC antigens on the basolateral cell membranes of enterocytes in the rat jejunum using the preembedding indirect immunoperoxidase technique. Tissue was fixed in 1% paraformaldehyde/0.125% glutaraldehyde. (a) Bases of enterocytes in the midvillus region, showing extensive interdigitations between adjacent cells and the presence of lateral spaces between cells. Basal membranes and the membranes bordering lateral spaces and surrounding interdigitating processes show clear reaction product associated with them. Staining is less evident or absent where the lateral cell membranes are more closely applied in the apical regions of the enterocytes. It is not clear whether the polarity of staining reflects barriers preventing diffusion of antibody to the unstained areas in fixed tissue or whether there are real differences in the expression of antigen between the apex and base of the enterocyte. (Magnification × 4000.) (b) High-power view of lateral cell membranes near the bases of two enterocytes. There is strong staining of most, but not all, interdigitations between the cells. Staining is strongest where there is a clear space between the adjacent membranes. The irregular cell borders, with the plane of section producing clefts of cell membrane apparently isolated in the cytoplasm, could create the appearance of intracellular antigen at the light microscopic level. (Magnification × 20,000.)

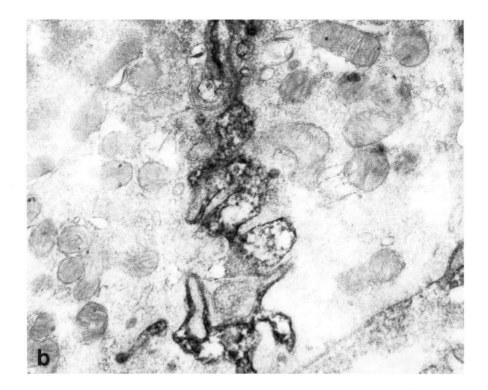

FIGURE 2b.

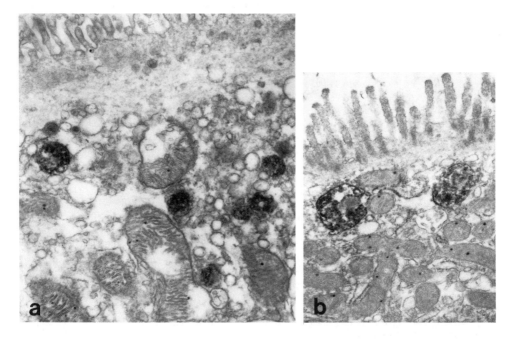

FIGURE 3. Localization of intracellular Class II MHC antigen by the indirect immunoperoxidase technique in the apical cytoplasm of jejunal enterocytes. Tissue was fixed with 1% paraformaldehyde/0.0125% glutaraldehyde. (a) Rat enterocyte, with brush border and terminal web at the top of the picture. Six bodies with multivesicular structure are stained strongly by the immunoperoxidase reaction product. (Magnification × 20,000.) (b) Human enterocyte, showing brush border, terminal web, and apical cytoplasm. Two stained multivesicular bodies lie below the terminal web. Staining is localized to the limiting membrane of the bodies and the membranes of the contained vesicles. (Magnification × 20,000.)

REFERENCES

1. **Hämmerling, G. J., Deak, B. D., Mauve, G., Hämmerling, U., and McDevitt, H. O.**, B lymphocyte alloantigens controlled by the I region of the major histocompatibility complex in mice, *Immunogenetics*, 1, 68, 1974.
2. **Hämmerling, G. J., Mauve, G., Goldberg, E., and McDevitt, H. O.**, Tissue distribution of Ia antigens: Ia on spermatozoa, macrophages and epidermal cells, *Immunogenetics*, 1, 428, 1975.
3. **Steinman, R. M. and Nussenzweig, M. C.**, Dendritic cells: features and functions, *Immunol. Rev.*, 53, 125, 1980.
4. **Hirschberg, H., Bergh, O. J., and Thorsby, E.**, Antigen-presenting properties of human vascular endothelial cells, *J. Exp. Med.*, 152, 249s, 1980.
5. **Wiman, K., Curman, B., Forsum, U., Klareskog, L., Malmnaäs-Tjernlund, U., Rask, L., Trägårdh, L., and Peterson, P. A.**, Occurrence of Ia antigen on tissues of non-lymphoid origin, *Nature (London)*, 276, 771, 1978.
6. **Klareskog, L., Forsum, U., and Peterson, P. A.**, Hormonal regulation of the expression of Ia antigens on mammary gland epithelium, *Eur. J. Immunol.*, 10, 958, 1980.
7. **Hart, D. N. J. and Fabre, J. W.**, Endogenously produced Ia antigens within cells of convoluted tubules of rat kidney, *J. Immunol.*, 126, 2109, 1981.
8. **Mayrhofer, G., Pugh, C. W., and Barclay, A. N.**, The distribution, ontogeny and origin in the rat of Ia-positive cells with dendritic morphology and of Ia antigen in epithelia, with special reference to the intestine, *Eur. J. Immunol.*, 13, 112, 1983.
9. **Matis, L. A., Jones, P. P., Murphy, D. B., Hedrick, S. M., Lerner, E. A., Janeway, C. A., McNicholas, J. M., and Schwartz, R. H.**, Immune response gene function correlates with the expression of an Ia antigen. II. A quantitative deficiency in A_e:A_α complex expression causes a corresponding defect in antigen-presenting cell function, *J. Exp. Med.*, 155, 508, 1982.
10. **Guillet, J.-G., Lai, M. Z., Thomas, J. B., Buus, S., Sette, A., Grey, H. M., Smith, J. A., and Gefter, M. L.**, Immunological self, nonself discrimination, *Science*, 235, 865, 1987.
11. **Germain, R. N.**, The ins and outs of antigen processing, *Nature (London)*, 322, 687, 1986.
12. **Werdelin, O.**, Determinant protection. A hypothesis for the activity of immune response genes in the processing and presentation of antigen by macrophages, *Scand. J. Immunol.*, 24, 625, 1986.
13. **von Boehmer, H.**, The developmental biology of T lymphocytes, *Annu. Rev. Immunol.*, 6, 309, 1988.
14. **Parr, E. L. and McKenzie, I. F. C.**, Demonstration of Ia antigens on mouse intestinal epithelial cells by immunoferritin labelling, *Immunogenetics*, 8, 499, 1979.
15. **Hirata, I., Austin, W. H., Blackwell, W. H., Weber, J. R., and Dobbins, W. O.**, Immunoelectronmicroscopic localization of HLA-DR antigens in control small intestine and colon and in inflammatory bowel disease, *Dig. Dis. Sci.*, 31, 1317, 1986.
16. **Cresswell, P.**, Intracellular Class II HLA antigens are accessible to transferrin-neuraminidase conjugates internalized by receptor-mediated endocytosis, *Proc. Natl. Acad. Sci. U.S.A.*, 82, 8188, 1985.
17. **Pernis, B.**, Internalization of lymphocyte membrane components, *Immunol. Today*, 6, 45, 1985.
18. **McMaster, W. R. and Williams, A. F.**, Identification of Ia glycoproteins in rat thymus and purification from rat spleen, *Eur. J. Immunol.*, 9, 426, 1979.
19. **Beckman, I. G. R., Bradley, J., Brooks, D., and Zola, H.**, Delineation of serologically distinct monomorphic determinants of human MHC Class II antigens: evidence of heterogeneity in their topographical distribution, *Mol. Immunol.*, 21, 205, 1983.
20. **Ey, P. L., Prowse, S. J., and Jenkin, C. R.**, Isolation of pure IgG1, IgG2a IgG2b immunoglobulins from mouse serum using protein A-Sepharose, *Immunochemistry*, 15, 429, 1978.
21. **Farr, A. G. and Nakane, P. K.**, Immunohistochemistry with enzyme labelled antibodies: a brief review, *J. Immunol. Methods*, 47, 129, 1981.
22. **Bulman, A. S. and Heyderman, E.**, Alkaline phosphatase for immunocytochemical labelling: problems with endogenous enzyme activity, *J. Clin. Pathol.*, 34, 1349, 1981.
23. **Hermanns, W., Colbatzky, F., Günther, A., and Steiniger, B.**, Ia antigens in plastic-embedded tissues: a post-embedding immunohistochemical study, *J. Histochem. Cytochem.*, 34, 827, 1986.
24. **Hodson, S. A.**, Light-microscopic autoradiography utilizing araldite sections, *J. Microsc. (Oxford)*, 89, 113, 1969.
25. **Reynolds, E. S.**, The use of lead citrate at high pH as an electron-opaque stain in electron microscopy, *J. Cell Biol.*, 17, 208, 1963.
26. **Mayrhofer, G. and Schon-Hegrad, M. A.**, Ia antigens in rat kidney, with special reference to their expression in tubular epithelium, *J. Exp. Med.*, 157, 2097, 1983.
27. **McLean, I. W. and Nakane, P. K.**, Periodate-lysine-paraformaldehyde fixative: a new fixative for immunoelectron microscopy, *J. Histochem. Cytochem.*, 22, 1077, 1974.

Chapter 6

A MICROPLATE ELISA FOR THE DETECTION OF INTERNAL AND EXTERNAL CELLULAR ANTIGENS

Raffaela M. Comacchio, Arthur W. Hohmann, Nigel Quadros, and John Bradley

TABLE OF CONTENTS

I. INTRODUCTION

The microplate enzyme-linked immunosorbent assay (ELISA) is one of the most wide-spread and adaptable methods for the detection and quantitation of antigens and antibodies. The original and still most common application uses antigens or antibodies which are soluble in aqueous solutions and can be coated easily onto the surface of the plastic ELISA wells. These ELISA methods have played a vital role in the production and characterization of monoclonal antibodies, the latter being a method which depends critically on the integrity and rapidity of screening assays. As monoclonal antibodies have started to unravel the complexity of cellular antigens, renewed interest has been generated in the antigens present both on the cell surface and in the cytoplasm. Immunohistochemical methods, although essential for complete analysis of these antigens, are relatively slow and laborious and not suited for large-scale screening procedures. A number of publications have described the use of ELISA methods for the detection of antibodies which react with cellular antigens,[1-7] (see also Chapter 3, Volume I) but ELISA is a tool which has not been widely used. There are clearly more difficulties involved in this form of ELISA compared to those which use soluble antigens, not the least of which is the limited density and lability of many cellular antigens.

We have developed a versatile microplate-cell ELISA using whole cells immobilized in the wells. We have used this assay extensively in the detection of human and mouse monoclonal antibodies as well as serum antibodies from patients. Depending on the fixation procedure used, external or internal cellular antigens can be identified. The procedures used in this method are detailed below, and results which demonstrate the versatility of this technique are presented.

II. MATERIALS

A. CELLS AND CELL LINES

1. Lymphocytes were prepared from peripheral blood of healthy volunteers and purified by density centrifugation on Ficoll-Hypaque.
2. Endothelial cells were obtained from human umbilical vein by collagenase digestion according to methods elaborated elsewhere.[8,9] Primary cultures were established in 25-cm^2 flasks precoated with 1% gelatin. The cultures were fed daily and confluent monolayers usually formed in 3 to 5 d.
3. MRC-5, a human fibroblast cell line, was obtained from Dr. Peter Hallsworth at Flinders Medical Centre.
4. RAJI and other lymphoblastoid cell lines were from stocks used routinely in this department.

B. MEDIA

RPMI-1640 (Flow Labs, McLean, VA) was used for all cell cultures and was supplemented with 10% fetal calf serum (Flow Labs), 2 m*M* glutamine (Flow Labs), and gentamicin at 100 μg/ml. Primary endothelial cell cultures were grown in the same media with the additional supplements of amphotericin B (2.5 μg/ml; Flow Labs) and insulin (0.02 U/ml; Sigma Chemical Company, St. Louis, catalog no. I1882).

C. TRYPSIN

Trypsin was used as a commercially available 0.05% sterile solution containing 0.02% ethylenediaminetetraacetic acid (EDTA) (Flow Labs).

D. PRIMARY ANTIBODIES
1. Mouse Monoclonal Antibodies

Mouse monoclonal antibodies to tubulin and actin were obtained from Amersham (Amersham, U.K.) and to vimentin from Biogenex (Dublin, CA). Mouse monoclonal antibodies to β-2 microglobulin (FMC 16), to a monomorphic determinant on the beta chain of human class II antigens (FMC 4), and to human B cells (FMC 1) were produced in this department and are described in other publications.[10,11] Leu4 was obtained from Becton Dickinson (Mountain View, CA), and OKT8 was obtained from Ortho Diagnostics (Raritan, NJ). Mouse monoclonal antibody to human IgM (HB57) was obtained from the American Type Culture Collection (Rockville, MD).

2. Human Monoclonal Antibodies

Human monoclonal antibodies were produced in this laboratory by procedures described elsewhere.[12] FMC H1 was derived from a patient with systemic lupus erythematosus and IK10 was derived from spleen cells from a patient with Hodgkin's lymphoma.

3. Human Serum Antibodies

Sera from patients with rheumatoid arthritis Felty's syndrome, and normal individuals were obtained from outpatient clinics at Flinders Medical Centre. These sera were fractionated into IgG and IgM-rich fractions by Sephadex G200 column chromatography.

E. SECONDARY ANTIBODIES

Rabbit anti-human IgG and IgM (DAKO, Glostrup, Denmark) and goat anti-mouse immunoglobulin prepared in this laboratory were labeled with alkaline phosphatase following conventional procedures.[13] The anti-human immunoglobulin preparations were tested and used at dilutions which detected equal amounts of IgG and IgM, respectively. Mouse monoclonal antibody against alkaline phosphatase (FMC 55) was prepared in this laboratory and was used to produce an alkaline phosphatase-anti-alkaline phosphatase complex (APAAP).[14]

The bridging antibody used in this method was the unlabeled goat anti-mouse immunoglobulin described above. Horseradish peroxidase (HRP)-labeled goat anti-human IgG (catalog no. 3201-0111) and IgM (catalog no 3201-0201) were from Cappel Labs (Cochranville, PA) and were used at 1/3000 and 1/4000 dilutions, respectively.

F. ELISA PLATES

Several types of ELISA plates have been used. Costar vinyl plates (catalog no. 2595) were advantageous in some situations since their thin walls allowed use of high-magnification microscope objectives when examining stained cells. The plates do, however, have a high-binding characteristic which makes them unsuitable for some applications such as with high concentrations of human sera. Flat-bottom Dynatech MicroELISA plates (Dynatech, Alexandria, VA) were used where nonsterile, flat-bottom wells were required. For culturing endothelial cells, sterile tissue-grade plates were used (Linbro, Flow Labs, McLean, VA; catalog no. 76-032-05).

G. BUFFERS
1. Phosphate Buffered Saline
Phosphate buffered saline (PBS) was 0.01 M phosphate, 0.15 M NaCl, and had a pH between 7.2 and 7.4.

2. Dilution Buffer
PBS containing 0.05% Tween 20 and 0.25% bovine serum albumin (BSA) (Sigma A-4503) (PBS-T-BSA) was used for diluting monoclonal antibodies, human sera, and enzyme-labeled reagents.

3. Washing Buffer
This was PBS containing 0.05% Tween-20 (PBS-T) and was used for all washing procedures of ELISA plates.

4. Blocking Buffer
This was used on glutaraldehyde-fixed cells and was 0.25% BSA, 0.1 M glycine in PBS.

H. ENZYME SUBSTRATES AND SUBSTRATE BUFFERS
1. Alkaline Phosphatase
a. Soluble Substrate (p-Nitrophenyl Phosphate)
Ninety-seven milliliters of diethanolamine were added to 800 ml of H_2O. To this was added 1 ml of 0.5 M $MgCl_2$ and the pH adjusted to 9.8 with 1 M HCl. The total volume was made to 1 l and the buffer stored at 4°C and used within 1 month. Substrate was prepared in this buffer by dissolving one 5-mg tablet (Sigma 104-105) of *p*-nitrophenyl phosphate in 5 ml just prior to use.

b. Insoluble Substrate (Naphthol ASMX-Fast Red)
Five milligrams of Naphthol AS-MX phosphate (Sigma N4875) were added to 10 ml of 0.1 M Tris pH 8.0 buffer. This was allowed to dissolve and 10 mg of Fast Red TR salt (Sigma F1500) were added, mixed vigorously for 15 s, filtered through Whatman No.1 paper, and used immediately. If necessary, 1 mM Levamisole (Sigma L9756) can be added as an inhibitor of endogenous alkaline phosphatase. It has little inhibitory activity on intestinal alkaline phosphatase.

2. Peroxidase — Soluble Substrate
This substrate buffer contains 7.3 g of citric acid and 9.47 g of Na_2HPO_4 (anhydrous) in 1 l of H_2O. The pH should be 5.0 without further adjustment. To prepare the substrate, 4 mg of *o*-phenylene diamine (Sigma P8787, 4-mg tablets) was added to 10 ml of the buffer, and 30 µl of a 3% solution of H_2O_2 were added just before use.

I. ENZYME STOP SOLUTIONS
These were used to stop the enzyme reactions with the soluble substrates and were 0.2 M EDTA for alkaline phosphatase and 2.5 N HCl for HRP.

J. FIXATIVES
1. Glutaraldehyde
Glutaraldehyde was obtained as the electron microscopy (EM) grade 25% solution from

TAAB (Reading, Berkshire, England). For use it was diluted to the desired concentration (v/v) in PBS.

2. Formaldehyde-Digitonin

Commercial grades of formaldehyde were received as approximately 37% solutions. Digitonin (Sigma D5628) was received as an 80% solution.

III. METHODS

A. PREPARATION OF CELLS FOR ELISA
1. Peripheral Blood Lymphocytes

Human lymphocytes, from blood taken from normal individuals, were prepared by centrifugation through Ficoll-Hypaque. Cells were washed twice in PBS and adjusted to a concentration of 2×10^6 cells per milliliter.

2. Cell Lines

Lymphocyte cell lines were harvested from cultures, washed twice in PBS and adjusted to a density of 2×10^6 cells per milliliter. Cell lines which grow as attached cells were removed from their culture vessels by trypsinization. For this procedure all media were removed from the cells, and the monolayers were washed twice with PBS. A small volume (approximately 5 ml for a 25-cm^2 flask) of 0.1% trypsin-EDTA was added to the flask and briefly rocked across the monolayer. The flask was turned on end and the trypsin solution removed. The flask was placed at 37°C for a brief period until the cell monolayer showed signs of detachment (approximately 1 min). The cells were removed from the flask by washing the surface with PBS. The cells were washed twice in PBS by centrifugation at $200 \times g$ for 8 min and suspended to a density of 2×10^5/ml in RPMI. One hundred microliters of this suspension were added to each well of a sterile 96-well, flat-bottom tissue culture plate. For endothelial cell cultures, the wells were precoated with 1% gelatin prior to addition of the cells. These plates were incubated at 37°C in 5% CO_2 for several days until 50 to 75% confluent.

B. PREPARATION OF CELL LAYERS FOR FIXATION
1. Unattached Cells

One hundred microliters of cell suspension was added to each well of a 96-well ELISA plate. To sediment the cells, two procedures were used.

a. Plate Centrifugation Method

The plates were centrifuged at $200 \times g$ for 5 min in a Sorval RC3 using a rotor specifically designed to carry micro-titer plates (Dupont Instruments Rotor HL-4). For methanol or ethanol fixation, the supernatants from the wells were removed as completely as possible by aspirating the liquid above the cells. Plates cannot be flicked out at this stage or else cells will be lost. For aldehyde fixation the supernatant was left in the wells and an equal volume (100 μl) of double-strength fixative added to each well (see below).

b. Settling Method

The simpler and preferred method, after addition of the cells to the wells, was to allow the cells to settle on the bench for 1 h at room temperature. The supernatant can then be either removed carefully or treated as above for fixation. (Although more rapid, centrifugation tends to distribute the cells away from the center of the well, whereas a more even distribution is seen after settling at $1 \times g$).

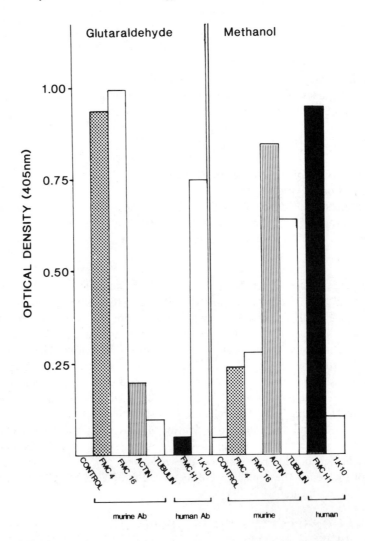

FIGURE 1.   Binding of mouse and human monoclonal antibodies to the human lymphoblastoid cell line, RAJI. RAJI cells were placed in 96-well trays, allowed to settle (at $1 \times g$), and fixed with 0.1% glutaraldehyde or 100% methanol. Monoclonal antibodies were applied and their binding detected by use of alkaline phosphatase-labeled anti-mouse or anti-human immunoglobulin and the soluble substrate *p*-nitrophenyl phosphate.

2. Attached Cells

Culture media were removed and the cell monolayer washed twice with PBS. The plates can be emptied by flicking out, although exposure to PBS prior to fixation should be for a minimal time.

C. FIXATION AND BLOCKING

Various fixatives and attachment factors (particularly poly-L-lysine) have been described for cell-ELISAs, and many of these have been used by us (Figures 1 to 4 and Table 1). We have included here only those which reliably yielded satisfactory results.

1. Aldehyde Fixation

Both glutaraldehyde and formaldehyde have been used. Glutaraldehyde was tested over a range of concentration from 0.025 to 0.25% final concentration. Concentrations below

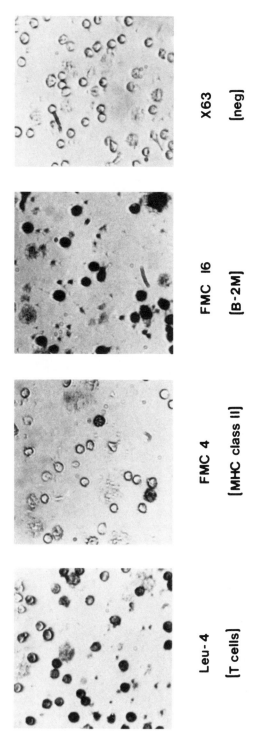

Leu-4   FMC 4   FMC 16   X63

[T cells]   [MHC class II]   [B-2M]   [neg]

FIGURE 2. Binding of mouse monoclonal antibodies to peripheral blood lymphocytes as revealed by Cell-ELISA. PBL were placed in flat-bottom ELISA trays, allowed to settle (at $1 \times g$), and fixed with 0.1% glutaraldehyde. Monoclonal antibodies to CD3 (Leu 4), MHC class II (FMC 4), and β-2-microglobulin were applied and their binding detected by the alkaline phosphatase-anti-alkaline phosphatase method and the insoluble substrate Naphthol ASMX and Fast Red.

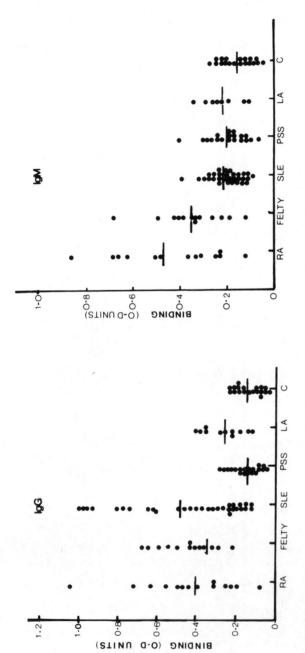

FIGURE 3.   Binding of human IgG and IgM to cultured endothelial cells as determined by ELISA. Human umbilical endothelial cells were grown in 96-well plates and fixed with 0.1% glutaraldehyde. These fixed cells were reacted with a 1/100 dilution of sera from patients, and the binding of IgG or IgM was revealed with horse-radish peroxidase labeled anti-human IgG or IgM and the substrate *o*-phenylene diamine. The horizontal lines represent the mean binding for each study group. Each dot is the mean result of quadruplicate determinations.

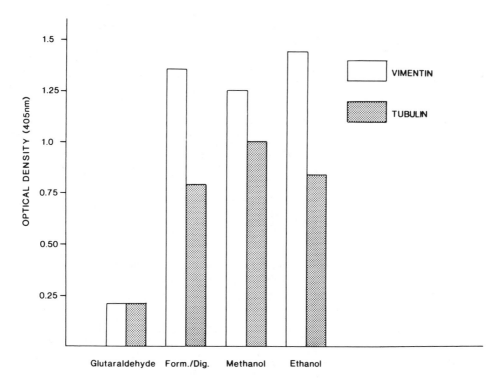

FIGURE 4.   Binding of mouse monoclonal antibodies against vimentin or tubulin to the MRC-5 human fibroblast cell line as shown by Cell-ELISA. MRC-5 cells were grown in 96-well plates and fixed with 0.1% glutaraldehyde, formaldehyde-digitonin, methanol, or ethanol as described in the text. Monoclonal antibodies were applied, and their binding revealed by alkaline phosphatase-labeled anti-mouse immunoglobulin and the soluble substrate, *p*-nitrophenyl phosphate.

TABLE 1

Comparison of the Binding of Mouse Monoclonal Antibodies to PBL As Determined by Cell-ELISA Using Soluble and Insoluble Enzyme Substrates[a]

Antibody	Specificity	% Positive	Optical density[b]
Leu4	T cells	67	0.48
OKT8	CD8	27	0.20
FMC4	MHC class II	10	0.40
FMC16	β-2-microglob.	96	0.87
FMC1	B cells	12	0.10
X63	neg.	1	0.01

[a]   PBL were allowed to sediment in 96-well plates and were fixed with 0.1% glutaraldehyde. Wells were reacted with monoclonal antibodies Leu4 (CD3), OKT8 (CD8), FMC 4 (MHC class II), FMC 16 (MHC class I), and FMC 1 (B-cells). The binding of these antibodies was revealed by immunoalkaline phosphatase and either the soluble substrate *p*-nitrophenyl phosphate (optical density, OD) or the insoluble substrate Naphthol ASMX-Fast Red (% positive).

[b]   Optical density at 405 nm.

0.1% resulted in incomplete attachment of cells to the wells, and, as a result, cells tended to be lost in subsequent washing procedures. The glutaraldehyde concentration of 0.1% was routinely used.

Glutaraldehyde was added to the wells to give a final concentration of 0.1% and was left for 15 min at room temperature. The contents of the plates were flicked out and washed twice by adding approximately 250 µl of PBS-BSA to each well for a few seconds each time. To block the remaining free aldehyde groups, 100 µl of 0.1 M glycine was added to each well and left at room temperature for 30 min. These plates were then ready for use.

2. Formaldehyde/Digitonin

Formaldehyde fixation was used prior to detergent permeabilization and was carried out in a manner similar to glutaraldehyde fixation. This procedure follows the method described by Oliver et al.[15] Formaldehyde was added to the wells to give a final concentration of 2%, and they were left at room temperature for 5 min. At this time an equal volume of a 0.1% solution of digitonin in PBS (i.e., 0.05% final concentration) was added and left for 2 min at 37°C. Plates were then washed and blocked, as described above, following glutaraldehyde fixation.

3. Methanol Fixation

Following deposition by growth, centrifugation, or settling of cells on the bottom of the wells, the supernatant was removed. With unattached cells such as peripheral blood lymphocytes (PBL) or lymphocyte cell lines the removal of supernatant was done carefully lest the cell layer be removed as well. To avoid this, a small volume (approximately 10 to 20 µl) of PBS was usually left in the wells. One hundred microliters of cold (-15°C) 100% methanol was added to each well and the plates left for 15 min at -15°C. The plates were then flicked out and allowed to dry. Plates were blocked with PBS-BSA for 30 min at room temperature. The plates were then ready for use.

D. ASSAY PROCEDURE

Both directly labeled second antibodies and a more sensitive enzyme-anti-enzyme complex were used with success with the soluble enzyme substrate(s). However, for detection of the deposition of reaction products on the cell surface the increased sensitivity of the enzyme-anti-enzyme complex was required.

Antibodies were added directly to the wells, either as undiluted culture supernatant or as a dilution of ascites or purified antibody diluted in PBS-T-BSA. Optimal dilution of the latter was generally 1/50 to 1/100, although some cytoskeletal antibodies reacted well at 1/5000. Serum antibodies from patients and normals were usually used at a 1/100 dilution. These antibodies were left in the plates for 1 to 2 h, although in cases where weak reactivity was expected (e.g., uncloned hybridoma supernatants), the antibodies were left overnight at 4°C. Following application of the primary antibody, the plates were washed three times with PBS-T. The washing was done by flicking out the plates, filling the wells with PBS-T, and leaving them for 3 min. This process was repeated twice, at which point the plates were emptied and tapped out on absorbent paper.

When the directly labeled antibodies were used, they were diluted in PBS-T-BSA; 100 µl were added to each well and left for 2 h at 37°C. The above washing procedure was carried out; 100 µl of soluble substrate were added and left for 20 to 30 min at room temperature. The enzyme reaction was terminated by the addition of 25 µl per well of 0.2 M EDTA (for alkaline phosphatase) or 2.5 N HCl (for HRP). The optical density was read on an ELISA plate reader at 405 nm for alkaline phosphatase or 492 nm for HRP.

Using the APAAP procedure, the second antibody was a 1/300 dilution of unlabeled goat anti-mouse immunoglobulin. One hundred microliters of this was added to the washed wells containing the primary antibody and left for 30 min at room temperature. The plates were washed as above, and 100 µl of the APAAP complex, prepared 30 min before (see Reference 14 for details), were added to the wells and left for 30 min at room temperature. The plates were washed (as above) and the ASMX-Fast Red substrate added and left for 15 min at room temperature. The plates were then washed with distilled water, counterstained with methyl green (when desired), tapped out, and preserved by the addition of 50 μl of Glycergel (DAKO).

E. STORAGE OF PLATES

Although plates were generally used within 1 to 2 d of preparation, they could be stored for periods of up to 1 month with no apparent loss of reactivity. Glutaraldehyde- or formaldehyde-fixed plates were stored at 4°C with PBS-BSA in the wells. Methanol-fixed plates were left dry and stored at 4°C.

IV. DISCUSSION AND CONCLUDING REMARKS

The major problem in the cell-ELISA is selecting a fixative which (1) anchors cells firmly in the wells, (2) preserves the antigenicity of labile epitopes, (3) does not damage the plastic itself, and (4) presents only external or internal antigens, as desired. It is unlikely that all of these requirements can be completely satisfied. For example, while aldehydes denature many epitopes recognized by monoclonal antibodies, no alternative works as well in anchoring cells in the wells and at the same time maintains the integrity of the cell membrane. This last point means that the interior of the cell is not exposed, and, therefore, only antibodies which react with cell surface determinants will be bound.

One of the best fixatives for epitopes recognized by monoclonal antibodies is acetone, a solvent which cannot be used on plastic ELISA plates. In its place we have used methanol or ethanol and detergents such as digitonin to disrupt the cell membrane to the extent that antibodies easily penetrate and bind to cytoplasmic constituents. However, this treatment does not totally abolish the reactivity of antibodies that recognize cell surface antigens. This may be due to the binding of these antibodies to the residual membrane or to the antigens present internally as they cycle from the cytoplasmic compartment onto the membrane.[16]

To demonstrate the flexibility of the cell-ELISA we have shown that the results can be expressed either as an optical density (OD) score (using soluble substrate) or as the number of cells stained (using insoluble substrate). This latter is of particular value when examining a heterogeneous population such as peripheral blood lymphocytes. Although an OD score can at times reflect the percentage of the cells stained, this depends much more on the affinity of the particular antibody used, the reaction of the secondary antibodies with this primary antibody, and the density and accessibility of the cellular epitopes.

Despite limitations, which are very much a part of any immunohistochemical technique, the cell-ELISA is a rapid, powerful, and efficient tool in the analysis of cellular antigens and in the search for antibodies which react with those antigens.

REFERENCES

1. **Suter, L., Bruggen, J., and Sorg, C.,** Use of an enzyme-linked immunosorbent assay (ELISA) for screening of hybridoma antibodies against cell surface antigens, *J. Immunol. Methods,* 39, 407, 1980.
2. **Cobbold, S. P. and Waldmann, H.,** A rapid solid-phase enzyme-linked binding assay for screening monoclonal antibodies to cell surface antigens, *J. Immunol. Methods,* 44, 125, 1981.
3. **Posner, M. R., Antoniou, D., Griffin, J., Schlossman, S. F., and Lazarus, H.,** An enzyme-linked immunosorbent assay (ELISA) for the detection of monoclonal antibodies to cell surface antigens on viable cells, *J. Immunol. Methods,* 48, 23, 1982.
4. **Feit, C., Bartal, A. H., Tauber, G., Dymbort, G., and Hirshaut, Y.,** An enzyme-linked immunosorbent assay (ELISA) for the detection of monoclonal antibodies recognizing surface antigens expressed on viable cells, *J. Immunol. Methods,* 58, 301, 1983.

5. **Effros, R. B., Zeller, E., Dillard, L., and Walford, R. L.,** Detection of antibodies to cell surface antigens by a simplified cellular ELISA (CELISA), *Tissue Antigens,* 25, 204, 1985.

6. **Glassy, M. C. and Surh, C. D.,** Immunodetection of cell-bound antigens using both mouse and human monoclonal antibodies, *J. Immunol. Methods,* 81, 115, 1985.

7. **Epstein, S. L. and Lunney, J. K.,** A cell surface ELISA in the mouse using only poly-L-lysine as cell fixative, *J. Immunol. Methods,* 76, 63, 1985.

8. **Jaffe, E. A., Nachman, R. L., Becker, C. G., and Minck, C. R.,** Culture of human endothelial cells derived from umbilical veins. Identification by morphologic and immunologic criteria, *J. Clin. Invest.,* 52, 2745, 1973.

9. **Gimbrone, M. A., Cotran, R. S., and Folkman, J.,** Human vascular endothelial cells in culture. Growth and DNA synthesis, *J. Cell. Biol.,* 60, 673, 1974.

10. **Beckman, I. G. R., Bradley, J., Brooks, D. A., Kupa, A., McNamara, P. J., Thomas, M., and Zola, H.,** Human lymphocyte markers defined by antibodies derived from somatic cell hybrids. II. A hybridoma secreting antibody against an antigen expressed by human B and null lymphocytes, *Clin. Exp. Immunol.,* 40, 593, 1980.

11. **Zola, H., McNamara, P. J., Moore, H. A., Smart, I. J., Brooks, D. A., Beckman, I. G. R. and Bradley, J.,** Maturation of human B lymphocytes. Studies with a panel of monoclonal antibodies against membrane antigens, *Clin. Exp. Immunol.,* 52, 655, 1983.

12. **Hohmann, A. W., Coleman, M., Comacchio, R. M., Skinner, J. M., and Bradley, J.,** Production of a polyspecific human monoclonal antibody reacting with an epidermal antigen, *Immunol. Cell Biol.,* 66, 239, 1988.

13. **Voller, A. and Bidwell, D. E.,** A simple method for detecting antibodies to Rubella, *Br. J. Exp. Pathol.,* 56, 338, 1975.

14. **Hohmann, A., Hodgson, A. J., Wang, D., Skinner, J. M., Bradley, J., and Zola, H.,** Monoclonal alkaline phosphatase anti-alkaline phosphatase (APAAP) complex: production of antibody, optimization of activity and use in immunostaining, *J. Histochem. Cytochem.,* 36, 137, 1988.

15. **Oliver, J. M., Senecal, J.-L., and Rothfield, N. L.,** Autoantibodies to the cytoskeleton in human sera, *Cell Muscle Motil.,* 6, 55, 1985.

16. **Pernis, B.,** Internalisation of lymphocyte membrane components, *Immunol. Today,* 6, 45, 1985.

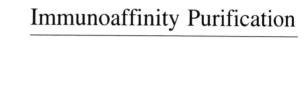

Immunoaffinity Purification

Chapter 7

THE USE OF IMMUNOAFFINITY CHROMATOGRAPHY IN THE PREPARATION OF THERAPEUTIC PRODUCTS

G. W. Jack

TABLE OF CONTENTS

I. INTRODUCTION

The concept of affinity as a property of biological molecules which may be exploited to assist in their purification has a long history of use. The basis of such affinity separations is the specific interaction between biological molecules. Thus, an enzyme will recognize its substrate or an inhibitor, a hormone its receptor, and an antibody its antigen. Affinity chromatography[1] has come to mean a separation system based on an immobilized molecule with a high specific affinity for the protein to be purified.

The principles of the method have often been described, but consist primarily[2] of contacting the crude protein solution with the affinity ligand immobilized convalently on an insoluble support matrix under conditions where the protein of interest will bind to the ligand. This binding must be sufficiently strong for the protein to remain bound while all the contaminating proteins are washed away. Only the insoluble support matrix bearing the ligand and bound protein in a buffer is left. By changing the conditions, the protein may be eluted off the insoluble support back into solution as a unique protein species. The immobilized ligand is thus free and unaffected, ready to be reequilibrated with buffer used for the initial contacting and binding.

The specificity of this technique depends on the nature of the immobilized ligand. If NAD or an NAD analog is used, then most dehydrogenases and other NAD-dependent enzymes will be bound, and similarly immobilized Concanavalin A will bind most glycoproteins, effecting the purification of classes of compounds rather than a single species. One of the most specific of biological interactions is that of an antigen with its antibody; hence, the degree of resolution of a chromatographic process based on this interaction will be very high. It has been found that protein purification protocols based on this technique, termed immunoaffinity chromatography (IAC), produce virtually pure protein in a single step. More conventional techniques such as ion-exchange and exclusion chromatography, precipitation, and ultrafiltration lack resolving power, and, hence, purification protocols will contain multiple steps. Even if each step is efficient, with recoveries greater than 80%, by the time several steps have been strung together, the overall yield of such a process is going to be poor.

Early examples of this methodology generally involved the use of immobilized antigen to purify polyclonal antibodies from animal sera.[3] The immobilization of one of the components of the interaction obviated the necessity of having to achieve a precise ratio of antigen to antibody in free solution to avoid solubilization of the complex by excess antigen.[4]

In the 1950s and 1960s reports appeared where the purification of antigen was postulated but the general interest of the time was the purification of antibody.[5] This, in part, was a natural consequence of the heterogeneity of antibodies present in the sera of animals even when immunized by a pure antigen. Such sera contained antibodies directed against the various epitopes of the antigen and with a range of affinities for each epitope. In particular, the presence of high-affinity antibodies resulted in a requirement for relatively severe conditions to dissociate the antibody-antigen complex which could inactivate the biological activity of a purified antigen. The impetus which the concept of IAC required to bring it to fruition as a usable tool for the protein chemist was the development by Kohler and Milstein[6] of monoclonal antibodies (MAbs). The growth of cloned hybridoma cell lines, either in culture or as ascites in mice or rats, provides the biochemist with a source of a unique antibody, the specificity and affinity of which may be determined. Some 10 years previous to this a suitably rigid and porous support matrix had been developed[7] based on agarose. By the 1970s IAC was being performed with polyclonal antibodies on agarose supports,[8] but it was not until 1980 that the first application of MAbs to the purification of a protein was reported.[9] Appropriately enough, the protein purified was α_2-interferon, for which at the time there was no known substrate or receptor; hence, only an antibody could be used as an affinity ligand.

This example shows one of the major advantages of MAb-based IAC over most other preparative techniques, namely, its high selectivity. Cultures producing interferon contain only minute amounts of the protein; yet, IAC is capable, in virtually a single step, of concentrating the interferon and purifying it to virtual homogeneity. It is this high resolving power of the technique that makes it attractive as a method in the production of therapeutic agents where very high degrees of purity must be attained to avoid unwanted side reactions.[10]

II. CHROMATOGRAPHIC MEDIA

A. GELS

The ideal properties of a gel for use as a support matrix for IAC include a high mechanical rigidity to withstand gel bead compression due to high hydrodynamic pressure resulting from high flow rates in column configurations. While the method will work in batch systems, fixed-bed column configurations offer the most efficient means of washing and eluting the gels as well as protecing the gel to minimize contamination during sterile operation. The gel should also be sufficiently porous to provide a large surface area per unit volume so as to allow sufficient antibody to be coupled. The pores should also be of sufficient size to allow ready diffusion even of large molecules to binding sites within the gel beads. The gel matrix should be readily derivatized by chemistries that form stable covalent bonds to minimize leakage of antibody from the matrix. Spherical beads have generally been used of a limited particle size distribution to assist in the formation of uniformly packed columns.

Some of the earliest examples of IAC employed fibrous support matrices such as cellulose,[3] but currently agarose[7] and a variety of synthetic and semisynthetic gels are the media of choice. Probably the most commonly used medium is Sepharose 4B, a 4% agarose gel, or its cross-linked derivative. The cross-linking is achieved by reacting agarose with 2,3-dibromopropanol followed by desulfation using alkaline borohydride to give a gel with a low content of ionizable groups.[11] Such gels, while exhibiting greatly improved rigidity and chemical stability, have fewer derivatizable groups and, hence, may not be as heavily loaded with MAb as the noncross-linked gel. Similar media are the Affi-Gel series produced by Biorad while Pharmacia-LKB's Sephacryl is an agarose-acrylamide copolymer. I.B.F. produces Trisacryl, a synthetic poly-*N*-acryloyl-2-amino-2-hydroxymethyl-1,3-propanediol, while Merck markets the Fractogel TSK series, a range of synthetic hydrophilic vinyl polymers.

Polyacrylamide gels may be derivatized in a variety of ways by conversion of carboxamide side groups to hydrazides,[12] but are largely unsuitable as IAC support matrices since gels with sufficiently large pores lack the mechanical rigidity to withstand even moderate flow rates. Copolymers of 2-hydroxyethyl methacrylate with ethylene dimethacrylate[13] (Spheron) have been used as IAC supports as have controlled pore glass beads[8] after suitable derivatization to mask their high nonspecific absorption of biomolecules in aqueous solution. Solid glass beads lack the high surface area-to-volume ratio desirable in an IAC support matrix. A variety of other inorganic supports have also been used including silica, titanium oxide, and alumina, but suffer from a number of drawbacks including pH instability and poor flow rates when packed as fixed-bed columns. Among the most recent developments, particularly for large-scale applications, has been the introduction of the "affinity filter". Such devices, typified by Domnick Hunter's Memsep, contain derivatizable polymers presented as sheets or fibers in a housing capable of operating at high pressures. On more traditional lines, duPont has introduced Perflex, a perfluorocarbon-based chemistry for which great chemical stability is claimed. For a more comprehensive description of available support matrices see Chapter 2.

B. COUPLING CHEMISTRIES

The attachment of MAbs to insoluble supports is a logical development of the methods developed for the immobilization of enzymes.[14] Such couplings are generally carried out through free amino groups on the surface of the antibody. A variety of chemistries are available to perform this reaction, and some are shown in Table 1, as are some of the chemistries linking through free carboxyl, hydroxyl, or sulfhydryl groups.

In small ligand affinity chromatography a spacer arm is frequently employed to place the ligand at some distance from the matrix backbone in an attempt to minimize steric hindrance of the interaction of ligand and target. In the case of antibody immobilization this is generally not thought to be a limiting factor in the interaction of the MAb and antigen. Spacer arms tend only to be introduced when they are part of the immobilization chemistry, as in the case of bisoxirane coupling.

When choosing a coupling method, a number of factors are likely to determine the method chosen. If only small columns for analytical purposes or small-scale purification are required, then it is likely to be more cost effective to buy in one of the commercially available activated matrices such as the Affi-Gels or CNBr or Tresyl-activated Sepharose rather than learn to activate gels in a reproducible manner. However, when large amounts of gel are required, it may be more economical to produce the activated gels in house, although the toxicity of some of the activating chemicals is such that the facilities of a well-equipped organic chemistry laboratory are required. Cyanogen bromide, hydrazine, and divinyl sulfone are all highly toxic while epichlorhydrin, tresyl chloride, and bisoxiranes are toxic and should only be handled in a well-ventilated fume cupboard. No guidance can really be given as to an immobilization method of choice. In our laboratory, immobilization chemistries have been used which have successfully coupled the antibody to the matrix only for the antibody then to fail to bind its antigen. This may result from multipoint attachment of the antibody to the matrix which, while producing a highly stable coupling, may constrain the antibody in such a way that its binding site is no longer active. In general, cyanogen bromide coupling is effective in producing an active immunosorbent as is tresyl activation. Both CNBr and divinylsulfone linkages are susceptible to alkaline hydrolysis above pH 10, while tresyl couplings withstand dilute alkali as do the relatively stable ether linkages produced by bisoxirane coupling. All linkages described appear to be stable in acid conditions.

C. THE ANTIBODY

One of the major advantages of using MAbs rather than polyclonal antibodies in IAC is that one can select the affinity of the immobilized antibody. Antibodies, in general, have affinities for their antigen in a range from 10^{-4} to $10^{-10}M$, and polyclonal sera will contain examples of antibodies with this range of affinities.[23] The antibodies will also be directed at all the epitopes on the antigen, which may be undesirable, as in the case, to be described more fully later, of the dimeric human pituitary glycoprotein hormones which share a common subunit.[25] Polyclonal antibodies raised against one of the hormones will cross-react with all the hormones and, if used in IAC, would result in the purification of a group of molecules rather than an individual molecular species.

During the elution of the antigen from the sorbent the conditions employed reduce the affinity of the antibody for its antigen. If an antibody is used with a very high affinity, of the order of $10^{-10}M$, then the conditions required to reduce the affinity sufficiently to achieve elution of the antigen may be so harsh as to destroy the biological activity of the antigen. The use of MAbs allows one to select an antibody of sufficient affinity to bind the antigen initially, yet achieve elution under sufficiently mild conditions as to retain the biological activity of the antigen. To this end, MAbs with affinity constants of the order of 10^{-7} to $10^{-8}M$ appear suitable.

When immobilizing MAbs, it is preferable to use purified antibody. In murine ascites fluid, although the antibody may be the predominant protein species, a range of other proteins is present. If the other proteins are immobilized along with the Mab, they will reduce the capacity of the sorbent for antigen and by virtue of the amphoteric nature of protein may act as ion exchangers, resulting in nonspecific binding to the sorbent. The purification of mouse IgG is conveniently achieved on columns of immobilized staphylococcal protein A[26] or by anion[27] or cation[28] exchange chromatography. Since antibodies have a range of isoelectric points, in our laboratory a method has been devised using both anion and cation exchange chromatography to purify MAbs from mice and rats irrespective of their IgG subclass.[82]

An extension of the use of protein A for antibody purification which does not appear to have been used extensively involves the covalent cross-linking of MAb bound to immobilized protein A.[29] Since protein A binds IgG through the Fc region, the antigen binding sites are left free to interact with antigen. The technique neatly combines the purification of the MAb and its immobilization.

The efficiency of operation of IAC sorbents is dependent on the level of antibody loading. Each antibody molecule has two antigen-binding sites so theoretically should bind 2 mol of antigen per mole of antibody. In practice this figure is never reached since, due to the random nature of coupling, some antibody molecules will be attached via an antigen-binding site, or a binding site may be masked by the close proximity of another antibody molecule. Perhaps the classic study is that by Eveleigh and Levy,[8] who showed that for antibody coupled to Sepharose by the CNBr method the maximum capacity obtained was 0.54 mol antigen per mole of antibody when the Sepharose was coupled at about 1 mg antibody per milliliter of gel. With porous glass they obtained higher efficiencies up to 1.5 mol of antigen per mole of antibody, but, again, at coupling densities of less than 1 mg per milliliter. Such lightly substituted gels make highly efficient use of the coupled MAb but have low capacity for antigen. More highly substituted gels will bind more antigen, but coupling more than about 10 mg MAb per milliliter of gel does not appear to increase the capacity of the gel for antigen. Coupling density has been shown to affect the distribution of antibody on the matrix.[30] At low levels of substitution, the antibody is found mainly on the surface of the gel beads. This may explain the high efficiency of such sorbents by minimizing the distance of diffusion from the feed solution to the absorption site. At high protein loadings, antibody was found to be more evenly distributed throughout the gel beads, requiring the antigen to diffuse into the bead pores to find a binding site.

D. LIGAND LEAKAGE

It is now admitted by all immunoaffinity chromatographers that ligand leakage occurs. Particularly in the field of therapeutic products researchers are being driven by the regulatory authorities to devise systems in which leakage is either abolished or, at least, greatly minimized. Those who say their gels don't leak should perhaps review their analytical techniques! A recent report[31] suggested that leakage from cyanogen bromide-activated gels only occurs when the sorbent is loaded with homologous antigen. Mere passage of nonantigen protein through the gel did not result in significant ligand leakage, and that leakage could be greatly reduced by using an *N*-hydroxysuccinimide-activated ester gel giving linkages similar to those obtained by bisoxirane coupling. One disadvantage of the method is the reported loss of antigen-binding capacity of MAbs coupled in this manner.[32] In contrast, in this laboratory using radioimmune assays to detect mouse and rat IgG the bulk of MAb leakage was found to coincide with the loading of a sample onto gels. Leakage of MAb from cyanogen bromide-coupled sorbents was found to be of the order 10 to 100 ng/ml of column effluent.

Leakage can be minimized by extended prewashing of sorbents, treating with chaotropic agents, or heating. An alternative strategy is the cross-linking of the coupled MAb to improve stability. This can be achieved by the use of either glutaraldehyde[33] or 2-hydroxy-5-nitro-benzyl bromide.[34] The result of ligand leakage, namely, MAb or antibody fragment present in the product of a purification, can be reduced subsequently by performing a gel-filtration or ion-exchange step after the IAC step. A logical step would be the use of a protein A

TABLE 1
Coupling of Ligands to Gels Activated by Various Chemistries

Ref.

Cyanogen Bromide

12, 15

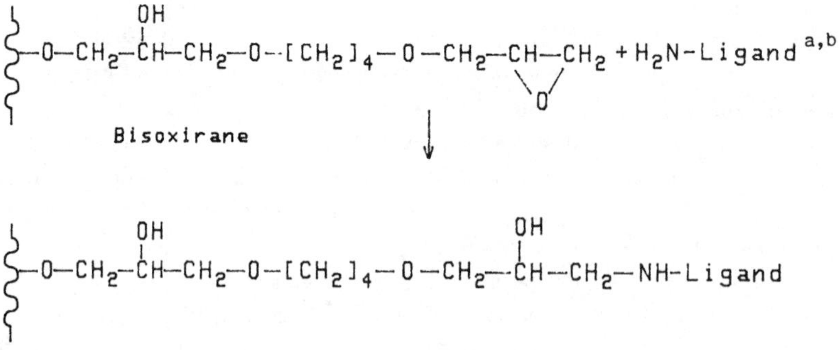

Bisoxirane

a,b

16

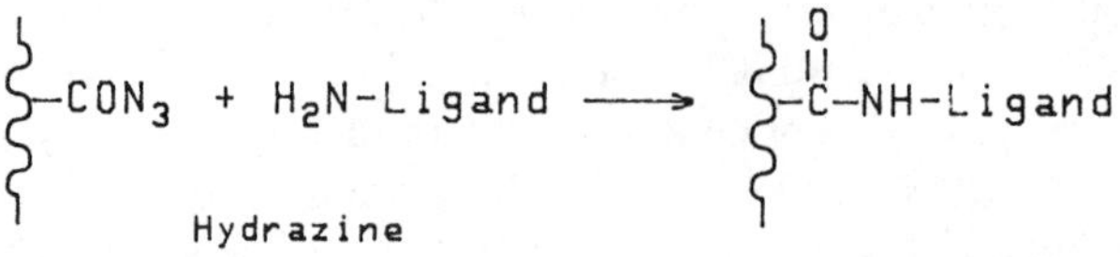

Hydrazine

17

Divinyl Sulphone

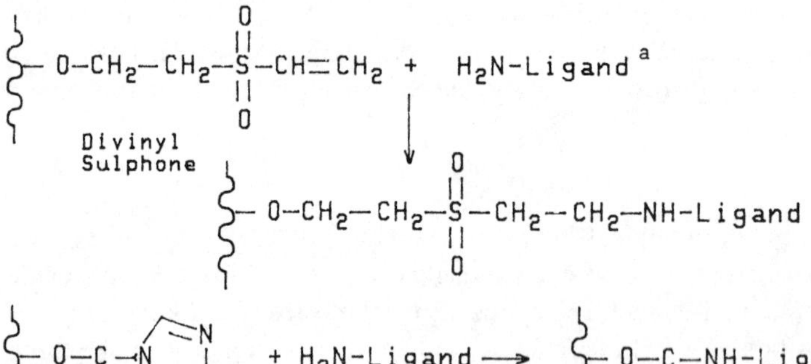

a

18

Carbonyldiimidazole

19

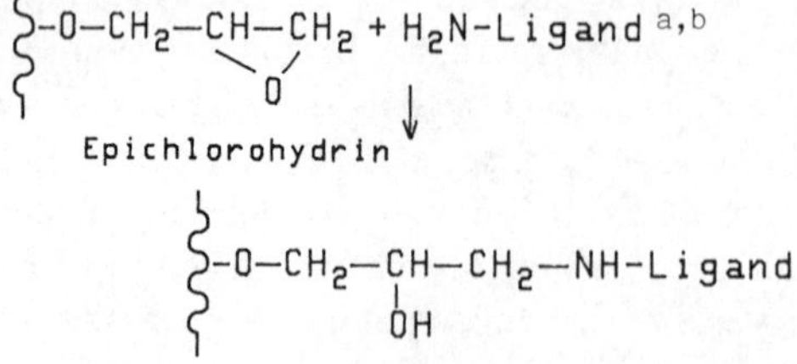

Epichlorohydrin

a,b

20

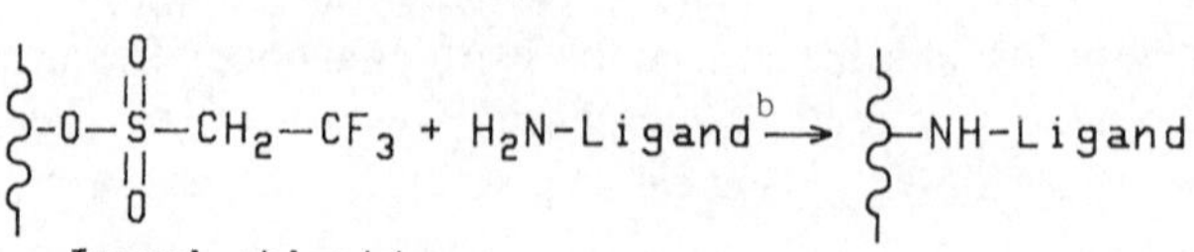

Tresyl chloride

21

TABLE 1 (continued)
Coupling of Ligands to Gels Activated by Various Chemistries

Ref.

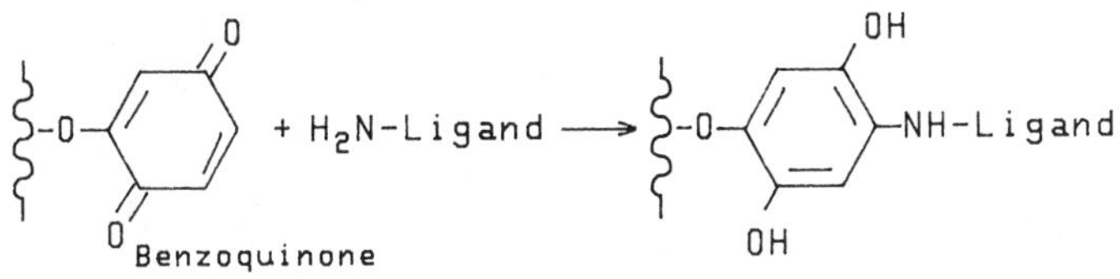

22

23

^a In these reactions coupling may occur through either the primary amine shown or a hydroxyl group.
^b This reaction may also induce coupling through free sulfhydryl groups.

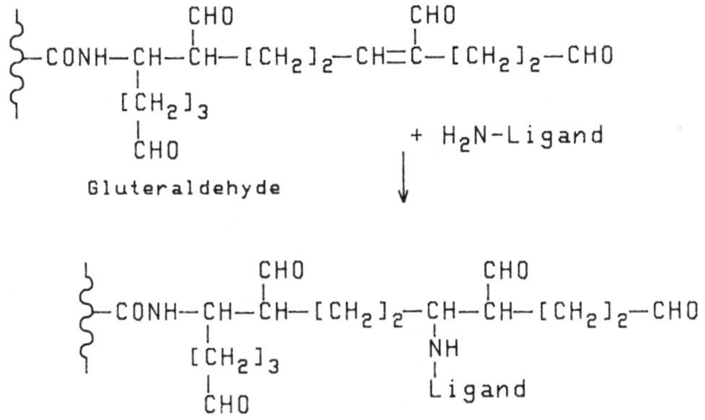

a   In these reactions coupling may occur through either the primary amine shown or a hydroxyl group.
b   This reaction may also induce coupling through free sulfhydryl groups.

column to remove IgG, but one is again faced with the problem of protein A leakage from the gel. Decisions on the most suitable matrices and coupling chemistries to minimize ligand leakage into the product probably still await the development of detection systems in the picogram range for IgG.

III. CHROMATOGRAPHIC PROCEDURES

No rules may be offered to aid in the design of an IAC purification protocol. Each case must be treated as a new and individual problem, dependent both on the MAb in use and the nature of the antigen being purified. This is particularly true of the conditions employed in the elution step where the stability of the antigen may place severe constraints on the nature of the eluant. With very robust, stable molecules, extremely denaturing conditions may be employed to effect elution as in the case of human growth hormone[35] where $8M$ urea in $0.1 M$ acetic acid was used while still producing a biologically active product. The design of these procedures including a mathematical treatment of the problem has been reviewed by Chase.[36]

A. ABSORPTION

The absorption stage of the chromatographic process is carried out under conditions which encourage the formation of the antibody-antigen complex. In general, this involves the use of innocuous buffer solutions at about neutral pH. Complex formation can be encouraged by the use of low flow rates to increase the time the antigen is in contact with the antibody, but using high concentrations of antigen and a high antigen-binding capacity. The use of low flow rates may be self-defeating in that it increases the time taken for each separation, while a higher flow rate may allow more cycles of operation per day. To achieve a high antigen concentration may require a prior time-consuming concentration step and ignores the ability of IAC to abstract from solution very dilute antigens.

During the absorption stage care must be taken to minimize nonspecific binding to the sorbent. Eveleigh and Levy[8] suggested the inclusion of 0.5 *M* NaCl and detergent in the buffer. While this step is undoubtedly successful in reducing nonspecific binding, the presence of a detergent may compromise a product for human therapeutic use due to the difficulties inherent in removing detergents from protein solutions. The precise nature of the buffer used for the binding stage can affect the extent of nonspecific interaction. Jack and Blazek[37] showed that changing from a Tris to a borate buffer reduced contamination levels during IAC purification of thyrotropin and that the subsequent inclusion of 0.5 *M* NaCl reduced them even further, to less than 0.5% by weight.

B. WASHING

The purpose of the washing step during IAC is to remove the material retained by the gel matrix nonspecifically. With highly porous matrices, small molecules in particular may be retained within the gel beads and require washing out. Nonspecific binding to the sorbent takes place either (1) by ionic interactions between the feed proteins and the ligand or any changed groups introduced into the matrix during coupling or (2) by hydrophobic interaction with the ligand. The use of high ionic-strength buffers during the absorption stage will minimize ionic binding, while washing with such buffers will elute any protein species lightly bound ionically. A subsequent wash with low ionic strength buffers, or even water, will help remove proteins bound by hydrophobic interactions. Such washing stages must be carried out, since elution of the specific antigen will be performed under conditions which will coelute nonspecifically bound protein.

C. ELUTION

Elution of the antigen from a MAb/antigen complex may be achieved by a wide variety of means including changes in pH, polarity-reducing agents, chaotropic ions, and denaturing/dissociating agents. Following immobilization the stability of IgG is improved drastically, enabling it to withstand conditions which would denature or inactivate it in free solution. The author's experience, however, is that murine MAbs will not withstand repeated exposure to pHs in excess of 10. The determination of conditions for elution can be a tedious and uncertain process, but a method[38] has been reported to screen elution methods rapidly by polyacrylamide gel electrophoresis.

Acid elution of antigen has usually been performed with acetic or propionic acids or low pH glycine buffers.[39] Where the MAb/antigen complex is stabilized by hydrophobic interactions, polarity-reducing substances may promote elution. To this end, dioxane and ethylene glycol[40] have been used successfully. Chaotropic ions are successful antigen eluants by virtue of their ability to disrupt the structure of water, thereby reducing hydrophobic interactions. In increasing chaotropic effect the ions may be ordered as follows:

$$Cl^- < I^- < CPO_4^- < CF_3COO^- < SCN^- \leqslant CCl_3COO^-$$

They are generally used at neutral pH in concentrations up to 3 M[39] with I^- and SCN^- the most popular ions in use. The denaturing/dissociating agents urea and guanidine hydrochloride are generally used at low pH.[35] Their inactivating effect on many proteins requires them to be removed rapidly from the purified protein. Now that the generation of MAbs is becoming a more routine laboratory practice, panels of antibodies are often available for an antigen of interest. Therefore, it is now possible to select a MAb of appropriate affinity to allow elution of the antigen using conditions least likely to be damaging to the antigen.

A number of novel approaches to the elution problem have been described recently.

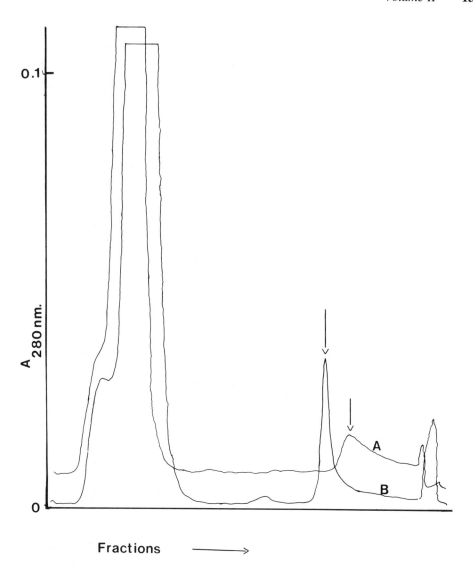

FIGURE 1. The IAC purification of human pituitary TSH. The column, 300 ml, was run at room temperature at a flow rate of 35 cm/h with (A) unidirectional flow or (B) reverse flow hormone elution. Arrows mark the TSH peaks.

Proteins which have cofactor or metal ion requirements may undergo profound conformational changes in the absence of the cofactor to the extent of abolishing epitopes present on the surface of the apoprotein. Prothrombin and factor IX have been purified on MAbs which recognize only the Ca^{2+}-stabilized forms of these proteins.[41] Elution from the antibody/antigen complex is performed by EDTA after binding antigen to MAb in the presence of Ca^{2+}. Although elegant, the concept is restricted mainly to metal-requiring proteins which are stable as both apo and holoprotein. A second approach has been the description of MAbs with a positive temperature-dependent affinity.[42] As a result, complexes formed at room temperature could be readily disrupted at 4°C.

Irrespective of the precise method of antigen elution, it should always be performed in the reverse direction to that of sample application.[39] The effect is illustrated in Figure 1. Columns tend to be loaded below their maximum capacity since before that point is reached antibody appears in the column effluent. Although the top of a column may be saturated with antigen, further down the column antigen-binding sites will be available. Therefore, if elution is conducted in the normal direction of flow, antigen will exchange on and off unoccupied sites in the lower part of the column, resulting in peak broadening. With reverse flow, antigen is moving to parts of a column with fewer and fewer free antigen-binding sites, resulting, as shown in Figure 1, in a sharp peak of eluted antigen.

D. AUTOMATION

Due to the high cost of MAbs it is not economical to produce a column of sorbent large enough to process a batch of product in a single chromatographic cycle. Methods to automate the process have been around as long as the technique itself and allow repetitive recycling of the chromatographic process with unattended operation.[11]

The basis of such systems is a programmer and a set of switchable three-way solenoid valves. Early programmers were of the cam-operated switch type[43] and were difficult to set precisely or change rapidly. The advent of the microprocessor-based programmable sequence timer provided a versatile instrument capable of controlling numerous devices and adequate for routine purifications.[44-46]

Such systems can be protected by the use of fluid sensors, level sensors, and pressure sensors to detect line breakages, low buffer levels, column blockage, or valve malfunction.

The latest development in the field is the use of a computer to control the sequence. This has the advantage of being an interactive system[36] in that the computer can take information from a UV-monitor, pH or conductivity meter and optimize each step of a separation to ensure a maintenance of product quality throughout many cycles of operation. Such a system has recently been marketed by Oros Systems Ltd. (Slough, U.K.), although it is dedicated to the purification of MAbs on immobilized protein A columns. This use of a computer has further advantages in the production of therapeutics in that it will record the data from each run for quality control purposes.

IV. PRODUCTION OF PARENTERAL PRODUCTS

In any IAC system one of the prime requirements is to maintain the sterility of the sorbent. The proteinaceous nature of the ligand renders it susceptible to proteolytic degradation, and any bacterial contamination of the gel bed will rapidly eliminate its antigen-binding capacity. In the author's experience, running IAC columns in the open laboratory, even with sterilized buffers and process material, results in column contamination within a matter of weeks. While this performance could possibly be improved by the use of in-line sterilizing filters, shifting the site of operation to a clean-room environment dramatically improves the lifetime of the sorbent.

A. ANTIBODY PRODUCTION

It has been known for some time that populations of rats and mice can carry viruses capable of infecting man. Some rodent viruses are capable of causing zoonotic disease, so it may be assumed that they, and possibly others, would cause infection if administered to humans as a contaminant of MAb-prepared parenteral. Among these viruses are Hantaan virus, lymphocytic choriomeningitis virus, rat rotavirus-like agent, and Sendai virus.[47] A further large group of rodent viruses are known to replicate in human cells *in vitro*. Thus, there is a possibility of MAbs produced in ascites fluid containing virus, while even the hybridoma cell line, derived from rodent cells, may carry a viral infection and shed viral particles when grown in culture.

This possibility has prompted national health authorities to draft guidelines for the pharmaceutical manufacture of MAbs. These guidelines are aimed mainly at the manufacturer of MAbs for direct therapeutic use, but the European Economic Communities document[48]

TABLE 2
Rodent Viruses of Known or Potential Threat to
Humans Which Should Not Be Present in MAb-
Producing Hydridoma

Virus	Virus group
Hantaan	Bunya
Sendai	Paramyxo
Lymphocytic choriomeningitis	Arena
Reovirus type 3	Reo
Rat rotavirus	Rota
Ectromelia	Orthopox
Epizootic diarrhoea	Rota
Lactate dehydrogenase	Toga
Minute	Parvo
Mouse adenovirus	Adeno
Mouse cytomegalovirus	Herpes
Mouse polio virus (GD-7)	Picorna
Mouse hepatitis	Corona
Mouse pneumonitis (K)	Papova
Pneumonia of mice	Paramyxo
Polyoma	Papova
Thymic	Herpes
Kilham rat virus	Parvo
Sialodacryoadenitis	Corona
Rat coronavirus	Corona
Toolan	Parvo

makes no distinction between MAbs for therapy and those to be used for the production of therapeutics. The U.S. guidelines issued by the Food and Drug Administration, (FDA)[49] are concerned both with murine and human monoclonals and are helpful in describing techniques used in reducing contamination. They also tackle the distinction between an agent used in the therapy of a life-threatening disease and one used to treat relatively healthy individuals, the infertile, for example. The Australian guidelines[50] follow the American lead of relating explicitly to injectable MAbs. The guidelines also contain practical details on viral decontamination using β-propiolactone which could be adopted for use with immobilized MAbs if immobilization improves their stability as it does with many proteins.[51]

The view taken by all the authorities, and commercial organizations which will produce MAbs on a contract basis, is that the hybridoma cell line should be well characterized and free of the viruses listed in Table 2 and of mycoplasmas. The regulations, then, describe the setting up of a master cell bank and a manufacturer's working cell bank. In production, finite passage of the cell line with an upper limit on the number of cell doublings is defined. If continuous culture of the hybridoma is considered for production, additional information on hybridoma stability is required. Purification of the antibody must be carried out to remove contaminants, including unwanted immunoglobulins, DNA, viruses, and any irritants, e.g., pristane, used in the generation of ascites. The purification procedure must be shown to be capable of removing specific contaminants by performing separations in which the crude preparation is spiked with, say, specific viruses or radio-labeled DNA, as was done in the case of the purification of human interferon from Namalwa tumor cells.[52] The commonest way of achieving this end is a combination of protein A and ion-exchange chromatography, carried out under the sorts of conditions used subsequently for the production of the parenteral product.

B. PRODUCTION FACILITIES

The production of parenteral therapeutic products for human use must be carried out in

TABLE 3
Environmental Requirements of Clean Rooms of Various Grades

	Max. particles (no./m³ of diameters)		Viable organisms (no./m³)	Standard classifications	
Grade	0.5 μ	5 μ		BS 5295	US Fed. Std.209B
1A (Work station)	3,000	0	< 1	1	100
1B	3,000	0	< 5	1	100
2	300,000	2,000	< 100	2	10,000

a clean environment under the regulations laid down by the local regulatory authority. In the U.K. the *Guide to Good Pharmaceutical Manufacturing Practice*,[53] the "orange guide", was compiled by the Medicines Inspectorate but has no statutory force and is merely a set of recommendations for manufacturers to follow. The guide interprets the statutory requirements of the Medicines Act 1968.[54]

Manufacture must be carried out in a clean area, a suite of rooms with environmental control of microbial and particulate contamination. The rooms should be constructed in such a way as to minimize surfaces for dust collection and with walls, floors, and ceilings of a readily cleaned material which is impervious and resists shedding of particles. The suite of rooms should be supplied with air at positive pressure appropriately filtered through high efficiency particle (HEPA) filters to the standards laid down by British Standard BS 5295[55] as shown in Table 3. Rooms in which aseptic processing is carried out should conform to Grade 1B standards, but Grade 1A conditions should be maintained under laminar flow work stations where the product is exposed and aseptic processes are carried out, such as the final filling of a sterile product prior to capping or freeze-drying. The status of the room should be monitored routinely with both a particle counter and an air sampler.[56] Grade 2 rooms can be used for the preparation of some parenterals providing that a pyrogen-free product with low microbial and particulate contamination is obtained which can subsequently be sterilized by filtration. The design and construction of aseptic areas is a specialized science[57] and best left to the experts.

The dirtiest things ever allowed into a suite of clean rooms are the human personnel! High standards of personal hygiene and cleanliness are required of the operators. On entering aseptic areas, personnel should change into all-enveloping, dedicated, sterile clean-room clothing which should shed virtually no particles and retain particles shed by the body. The clothing, including head and footwear, should be supplemented with powder-free rubber or plastic gloves and a nonlinting face mask. Such clothing can be rather hot to wear, highlighting the need for environmental control and drier, cooler working conditions than might be tolerated elsewhere.

All buffers and reagents used in a clean-room facility should be pyrogen free and sterile. Sterility may be achieved either by autoclaving, attaining a temperature of 121°C in the solution for 15 min, or by filtration through a 0.22 filter. Filtration has the advantage over autoclaving of also rendering the solution virtually particle free. Nonpyrogenicity can generally be achieved by the use of high-grade reagents and pyrogen-free water. The pyrogenicity of solutions is most easily checked by the limulus amoebocyte lysate assay.[58] Glassware may be sterilized by dry heat at 160°C for 1 h or at 180°C for 3 h, a procedure which has the added benefit of depyrogenating the glassware. All crude biological material should be filter sterilized prior to introduction into a clean-room environment. When potentially hazardous material such as human tissues or known pathogens is handled, it should be done in contained microbiological safety cabinets, operating at negative pressure to the room.

Of prime importance to the successful running of a clean-room suite are the training of the personnel and the validation of equipment and processes. Training, besides instruction on carrying out the manufacturing processes, should include reference to hygiene and basic microbiology so that staff are aware of the hazard they present to product integrity. Continual checks on performance[59] and validation of equipment, particularly air filters, stills, and sterilizers, is essential together with a maintenance of records of such operations. Water sources in particular should be checked regularly for microbial, pyrogen, and chemical contaminants. All operations in the manufacture of a particular product have to be carried out to a specific set of instructions contained in that product's Master Formula and Method. All manipulations and processes used in the manufacture of a batch of product must be recorded and countersigned in a Batch Manufacturing Record.

The regulations for the production and purification of MAbs are strict as are the guidelines relating to the manufacture of therapeutic products. It is, however, only by strict adherence to the concept of good manufacturing practice that safe products may be routinely produced.

V. PURIFICATION OF HUMAN PITUITARY GLYCOPROTEIN HORMONES

Until 1985, this laboratory produced the U.K. supply of human growth hormone (hGH) for the treatment of pituitary dwarfism. At that time a number of cases were reported[60] of Creutzfeldt-Jakob disease among patients who had received therapy with pituitary hGH preparations, and its use was discontinued in preference to a genetically engineered product. In the course of hGH production a side fraction is produced rich in the glycoprotein hormones thyrotropin (TSH), follitropin (FSH), and leutropin (LH). Ethanol fractionation of this side product gave a glycoprotein fraction suitable as starting material for the IAC purification of TSH, FSH, and LH.

The initial extraction and fractionation of the pituitaries was carried out in contained microbiological safety cabinets in Grade 2 rooms (Table 3). The hazard from Creutzfeldt-Jakob disease was felt to be minimal since pituitaries were taken at autopsy only from people who had shown no neurological symptoms, but the Hepatitis B status of the bodies was unknown. The subsequent IAC purification of the crude glycoproteins was carried out under Grade 2 conditions with the final filling and freeze-drying of a sterilized product taking place under Grade 1A conditions.

The initial intention of the development of this method was to produce clinically acceptable TSH, FSH, and LH for the treatment of thyroid cancer and some forms of infertility. Since the ban on the clinical use of human pituitary-derived material, the hormones are produced as standards for diagnostic kits or for research purposes. Prior to the ban, some TSH was used in humans[61] to determine its effectiveness in stimulating the activity of the thyroid. This work has been described in detail elsewhere.[37,46]

The MAbs, designated WCG-73 for TSH (from Wellcome Research Laboratories), ES13 for FSH and ES17 for LH (from Dr. K. James, Dept. of Surgery, Edinburgh University) were immobilized on CNBr-activated Sepharose CL4B at 2 mg/ml of gel. The freeze-dried gel was first swollen and washed in 1 mM HCl, then coupled in 0.2 M bicarbonate buffer pH 8.6 containing a 0.5 M NaCl for 4 h at room temperature, then 16 h at 4°C. The gel then was washed with coupling buffer followed by 0.5 M Tris/HCl pH 8 for 1 h. The sorbent was then washed alternately with 0.1 M acetate buffer pH 4.0 and 0.05 M borate buffer pH 8.5, both containing 0.5 M NaCl through six cycles. All washing procedures were carried out in a scintered-glass funnel under suction. The gels, each 300 ml, were then packed in 5-cm diameter columns and washed further with the acetate and borate buffers. Prior to use, the columns were treated with the sequence of buffers used in the preparative protocol at a flow rate of 35 cm/h.

The process was automated using a microprocessor-based programmable sequence timer

TABLE 4
The Binding Capacity of MA-Based Immunosorbents for Their Antigen

Antigen	Antibody designation	Affinity constant	Capacity IU antigen/ml gel
TSH	WGC-73	$5 \times 10^{-7}M$	0.5
FSH	ES 13	$10^{-8}M$	360
	ES 14	—	220
LH	ES 17	$10^{-8}M$	980
	44/3.4	—	480

Note: All antibodies were immobilized on CNBr-activated Sepharose 4B at 4 mg/ml of gel.

(PSC II, Tempatron Ltd., 6 Portman Road, Battle Farm Estate, Reading, U.K.), PSV-100 three-way solenoid valves of 24 V DC operation and a FRAC-300 programmable fraction collector with a UV-absorbance-driven level sensor (Pharmacia). This arrangement[46] automatically applies crude material to the column followed by the washing and eluting buffer, changes the direction of flow through the column for antigen elution, and collects the column effluent as either a waste or fractions to be retained.

Both crude material and the columns were equilibrated at room temperature with 50 mM borate buffer pH 8.5 (12 g/l boric acid and 2 g/l NaOH) containing 0.5 M NaCl. Column capacities for the hormones are given in Table 4. After the crude material was applied to the column, it was washed with two column volumes of equilibration buffer followed by one column volume of 50 mM borate buffer pH 8.5. The direction of flow through the column was then reversed while 1.5 column volumes of the antigen eluting buffer, 0.1 M glycine/HCl containing 0.5 M NaCl pH 3.5, was passed through the column. This was followed by half a column volume of 8 M urea in 0.1 M acetate buffer pH 4.0 as a sanitizing agent and a column volume of equilibration buffer. Flow direction through the column was again reversed prior to a final reequilibration with two volumes of the starting buffer.

The column effluent was sent to waste by the fraction collector other than when the unabsorbed protein emerged from the column and during elution of the antigen. Since the crude material contained more than one protein of interest, the columns were effectively run in a cascade system.[2] The unabsorbed protein fraction from the first column, usually anti-TSH, still contained FSH and LH in the appropriate buffer for binding to their specific absorbent. The purified hormone fractions from a number of cycles were pooled and concentrated by ultrafiltration over an Amicon PM-10 membrane and subjected to gel filtration through a column 2.6 cm in diameter by 80 cm long of Sephadex G-100 equilibrated against 50 mM Tris/HCl pH 8.5. The hormone-containing fractions were again concentrated to ten times the concentration required in the finished product, then diluted with 9 volumes of 1.11% w/v lactose monohydrate in water, sterile filtered, filled in vials as 1-ml lots, and freeze-dried. These hormones are rather "sticky" and should only be filtered through membranes of cellulose acetate, polycarbonate, or polysulfone.

In general, we have now stopped using Sephadex gel filtration media in favor of the Sephacryl series since they may be sterilized and depyrogenated even when packed as a column with 0.5 N NaOH. When operated in a clean room environment with sterile buffers and crude material, the MAb sorbents are very stable. Columns were used for up to 50 cycles of operation spread over many months and produced hormones of very high quality (Table 5).

The rationale behind the use of MAbs for these purifications lies in the similarity of all the hormones. Each is a dimeric protein with one subunit common to all three hormones while the unique subunits show considerable sequence homology.[25] Each hormone is glycosylated to varying degrees[62] resulting in each hormone displaying microheterogeneity with a wide spread of isoelectric point. These spreads of isoelectric point overlap, and since they

TABLE 5
Specification of Human Pituitary Glycoprotein Hormones Produced
by MAb-Based IAC

Hormone	TSH	FSH	LH
Specific activity[a]	16.6 IU/mg	10,000 IU/mg	25,000 IU/mg
Biological activity[b]	0.8	1.2	0.9
Contaminants[c]			
TSH	—	0.07%	0.02%
FSH	0.01%	—	0.04%
LH	0.03%	0.09%	—
hGH	0.06%	0.009%	<0.09%
MAb	0.01%	<0.15%	<0.6%
Bacteria	<5 c.f.u./ml	<5 c.f.u./ml	<5 c.f.u./ml
Pyrogenicity	6—12 EU/ml	4—8 EU/ml	1.25—2.5 EU/ml

[a] Specific activity is defined as the activity measure by radioimmune assay against an International Reference Preparation per milligram of protein.
[b] Biological activity was determined *in vivo* and expressed as a fraction of the radioimmune assay value.
[c] Where appropriate the contaminants are expressed as % w/w.

TABLE 6
Stability of Freeze-Dried Human Glycoprotein Hormones
Produced by MAb-Based IAC

	Time in weeks for 5% decrease in bioactivity		
	TSH	FSH	LH
−20°C	967	234	264
4°C	54	38	31
20°C	10	13	9.4

all have very similar molecular sizes, purification methods based on size and charge are inefficient in producing pure hormone in high yield. The IAC-based methods, in contrast, give yields of the order of 50% of material which is still biologically active and, after freeze-drying, very stable (Table 6).

VI. PRODUCTS

Since the development of MAb-based IAC a large number of proteins have been purified by the method although only a limited number of these proteins are of potential therapeutic interest. Alternatively, a number of proteins which have been purified by this method are produced commercially by alternative methods: such a protein is hGH which has been purified on an immobilized MAb[63] while the commercial product is obtained by other means.[64] Some of the proteins of potential therapeutic value which have been purified on MAb-based immunosorbents are listed in Table 7. Few of these products are as yet licensed pharmaceuticals, the majority still being classified in the U.S. as investigational new drugs and in the U.K. for clinical trials only. Among the exceptions is a factor VIII preparation (see Chapter 8) for the treatment of hemophilia, prepared from human blood and rendered free of hepatitis and AIDS viruses by virtue of the method of preparation. Urokinase is also being produced commercially by IAC while there is also interest in the purification of viral coat proteins[79] for new vaccines. A further application in the therapeutic field may lie in the use of IAC to deplete serum of specific hemostatic factors[80] present in excess and causing life-threatening illness.

TABLE 7
Proteins of Potential Therapeutic Value Which Have Been
Purified by MAb-Based IAC

Protein	Use	Ref.
Interferon	Cancer	65
Interferon-alpha C	Cancer	66
Interferon-gamma	Cancer	67
Interleukin-2	Cancer	68
Erythropoietin	Anemia	69
Tissue plasminogen activator	Dissolution of blood clots	70
Human growth hormone	Dwarfism	63
Somatomedin C	Tissue stimulation	71
Urokinase	Heart attack	72
Factor VIII	Hemophilia	73
Factors VII, IX, X; proteins C and 5	Hemostasis	74,75
Factor XII	Hemostasis	76
Factor XIII	Hemostasis	77
Glucocerebrosidase	Gaucher's disease	78

VII. CONCLUSIONS

Monoclonal antibody-based immunoaffinity chromatography provides a rapid and highly efficient means of purifying complex biological molecules to near homogeneity in a single step. The method is well established as a laboratory-scale procedure for purifying small amounts of protein for characterization and investigation, but the method is open to scale up for commercial production of high-value products. The method has similar requirements for scaleup as gel filtration and ion-exchange, and these processes are now well defined for large-scale operation.[81] The process may readily be automated and the columns recycled repetitively over an extended period to improve the economics of operation. However, due to the high cost of producing MAbs and validating their purification for use as ligands, IAC is likely to remain a purification technique applicable only to very high-value products.

One of the major advantages of IAC, namely its ability to remove from solution antigens present only in very low concentrations, is being eroded by the application of cloning techniques which result in high levels of expression of commercially important proteins. These may then be purified adequately by more conventional chromatographic techniques. For highly labile proteins, however, such as the factor VIII/von Willebrand complex even when produced by recombinant DNA technology, IAC appears to be the method of choice for their purification. The method also has the power to resolve closely related forms of a natural product as in the case of the family of proteins which is hGH.[63]. The selectivity of IAC, given the appropriate MAbs, is further illustrated by the resolution of the human pituitary glycoprotein hormones.[46]

The extent of commercial use of IAC for the production of therapeutics will depend in part on the cost of large-scale production of MAbs and the ease with which the requirements of regulatory authorities can be met. Otherwise, the main problem associated with IAC is leakage of ligand into the product. This may be overcome with the development of stable matrices and coupling chemistries. The challenge to IAC as the technique with the greatest resolving power is being posed by the manufacturers of ion-exchange matrices producing matrices of smaller size and tighter particle size distribution. There will remain occasions when IAC is the method of choice for some purification problems and could only be superseded by chromatography using as ligand a receptor molecule for the protein of interest which only binds the protein when still in a biologically active conformation.

ACKNOWLEDGMENT

The author wishes to thank Miss S. M. Hill for her assistance in the preparation of this manuscript.

REFERENCES

1. **Cuatrecasas, P.,** Affinity chromatography of macromolecules, *Adv. Enzymol.,* 36, 29, 1972.
2. **Jack, G. W. and Wade, H. E.,** Immunoaffinity chromatography of clinical products, *Tibtech,* 5, 91, 1987.
3. **Campbell, D. H., Luescher, E., and Lermon, L. S.,** Immunologic adsorbents. I. Isolation of antibody by means of a cellulose-protein antigen, *Proc. Natl. Acad. Sci. U.S.A.,* 37, 575, 1951.
4. **Heidelberger, M.,** Quantitative absolute methods in the study of antigen-antibody reactions, *Bacteriol. Rev.,* 3, 49, 1939.
5. **Tozer, B. T., Cammack, K. A., and Smith, H.,** Separation of antigens by immunological specificity. II. Release of antigen and antibody from their complex by aqueous carbon dioxide, *Biochem. J.,* 84, 80, 1962.
6. **Kohler, G. and Milstein, C.,** Continuous culture of fused cells secreting antibody of predefined specificity, *Nature, London,* 256, 495, 1975.
7. **Hjerten, S.,** The preparation of agarose spheres for chromatography of molecules particles, *Biochim. Biophys. Acta,* 79, 393, 1964.
8. **Eveleigh, J. W. and Levy, D. E.,** Immunochemical characteristics and preparative application of agarose-based immunosorbents, *J. Solid Phase Biochem.,* 2, 45, 1977.
9. **Secher, D. S. and Burke, D. C.,** A monoclonal antibody for large-scale purification of human leukocyte interferon, *Nature, London,* 285, 446, 1980.
10. **Fryklund, L., Skoog, B., Friberg, J., Brandt, J., and Fholenhag, K.,** Methionyl human growth hormone: characteristics and analytical criteria for human pharmaceutics, in *Quo Vadis? Therapeutic Agents Produced by Genetic Engineering,* Joueaux, A., Leygue, G., Moore, M., Roncucci, R., and Schmelck, P. H., Eds., *Quo Vadis?* Symposium, Sanofi Group, Toulouse-Labege, 1985, 131.
11. **Porath, J., Janson, J. C., and Laas, T.,** Agar derivatives for chromatography, electrophoresis and gel-bound enzymes. I. Desulphated and reduced cross-linked agar and agarose in spherical bead form, *J. Chromatogr.,* 60, 167, 1971.
12. **Cuatrecasas, P.,** Protein purification by affinity chromatography-derivatization of agarose and polyacrylamide beads, *J. Biol. Chem.,* 245, 3059, 1970.
13. **Coupek, J.,** Macroporous spherical hydroxyethyl methacrylate copolymers, their properties, activation and use in high performance affinity chromatography, *Anal. Chem. Symp. Ser.,* 9, 105, 1982.
14. **Zaborsky, D. R.,** *Immobilised Enzymes,* CRC Press, Boca Raton, FL, 1973.
15. **Porath, J., Axen, R., and Ernback, S.,** Chemical coupling of proteins to agarose, *Nature, London,* 215, 1491, 1967.
16. **Porath, J.,** General methods and coupling procedures, *Methods Enzymol.,* 34, 13, 1974.
17. **Inman, J. K.,** Covalent linkage of functional groups, ligands and proteins to polyacrylamide beads, *Methods Enzymol.,* 34, 30, 1974.
18. **Fornstedt, N. and Porath, J.,** Characterization studies on a new lectin found in seeds of *Vicia ervilia,* *FEBS Lett.,* 57, 187, 1975.
19. **Bethell, G. S., Ayers, J., Hancock, W. S., and Hearn, M. T. W.,** A novel method of activation of cross-linked agaroses with 1, 1-carbonyldiimidazale which gives a matrix for affinity chromatography devoid of additional changed groups, *J. Biol. Chem.,* 254, 2592, 1979.
20. **Porath, J. and Fornstedt, N.,** Group fractionation of plasma proteins on dipolar ion exchangers, *J. Chromatogr.,* 51, 479, 1970.
21. **Nillson, K. and Mosbach, K.,** Immobilization of enzymes and affinity ligands to various hydroxyl group carrying supports using highly reactive sulphonyl chlorides, *Biochem. Biophys. Res. Commun.,* 102, 449, 1981.
22. **Brandt, J., Andersson, L. D., and Porath, J.,** Covalent attachment of proteins to polysaccharide carriers by means of benzoquinone, *Biochim. Biophys. Acta,* 386, 196, 1975.
23. **Weston, P. D. and Avrameas, S.,** Proteins coupled to polycacrylamide beads using glutaraldehyde, *Biochem. Biophys. Res. Commun.,* 45, 1574, 1971.
24. **Goding, J. W.,** *Monoclonal Antibodies: Principles and Practice,* Academic Press, London, 1983, chap. 2.

25. **Pierce, J. G. and Parsons, T. F.,** Glycoprotein hormones: structure and function, *Annu. Rev. Biochem.,* 50, 456, 1981.
26. **Goding, J. W.,** Use of staphylococcal protein A as an immunological reagent, *J. Immunol. Methods,* 20, 241, 1978.
27. **Burchiel, S. W., Billman, J. R., and Alber, T. R.,** Rapid and efficient purification of mouse monoclonal antibodies from ascites fluid using high performance liquid chromatography, *J. Immunol. Methods,* 69, 33, 1984.
28. **Carlsson, M., Hedin, A., Inganas, M., Harfast, B., and Blomberg, F.,** Purification of *in vitro* produced mouse monoclonal antibodies. A two-step procedure utilising cation exchange chromatography and gel filtration, *J. Immunol. Methods,* 79, 89, 1985.
29. **Gersten, D. H. and Marchalonis, J. J.,** A rapid novel method for the solid-phase derivatization of IgG antibodies for immunoaffinity chromatography, *J. Immunol. Methods,* 24, 305, 1978.
30. **Chase, H. A., Horstmann, B. J., and Fowell, S. L.,** The performance of affinity separations using immobilised antibodies, *J. Chem. Technol. Biotechnol.,* in press.
31. **Peng, L., Calton, G. J., and Burnett, J. W.,** Stability of antibody attachment in immunosorbent chromatography, *Enzyme Microb. Technol.,* 8, 681, 1986.
32. **Pfeiffer, N. E., Wylie, D. E., and Schuster, S. M.,** Immunoaffinity chromatography utilising monoclonal antibodies: factors which influence antigen-binding capacity, *J. Immunol. Methods,* 97, 1, 1987.
33. **Van Wezel, A. L. and Van Der Marel, P.,** The application of immunoadsorption on immobilised antibodies for large scale concentration and purification of vaccines, *Anal. Chem. Symp. Ser.,* 9, 283, 1982.
34. **Murphy, R. F., Iman, A., Hughes, A. E., McGucken, M. J., Buchanan, K. D., Conlan, J. M., and Elmore, D. T.,** Avoidance of strongly chaotropic eluents for immunoaffinity chromatography by chemical modification of immobilised ligands, *Biochim. Biophys. Acta.* 420, 87 1976.
35. **Jack, G. W. and Gilbert, H. J.,** The purification by immunoaffinity chromatography of bacterial methionyl-(human somatotropin), *Biochem. Soc. Trans.,* 12, 246, 1984.
36. **Chase, H. A.,** Affinity separations utilising immobilised monoclonal antibodies — a new tool for the biochemical engineer, *Chem. Eng. Sci.,* 39, 1099, 1984.
37. **Jack, G. W. and Blazek, R.,** The purification of human thyroid-stimulating hormone by immunoaffinity chromatography, *J. Chem. Technol. Biotechnol.,* 39, 1, 1987.
38. **Janatova, J. and Gobel, R. J.,** Rapid optimisation of immunoadsorbent characteristics, *Biochem. J.,* 221, 113, 1984.
39. **Kristiansen, T.,** Matrix-bound antigens and antibodies, in *Affinity Chromatography,* Hoffman-Ostenhof, O., Ed., Pergamon Press, Oxford, 1978, 191.
40. **Andersson, K. K., Benjamin, Y., Douzou, P., and Balny, C.,** The effects of organic solvents and temperature on desorption of yeast 3-phosphoglycerate kinase from immunosorbent, *J. Immunol. Methods,* 25, 375, 1979.
41. **Liebman, H. A., Limentani, S. A., Furie, B. C., and Furie, B.,** Immunoaffinity purification of factor IX (Christmas factor) by using conformation-specific antibodies directed against the factor IX-metal complex, *Proc. Natl. Acad. Sci. U.S.A.,* 82, 3879, 1985.
42. **Dawes, J.,** Immunopurification process. Int. Patent WO 86/00910, 1986.
43. **Anderson, N. G., Willis, D. D., Holladay, D. W., Caton, J. E., Holleman, J. W., Eveleigh, J. W., Attrill, J. E., Ball, F. L., and Anderson, N. L.,** Analytical techniques for cell fractions. XIX. The Cyclum: an automatic system for cyclic chromatography, *Anal. Biochem.,* 66, 159, 1975.
44. **Eveleigh, J. W.,** Practical considerations in the use of immunosorbents and associated instrumentation, *Anal. Chem. Symp. Ser.,* 9, 293, 1982.
45. **Bazin, H. and Malache, J.-M.,** Rat (and mouse) monoclonal antibodies. V. A simple automated technique of antigen purification by immunoaffinity chromatography, *J. Immunol. Methods,* 88, 19, 1986.
46. **Jack, G. W., Blazek, R., James, K., Boyd, J. E., and Micklem, L. R.,** The automated production by immunoaffinity chromatography of the human pituitary glycoprotein hormones thyrotropin, follitropin and lutropin, *J. Chem. Technol. Biotechnol.,* 39, 45, 1987.
47. **Carthew, P.,** Is rodent virus contamination of monoclonal antibody preparations for use in human therapy a hazard? *J. Gen. Virol.,* 67, 963, 1986.
48. Commission of the European Communities — Committee for Propietary Medicinal Products, Guidelines on the production and quality control of monoclonal antibodies of murine origin intended for use in man, *Tibtech,* 6, G5, 1988.
49. Points to consider in the manufacture of injectable monoclonal antibodies products intended for human use in vivo, Office of Biologics Research and Review, Center for Drugs and Biologics, U.S. Food and Drug Administration, Bethesda, MD, 1985.
50. Commonwealth Department of Health, Guidelines for the Production of Monoclonal Antibodies Intended for Therapeutic Use, P.O. Box 100, Woden, ACT 2606, Canberra, Australia, 1985.
51. **Weetall, H. H. and Cooney, D. A.,** Immobilised therapeutic enzymes, in *Enzymes as Drugs,* Holcenberg, J. S. and Roberts, J., Eds., John Wiley & Sons, New York, 1981, 395.

52. **Finter, N. B. and Fantes, K. H.,** The purity and safety of interferons prepared for clinical use: the case for lymphoblastoid interferon, in *Interferon,* Vol. 2, Gresser, I., Ed., Academic Press, London, 1980, 65.
53. **Sharp, J. R., Ed.,** *Guide to Good Pharmaceutical Manufacturing Practice,* 1983, Her Majesty's Stationery Office, London.
54. The Medicines Act 1968, Her Majesty's Stationery Office, London, 1968.
55. BS 5295: Environmental Cleanliness in Enclosed Spaces, British Standards Institute, London, 1976.
56. **Tetzlaff, R. F.,** Regulatory aspects of aseptic processing, *Pharmaceutical Technology, 8,* 38, 1984.
57. **Stockdale, D. P.,** Design and operation of an aseptic filling system. *Pharmaceutical Technology, 9,* 38, 1985.
58. **Levin, J. and Bang, F. B.** Clottable proteins in limulus: its location and kinetics of its coagulation by endotoxin. *Thromb. Diath. Haemorrh.,* 19, 186, 1968.
59. **Mead, W. J.** Proper maintenance: a key to product quality. *Pharm. Technol.,* 9, 42, 1985.
60. **Powell-Jackson, J., Weller, R. D., Kennedy, P., Preece, M. A., Whitcombe, E. M., and Newsom-Davis, J.,** Creutzfeldt-Jacob disease after administration of human growth hormone, *Lancet,* 2, 244, 1985.
61. **Law, A., Jack, G. W., Tellez, M., and Edmonds, C. J.,** In-vivo studies of a human-thyrotropin preparation, *J. Endcrinol.,* 110, 375, 1986.
62. **Van Ginkel, L. A. and Loeber, J. G.,** Heterogeneity of human luteinizing hormone. Effect of neuraminidase treatment on biologically active hormone and α- and β- subunit, *Acta Endocrinol.,* 114, 577, 1987.
63. **Jonsdottir, I., Skoog, B., Ekre, H.-P. T., Pavlu, B., and Perlman, P.,** Purification of pituitary and biosynthetic human growth hormone, using monoclonal antibody immuno-adsorbent, *Mol. Cell. Endocrinol.,* 46, 131, 1986.
64. **Olson, K. C., Fenno, J., Lin, N., Harkins, R. N., Snider, C., Kohr, W. H., Ross, M. J., Fodge, D., Prender, G., and Stebbing, N.,** Purified human growh hormone from *E. coli* is biologically active, *Nature,* 293, 408, 1981.
65. **Staehelin, T., Hobbs, D. S., Kung, J.-F., Lai, C.-Y., and Pestka, S.,** Purification and characterisation of recombinant human leukocyte interferon (IFLrA) with monoclonal antibodies, *J. Biol. Chem.,* 256, 9750, 1981.
66. **Vaks, B., Mory, Y., Pederson, J. U., and Horovitz, O.,** A semi-continuous process for the production of human and interferon- C from *E. coli* using tangential-flow microfiltration and immunoaffinity chromatography, *Biotech. Lett.,* 6, 621, 1984.
67. **Le, J., Barrowclough, B. S., and Vilcek, J.,** Monoclonal antibodies to human immune interferon and their application for affinity chromatography, *J. Immunol. Methods,* 69, 61, 1984.
68. **Stadler, B. M., Berenstein, E. H., Siraganian, R. P., and Oppenheim, J. J.,** Monoclonal antibody against human interleukin 2. I. Purification of interleukin 2 for the production of monoclonal antibodies, *J. Immunol.,* 128, 1620, 1982.
69. **Yanagawa, S., Hirade, K., Ohnota, H., Sasaki, R., Chiba, H., Ueda, M., and Gotor, M.,** Isolation of human erythropoietin with monoclonal antibodies, *J. Biol. Chem.,* 259, 2707, 1984.
70. **Reagan, M. E., Robb, M., Bornstein, I., and Niday, E.,** Immunoaffinity purification of tissue plasminogen activator from serum-supplemented conditioned media using monoclonal antibody, *Thromb. Res.,* 40, 1, 1985.
71. **Chernausek, S. D., Chatelain, P. G., Svoboda, M. E., Underwood, L. E., and Van Wyk, J. J.,** Efficient purification of somatomedin C-insulin-like growth factor I using immunoaffinity chromatography, *Biochem. Biophys. Res. Commun.,* 1626, 282, 1985.
72. **Corti, A., Nolli, M. L., Suffientini, A., Guindani, A., Blasi, F., and Cassani, G.,** Purification of a single-chain urokinase precursor from A-431 human epidermoid carcinoma cells by monoclonal antibody immunoaffinity chromatographin, in *'Protides' of the Biological Fluids, Vol. 33,* Peeters, H., Ed., Pergamon Press, Oxford, 1985, 623.
73. **Zimmerman, T. S. and Fulcher, C. A.,** Ultrapurification of Factor VIII using monoclonal antibodies, U.S. Patent 4,361,509, 1982.
74. **Jenny, R., Church, W., Odegaard, B., Litwiller, R., and Mamm, K.,** Purification of six human vitamin K-dependant proteins in a single chromatographic step using immunoaffinity columns, *Prep. Biochem.,* 16, 227, 1986.
75. **Nakamura, S. and Sakata, Y.,** Immunoaffinity purification of protein C by using conformation-specific monoclonal antibodies to protein C- Calcium ion complex, *Biochim. Biophys. Acta,* 925, 85, 1987.
76. **Pixley, R. A., Procino, L. G., Silver, L., and Colman, R. W.,** Purification of functionally active Factor XII by immunoaffinity chromatography using a monclonal antibody to the Factor XII heavy chain region, *Clin. Res.,* 33, 550A, 1985.
77. **Gniewek, R. A., Schweinle, J. E., and Kurusky, A.,** Monoclonal antibodies specific for human plasma Factor XIIIB subunit and their use in the purification of human plasma Factor XIII by immunoaffinity chromatography, *Fed. Proc.,* 44, 1070, 1985,

78. **Aerts, J. M. F. G., Donker-Koopman, W. E., Murray, G. J., Barranger, J. A., Tager, J. M., and Schram, A. W.**, A procedure for the rapid purification in high yield of human glucocerebrosidase using immunoaffinity chromatography with monoclonal antibodies, *Anal. Biochem.*, 154, 655, 1986.
79. **Kikuchi, G. E., Baker, S. A., Merajver, S. D., Coligan, J. E., Levine, M., Glorioso, J. C., and Nairn, R.**, Purification and structural characterisation of *Herpes simplex* virus glycoprotein C, *Biochemistry*, 26, 424, 1987.
80. **Prowse, C. V.**, Immunoaffinity purification of haemostasis proteins, *J. Chem. Technol. Biotechnol.*, in press.
81. **Curling, J. M. and Cooney, J. M.**, Operation of large scale gel filtration and ion-exchange systems, *J. Parenter. Sci. Technol.*, 36, 59, 1982.
82. **Beer, D.**, unpublished.

Chapter 8

METHODOLOGY FOR THE IMMUNOPURIFICATION AND IMMUNODEPLETION OF COAGULATION FACTORS, WITH EMPHASIS ON FACTORS VIII AND IX

Christopher V. Prowse

TABLE OF CONTENTS

I. BACKGROUND

The intent of this review is to give some detail of the methodology and approach used in developing procedures for the purification and plasma-depletion of coagulation factors VIII and IX in my laboratory. These will be summarized in the context of corresponding methods developed elsewhere. While immunoaffinity chromatography can involve any antibody, most groups now use monoclonal antibodies (MAb) for this purpose, and only this aspect will be covered here.

Monoclonal antibodies (MAb) to virtually all the known coagulation factors (Table 1)[1-39] have now been described. These were mainly developed for studies on the structure and function of individual factors or to allow development of specific assays of these glycoproteins.[40] In many cases these MAb have subsequently been used for immunopurification purposes (Table 1). Since the MAb were not usually selected for purification, the elution conditions described are frequently severe, e.g., pH 2 or 3 M KSCN. Despite this, the exquisite specificity of MAb can allow high purifications in a single step. This is, perhaps, best illustrated by the use of a series of MAb columns to separate essentially pure factors II, VII, IX, and X, and protein C and S from a barium eluate of human plasma containing this homologous series of vitamin-K-dependent proteins.[5]

While purification of individual coagulation factors may be immensely useful in laboratory research, this is usually carried out on a fairly small scale. Currently, coagulation factors VIII and IX and plasminogen activators are the major hemostatic factors used therapeutically. While MAb-based purification methods have been developed for all these proteins, there are simpler and cheaper methods for the affinity purification of plasminogen activators. Immunoaffinity procedures developed for the purification of factors VIII and IX are summarized in Table 2.[5,14,16-20,25,26,41-43] Two of these procedures, both for factor VIII, have been used on a large scale to prepare therapeutic products for clinical use.

MAb columns may also be used to remove proteins from complex mixtures. In the case of factors VIII and IX this approach offers an alternative route to preparation of plasma deficient in specific coagulation factors (Table 3).[21,44-48] Such reagents are used in the assay of factors VIII and IX and are increasingly difficult and costly to obtain from the traditional source: patients with congenital deficiency, many of whom are now infected with the human immunodeficiency virus.

II. ANTIBODY PRODUCTION, IMMOBILIZATION, AND ELUTION

Nearly all described MAb to coagulation factors have been produced by conventional techniques using purified protein as immunogen and enzyme-linked or radioimmunoassays to detect the antibodies. In certain situations enzyme digested or denatured protein or synthetic peptides have been used to produce antibodies with specificities not easily obtained using the native protein.[49,50]

Both immunopurification and depletion usually require use of immobilized purified antibody. A wide variety of commercial systems are marketed which allow MAb purification. In this laboratory we start with ascitic fluid or, more usually, ten-fold concentrated culture supernatant (Amicon PM-10 membrane) and purify on columns of protein A Sepharose 4B using a gradient from 0.1 M phosphate-citrate pH 8.4, 2 M NaCl to 0.05 M phosphate-citrate pH 2.[51] The high salt content and pH of the loading buffer ensure maintenance of column IgG capacity, even for MAb of IgG1 isotype. In this gradient system we usually observe a small peak of bovine IgG, eluting at about pH7, when the source material contained fetal calf serum. Alternatively, immunoglobulins are partially purified by repeated precipitation with 18% w/w sodium sulfate at 20°C after addition of an equal volume of 0.2 M

TABLE 1
Monoclonal Antibodies to Blood Coagulation Factors: References

Factor	Description	Use in purification
Fibrinogen	1,2	—
Prothrombin	3,4	5
Tissue factor	6,7	6,7
Factor V	8,9	10
Factor VII	11	5,12
Factor VIII/von Willebrand factor	13 (registry)	14,15,16,17,18,19,20
Factor IX	21,22,23,24	5,25,26
Factor X	27	5,27
Factor XI	28,29	28
Factor XII	30,31	30,31
Factor XIII	32	32
Prekallikrein	33	—
High molecular weight kininogen	34	—
Protein C	35,36	5,37
Protein S	38,39	5,38,39

Note: This listing is illustrative rather than comprehensive. For a fuller review including antibodies to fibrinolytic enzymes and inhibitors of coagulation and fibrinolysis, see Reference 40.

phosphate pH 8.0. In either case we then dialyze the MAb, at a concentration of 1 to 2 mg/ ml, into 25 mM phosphate, 0.1 M NaCl pH 7.4.

For immobilization of MAb most studies to date have used, at least initially, either Pharmacia's cyanogen bromide activated Sepharose or Biorad's Affigel material. While there is now an increasing range of activated supports, both the above materials have a pore size that may restrict entry of IgG into the gel. This means that during the immobilization of MAb to such gels the amount of gel surface available for IgG coupling is reduced. This does not, in itself, matter, except in terms of the eventual capacity per milliliter of immunoaffinity gel. What does matter is that subsequent access of proteins to the immobilized MAb will be similarly restricted. Thus, the factor VIII complex, with a size in excess of 1 million Da, will only have access to a small proportion of MAb (M 150,000 Da) immobilized on 4% agarose. On the other hand, factor IX, with a molecular weight of about 55,000 Da, will not be restricted in this way. The other disadvantage of agarose-based gels is that they are relatively "soft" or compressible, i.e., they do not perform well at high flow rates. This is not usually important for initial laboratory studies but can be vital in developing larger-scale processes. One way around both these problems is to use small-pore "hard" gels, such as those based on coated silica, wherein reactions largely take place on (the relatively small area of) the surface of the beads. Our own experience has been largely with the alternative of using gels, such as Pharmacia's Sephacryl S-1000, which have a large enough pore size to allow free access by IgG and factor VIII and are reasonably noncompressible.

Using such gels, a method for MAb immobilization based on reductive amination after periodate oxidation of the Sephacryl gel was developed.[52] While this has worked well, giving high coupling yields at protein loads as high as 100 mg/ml gel, subsequent experience at loadings around 1 mg/ml has shown that, while there is no problem with immobilization, the resultant affinity gel is *in some cases* inactive. We suspect this is due to multipoint immobilization of the MAb resulting in conformational inactivation, i.e., overcoupling. For this reason our pilot studies are now usually carried out with MAb immobilized on Sephacryl S-1000 using a modification of Kohn and Wilchek's low temperature cyanogen bromide method[53] developed by Dr. Duncan Pepper (see Chapter 9 in this volume). This is suitable for all Sepharoses, Sephacryls, and Fractogels and has the advantage of using water-soluble solvents and resulting in an activated product that can be stored for months prior to use.

TABLE 2
Immunopurification of Human Factors VIII and IX

Name	MAb to	Ref	Scale (liters)	Product	Eluant	Specific activity (μ/mg)	Yield %
Monoclate*	vWf	14,41	>100	VIII	0.25 M CaCl$_2$	3500	—
Bourgouis 1987	vWf	18	10	VIII & vWf	pH 8.5	60	40
Hornsey 1987	vWf	19	<1	VIII & vWf	KI, Lysine	50	50
HemofilM/AHF.M*	VIII	42	>100	VIII	40% EG	1800	—
Rotblat 1985	VIII	16	20	VIII	20% EG,KI	5000	80
Croissant 1986	VIII	17	<1	VIII	50% EG	—	≤60
Ganz 1988	VIII	20	<1	VIII	50% EG, NaCl	5000	80
Jenny 1986	IX	5	<1	IX	3 M NaSCN	132	—
Bessos 1986	IX	25	<1	IX	50% EG, pH 10	180	80
Limentani 1987	IX	43	<1	IX	EDTA	~150	—
Smith 1988	IX	26	3—10	IX	EDTA	150	90

Note: MAb, EG, and EDTA denote monoclonal antibody, ethylene glycol, and ethylenediamine tetraacetic acid, respectively. Scale refers to the volume of material loaded onto the immobilized antibody. * denotes products at clinical trial.

TABLE 3
Immunoaffinity Depletion of Normal Human Plasma

Depletion	Antibody	Supplier	Ref.
Factor VIII	VIII(2) & vWf	Diagen/Porton	44
	VIII(2) & vWf	—	45
	NS	Mertz & Dade	—
	NS	Stago	—
	VIII	Organon Teknika	—
	VIII	—	46
Factor IX	IX	Diagen/Porton	21
	IX	Organon Teknika	—
	IX	—	47
	IX	—	48

Note: NS: not specified. Plasmas depleted of factor VII (Diagen), factor V, or factor X (Dade) are also available commercially.

The method requires use of a fume hood for all operations involving cyanogen bromide and of an internally spark-free freezer. Use of a Pyrex glass vessel with a rubber-lined screw cap, capable of centrifugation at $1000 \times g$ allows the same container to be used throughout activation and coupling:

1. Prior to activation the Sephacryl is autoclaved at 121°C for 30 min and then washed with distilled water. The autoclaving enhances reactivity with CNBr, increasing the activation yield from 5 to 25 μmol cyanate per milliliter gel.
2. In a 500-ml glass container, 100 ml of settled gel is mixed with 50 ml distilled water, using a magnetic stirrer, and 350 ml acetone slowly added with continuous mixing. This is allowed to cool to $-20°C$ by standing overnight in a $-20°C$ freezer (WHICH MUST BE INTERNALLY SPARK FREE), *after* aspiration of 350 ml supernatant fluid from the settled mixture.
3. Rapidly, so the temperature remains at $-20°C$, and with vigorous stirring, add 5 g of cyanogen bromide dissolved in a minimum volume of acetone (5 ml) and then 10 ml of triethylamine. The reaction vessel is then placed in cold ethanol at $-15°C$ in a 2-liter Dewar flask and the reaction stopped after 90 s by addition of 350 ml glacial acetic acid (precooled to $-15°C$) to give a final pH of 2.0 ± 0.5. Stand for 30 min at $-15°C$ and aspirate supernatant fluid from settled gel.
4. For use, the gel is washed with ice-cold 1 mM HCl. Alternatively, it may be stored in activated form for at least 6 months at $-40°C$ after washing and resuspension in acetone containing 10% v/v acetic acid.
5. For coupling, purified MAb in buffer pH 7.4 (see above) is added to an appropriate volume of activated moist gel (excess fluid being removed by suction on a sintered glass filter just before use) and the pH adjusted to between 7 and 9 with 0.1 M sodium bicarbonate and mixed end-over-end in a closed vessel overnight at room temperature. After blocking residual active groups by suspension in 1 M ethanolamine pH 9.0 for 2 h, the resultant immunoaffinity gels were usually preconditioned by heating in neutral buffer at 56°C for 30 min to remove weakly bound antibody and stored in 15 mM citrate, 150 mM NaCl pH 7.0. Depending on the study involved, gels were in some studies also precycled through pH 4 and pH 9 buffer cycles just prior to use.

The above procedure yields active MAb coupled to Sephacryl S-1000 with a coupling yield of more than 90% (or more than 80% after preconditioning) at MAb loadings between 1 and 3 mg/ml. In most studies we have found little benefit in coupling greater amounts of antibody than this.

The final requirement for immunoaffinity chromatography is the selection of appropriate loading and elution conditions. For coagulation factors the nature of the source material (human plasma or cell culture supernatant) usually requires that loading be carried out under conditions close to neutrality and physiological ionic strength. In an ideal world, elution conditions would be minimally different from loading conditions and result, for example, from small changes in temperature, pH, or metal ion concentration. In practice, MAb that allow such procedures are usually only obtained by careful selection at the time of fusion and cloning. At least to date, most studies have used preexisting MAbs and have had to resort to more severe elution conditions. Eluting buffers that have been used including those with extremes of ionic strength or pH, organic solvents, detergents, chaotropic buffers (such as thiocyanate, magnesium sulfate, urea), use of diols such as ethylene glycol to disrupt hydrophobic interactions, use of competitive ligands, and use of buffers to disrupt protein heterodimers after binding to the MAb (see factor VIII -von Willebrand factor [vWf] below). Our approach has been to screen panels of such buffers to determine those allowing retention of the appropriate activity and then determining which of these buffers allows elution of radiolabeled protein from MAb immobilized on Sephacryl, prior to small-scale column studies using activity assays on eluted proteins.[19] For the first and last of these stages the activity assays must be sufficiently sensitive to allow assessment at sample dilutions where the chosen buffers do not interfere with the assay itself. For the intermediate step, provided the appropriate protein is available in pure, radiolabeled form, our usual approach may be outlined as follows:

1. Radiolabeled protein ($\sim$10,000 dpm) in 0.9 ml loading buffer or, preferably, in the proposed feedstock material, e.g., plasma, is mixed for 2 h with 50 to 100 μg of MAb immobilized on 0.05 ml settled Sephacryl.
2. Following centrifugation the supernatant fluid is aspirated and its radioactive content determined. After washing with 1 ml loading buffer the gel is mixed with 0.5 ml control or elution buffer for 30 min. After centrifugation this buffer is aspirated and the gel washed with a further 0.5 ml elution buffer. The radioactivity in the two combined portions is determined and used to assess percentage elution.

We have found that conical 1.8-ml Eppendorf polyethylene tubes are convenient for such studies and that the suggested volumes are appropriate to avoid excessive dilution of elution buffers by fluid trapped in the internal space of gels. For some elution buffers, particularly those containing iodide, there may be considerable quenching of ^{125}I radioactivity, and this must be corrected for.

The above approach tests, as it should, the ability of immobilized antibody to bind and release protein in solution. It is, however, not easily applied to rapid screening of large numbers of hybridoma supernatants. To screen large numbers of crude antibodies the use of enzyme-linked immunosorbent assay (ELISA) methods can be more convenient. If pure antigen or enzyme-labeled specific antibody is available, then a multi-sandwich ELISA can be set up in which anti-mouse antibody on microplates is used to capture MAb from crude solution. After washing, this is then probed for ability to bind antigen (either directly with labeled antigen or indirectly with labeled specific antibody) in the presence of various buffers. This approach suffers from the disadvantage that multiple interactions are required for a positive signal, any one of which may be disrupted by the test elution buffer. Despite this, the approach has been used to select MAb which allow elution under mild conditions, such as absence of calcium ions.[26]

A more generally applicable ELISA screen involves coating microplates with the protein of interest, which for these purposes does not need to be highly purified. Crude MAb are then allowed to react with the bound protein in the presence of various buffers and proportion of MAb bound determined, after washing, using enzyme-labeled second antibody. The potential disadvantages of this approach, which in our experience allows screening of large numbers of MAb with a large buffer panel within 1 d, are that it assesses antibody elution from immobilized protein rather than the reverse and that certain buffers might be expected themselves to elute the protein antigen from the plate. In practice our limited experience suggests that even the most chaotropic buffer known to us, lithium diiodosalicylate, will not elute protein from standard ELISA microtrays once it has been coated on, as determined using radiolabeled vWf.

Once a MAb and a corresponding elution buffer have been selected, we usually proceed to small-scale conventional chromatography studies using 1-ml amounts of gel in small disposable plastic columns (Biorad). For purification procedures the activity yield at this stage should be in excess of 50%. This usually requires that more than 80 to 90% of the loaded activity is bound, with elution of more than 70% of the bound activity. To obtain such performance it is often necessary to preprocess the feedstock material. Thus, in our work on factor VIII, such performance was not easily obtained if human plasma was loaded directly on the column, but was readily achieved if plasma was first processed to cryoprecipitate to concentrate the factor VIII and reduce the protein load.[19] For depletion the initial approach is exactly the same, but it is usually necessary to exceed 99% removal of protein from the feedstock plasma rather than only 80 to 90%. On the other hand, for depletion it is only necessary to elute the bound protein, not to recover it in active form, during column regeneration.

For either purification or depletion, the high cost of MAb requires that the antibody column be used on multiple occasions. This is aided by avoidance of conditions that may precipitate proteins within the gel and of contamination, by prefiltration of all buffers and samples. In our experience we have found it helpful to include a prefilter column of Sephacryl, with no bound antibody, upstream of the antibody column, for this purpose. When using multiple gels it is preferable that they be in physically separate columns, rather than packed in a single container, as this reduces the risk of damaging the various (costly) antibody gels and allows simple separation of the gels should this be required.

III. PURIFICATION AND DEPLETION OF FACTOR VIII

A. Purification

Factor VIII circulates in blood plasma as a complex with vWf. The existence of this calcium-dependent complex in plasma, but not in recombinant sources of factor VIII until recently, has allowed three approaches to the immunopurification of VIII (Table 3). Most simply, MAb to VIII is used to bind VIII from plasma or, more usually, from more-processed materials. Elution of up to 80% of the bound material has been reported and when used as the terminal step in a purification scheme can result in essentially pure factor VIII. Curiously, elution buffers in the published reports using this approach all include ethylene glycol, suggesting that an immunodominant epitope on factor VIII involves a hydrophobic region. This approach is used in the preparation of Hemofil M prepared by Hyland Inc. and now being used therapeutically in patients.

The two alternative approaches involve the use of MAb to vWf to bind the vWf-VIII complex from plasma and plasma derivatives. In the approach patented by Zimmermann[41] and used in the preparation of the therapeutic product "Monoclate", factor VIII is then eluted by disruption of the vWf-VIII complex using excess calcium, leaving vWf bound to the antibody. Alternatively, both factor VIII and vWf may be eluted by conditions allowing release of vWf from the antibody. This was the approach taken in this laboratory, using a preexisting panel of MAb to vWf and aimed at making a product containing both active factor VIII and active vWf. As described above, we initially screened a large number of

TABLE 4
Some Potential Elution Buffers for Factor VIII and IX Chromatography

Buffer	Percent residual activity		
	VIII	vWf	IX
1 M KI, 1 M lysine	100	102	N.D.
20% EG, 1 M KI	72	48	N.D.
0.25 M CaCl$_2$	105	45	N.D.
1 M LiCl	84	75	N.D.
1 M lysine	81	95	N.D.
3 M urea (5 M urea)	88 (0)	77 (0)	100 (100)
1 M ethanolamine	100	95	N.D.
1 M diaminohexane	76	95	N.D.
50% dimethyl sulfoxide	77	90	N.D.
50% EG, 0.2 M glycine pH 10	—	—	100
0.1 M glycine pH 10	0	0	N.D.
0.1 M glycine pH 2.5	0	0	75
3 M KSCN pH 7.7	9	0	65

Note: For factor VIII.vWf all buffers were made up in 20 mM imidazole, 5 mM CaCl$_2$, 0.1M lysine pH 6.5 except the last three. EG: ethylene glycol. N.D. = not determined.

potential eluting buffers to see which allowed retention of activity. On the basis of previous studies in other laboratories we restricted buffers to the range pH 6.5 to 8, and included 5 mM calcium and 0.1 M lysine in the buffers as known stabilizers of factor VIII and vWf. Subsequently, the assumption of selecting nonalkaline buffers has proved erroneous since the group of Bourgois et al.[18] have described a MAb column allowing binding of VIII/vWf from plasma at neutral pH, with elution of the active complex by a simple change in pH to 8.5.

Table 4 lists some buffers that we found to allow retention of factor VIII and vWf (and factor IX) activity. Obviously, there are a whole range of less severe buffers that do not inactivate coagulation factors, but the list may be of use to those attempting to select appropriate buffers for immunopurification. It is worth mentioning that the microtray version of the chromogenic factor VIII assay, sold by Kabi, was very useful in these studies since it allowed screening of up to 40 samples within an hour and involves a large predilution step which reduces or avoids interference by many buffers.

In our studies we used plasma or cryoprecipitate as starting material and obtained a product containing about 50 U FVIII:C per milligram total protein, of which half was vWf. At this purity we had no problems with absorbtive losses, but for products of higher purity, factor VIII is easily lost, due to adsorption to the walls of the container. Such losses may be avoided through the use of high salt concentrations or addition of amino acids such as lysine or readdition of protein.

B. Depletion

For plasma depletion initial studies were carried out using MAb to factor VIII,[54] immobilized on Sephacryl S1000, and batch adsorption experiments similar to those described above, except that residual factor VIII activity rather than supernatant radioactivity was determined. The most promising MAb were then assessed in small-scale column experiments. While the results of the latter studies showed somewhat improved performance (faster and greater depletion) over the small-scale batch-mixing experiments, subsequent results suggest that adequate prediction of performance requires use of columns containing at least 5, and preferably 10, ml of gel. Smaller-scale studies can result in overestimation of the extent of depletion of plasma due to dilution of samples with equilibration or washing/recycling buffers.

Based on these assessments we found that our best single MAb allowed depletion of more than 90% of factor VIII from plasma, but did not allow the 99% or more depletion required for production of a suitable deficient plasma for use as a factor VIII assay reagent. To achieve this it was necessary to also use columns containing a second MAb to factor VIII and one to vWf.[45] The use of the latter MAb results in a plasma reagent that is both VIII and vWf depleted. However, this appears to perform well in most situations,[45] and corresponding results have been reported for another reagent prepared in a similar manner.[44]

Other factor VIII immunodepleted plasmas are deficient in VIII, but not vWf (Table 3) showing that, given an appropriate antibody, such a product can be produced. After the plasma depletion step it is necessary to elute bound material from the column(s). In our studies, using three serially connected MAb columns, this was achieved with the columns still connected together and using buffer containing high amounts of potassium iodide. While it was not rigidly proved that this resulted in complete elution of VIII/vWf from the affinity gels, they were used through at least ten cycles with no loss of capacity. From this we infer that either the majority of bound material is eluted or that only a fraction of column capacity was being used on each cycle.

Other points that may be helpful for those attempting to prepare deficient plasma by immunoaffinity chromatography are

1.   Prior to recycling with buffers containing KI it is essential to wash columns well, e.g., with 15 mM citrate, 150 mM NaCl pH7, otherwise residual plasma proteins may be precipitated in the gel.
2.   To be usable, the prepared deficient plasma must contain normal levels of other coagulation factors. Fibrinogen and contact factors may be lost by adsorption, particularly on new columns. All glass surfaces should be siliconized. Unfortunately, Sephacryl appears somewhat worse in respect to nonspecific adsorption from human plasma than are Sepharose gels.
3.   Factor V is inherently labile in citrated plasma. To avoid depletion to levels below 50% of those in normal plasma, processing should be carried out within 6 to 8 h. It is also feasible that fibrinogen and factor VIII may be lost by cryoprecipitation if processing were carried out at temperatures below 10°C, rather than 20°C.
4.   In our hands the flow rate of plasma through the antibody columns was critical. Adequate depletion was achievable at 1 column volume per hour, but not at 3, even with higher amounts of antibody than the usual 1 to 3 mg/ml of gel.
5.   Once prepared, the deficient plasma was stable for up to 6 months at −40°C. Beyond 6 months it tended to lose sufficient factor V activity to become unusable. Alternatively, the reagent could be aliquoted and freeze-dried. This allowed storage at 4°C or below, but again the reagent became unusable after storage at 20°C or 37°C due to loss of factor V and fibrinogen. These losses, associated with storage in the frozen or freeze-dried state are also seen for congenitally deficient plasmas.

IV. PURIFICATION AND DEPLETION OF FACTOR IX

Our approach to developing methods for the purification and depletion of factor IX was identical to that used for factor VIII except that it was found possible to use a single antibody for both procedures.[25,47] Our panel of MAb to factor IX is directed to the C-terminal portion of the molecule, which forms the heavy chain after activation. Such antibodies allowed purification of essentially pure factor IX from clinical concentrates using urea or alkaline ethylene glycol as eluants. The N-terminal light chain is subject to posttranslational carboxylation during factor IX synthesis. The resultant gamma-carboxy glutamic acid residues are involved in the binding of calcium required as part of the activity of factor IX. Antibodies to this portion of the molecule might thus be expected to exhibit divalent-metal ion-dependent binding of factor IX and to exhibit selective binding of the active carboxylated forms of factor IX.

Two groups have developed purification methods based on binding of factor IX to such MAb in the presence of millimolar concentrations of metal ions, with elution using EDTA.[26,43] These mild procedures allow purification of fully active factor IX separately from potentially contaminating virus[43] and exhibiting reduced side effects in animal models of thrombosis.[26]

Factor IX-deficient plasma prepared by immunoaffinity chromatography allows assay of factor IX by the conventional clotting time method and yields results equivalent to those obtained with congenitally deficient plasma, provided at least 99% of the normal plasma factor IX is removed.[21,47,48]

V. CONCLUDING REMARKS

Monoclonal antibodies are still expensive compared to other ligands that may be used in affinity chromatography. If they are to be used in such procedures they must be coupled to the chromatography matrix efficiently and then be usable repeatedly. This is achievable. For example, Sandberg et al. have recently reported the use of MAb columns for more than 100 runs without loss of capacity.[55]

Since MAb are produced from transformed animal cells, they are potentially infective, and there are extensive requirements relating to their use as therapeutic agents or for the purification of such agents.[56] Where MAb are used for purification, they may also contaminate the therapeutic product, by leakage from the affinity column, and result in an immune response in the recipient. Our experience is that, regardless of the chemistry used for coupling, leakage of MAb from affinity columns is usually reduced to nanograms per milliliter after a few cycles of use. Such levels are not immunogenic in man after single intravenous doses. If this is also true over a lifetime of repetitive dosing has yet to be determined, but is one aim of the current trials with MAb purified factor VIII.

The use of MAb for therapeutic production of coagulation factor VIII or IX offers the prospect of a high degree of purification in a single step. Such high-purity products may have benefits in terms of a reduced capacity to suppress the immune system (VIII or IX), a reduced potential to cause thrombotic side effects (IX), and a reduced risk of viral contamination (VIII or IX). None of these has been demonstrated in human patients, and the last is unlikely ever to be tested since regulatory authorities currently require viral inactivation steps such as heat or detergent treatment in addition to any immunoaffinity chromatography step.

If the current trials with factor VIII demonstrate a benefit in using highly purified products, it is likely that alternative purifications yielding equivalent products will eventually be developed. For the preparation of depleted normal plasma, MAb are likely to remain the easiest approach to preparing specifically deficient, noninfective laboratory reagents from normal, screened donor plasma.

ACKNOWLEDGMENTS

I would like to thank the many colleagues who have participated in the studies outlined above; in particular, V. Hornsey, H. Bessos, D. Pepper, K. James, L. Micklem, and D. B. L. McClelland.

REFERENCES

1. **Sola, B., Avner, P. R., Zilber, M. T., Connan, F., and Levy, D.,** Isolation and characterisation of monoclonal antibody specific for fibrinogen and fibrin of human origin, *Thromb. Res.,* 29, 643, 1983.
2. **Francis, S. E., Joshua, D. E., Exner, T., and Kronenberg, H.,** Some studies with a monoclonal antibody directed against human fibrinogen, *Am. J. Hematol.,* 18, 111, 1985.
3. **Malhotra, O. P. and Sudilovsky, O.,** Monoclonal antibodies to prothrombin, *Thromb. Res.,* 47, 501, 1987.
4. **Owens, J., Lewis, R. M., Cantor, A., Furie, B. C., and Furie, B.,** Monoclonal antibodies against human abnormal (des-gamma-carboxy) prothrombin specific for the calcium-free conformer of prothrombin, *J. Biol. Chem.,* 259, 13800, 1984.
5. **Jenny, R., Church, W., Odegaard, B., Litwiller, R., and Mann, K. G.,** Purification of six human vitamin K-dependent proteins in a single chromatographic step using immunoaffinity columns, *Prep. Biochem.,* 16, 227, 1986.
6. **Carson, S. D., Ross, S. E., Bach, R., and Guha, A.,** An inhibitory monoclonal antibody against human tissue factor, *Blood,* 70, 490, 1987.
7. **O'Brien, D. P., Giles, A. R., Tate, K. M., and Vehar, G. A.,** Factor VIII bypassing activity of bovine tissue factor using the canine hemophilic model, *J. Clin. Invest.,* 82, 206, 1988.
8. **Katzmann, J. A.,** Isolation of functional human coagulation factor V by using a hybridoma antibody, *Proc. Natl. Acad. Sci., U.S.A.,* 78, 162, 1978.
9. **Annamalai, A. E., Rao, A. K., Chiu, H. C., et al.,** Epitope mapping of functional domains of human factor Va with human and murine monoclonal antibodies: evidence for the interaction of heavy chain with factor Xa and calcium, *Blood,* 70, 139, 1987.
10. **Foster, W. B., Tucker, M., Katzmann, J. A., Miller, R. S., Nesheim, M. E., and Mann, K. G.,** Monoclonal antibodies to human coagulation factor V and factor Va, *Blood,* 61, 1060, 1983.
11. **Broze, G. J., Hickman, S., and Miletich, J. P.,** Monoclonal anti-human factor VII antibodies: detection in plasma of a second protein antigenically and genetically related to factor VII, *J. Clin. Invest.,* 76, 937, 1985.
12. **Cerkus, A. L., Ofosu, F. A., Birchall, K. J., et al.,** The immunodepletion of factor VII from human plasma using a monoclonal antibody, *Br. J. Haematol.,* 61, 467, 1985.
13. **Goodall, A. H. and Meyer, D.,** Registry of monoclonal antibodies to factor VIII and von Willebrand factor, *Thromb. Haemostas.,* 54, 878, 1985.
14. **Fulcher, C. A., Roberts, J. R., and Zimmerman, T. S.,** Thrombin proteolysis of purified factor VIII procoagulant protein, correlation of activation with generation of specific polypeptide, *Blood,* 61, 807, 1983.
15. **Fulcher, C. A., Roberts, J. R., Holland, L. Z., and Zimmerman, T. S.,** Human factor VIII procoagulant protein-monoclonal antibodies define precursor product relationships and functional epitopes, *J. Clin. Invest.,* 76, 117, 1985.
16. **Rotblat, F., O'Brien, D. P., O'Brien, F. J., Goodall, A. H., and Tuddenham, E. G.,** Purification of human factor VIII:C and its characterization by western blotting using monoclonal antibodies, *Biochemistry,* 24, 4294, 1985.
17. **Croissant, M. P., Vandepol, H., Lee, H. H., and Allain, J. P.,** Characterisation of four monoclonal antibodies to factor VIII coagulant protein and their use in immunopurification of factor VIII, *Thromb. Haemostas.,* 56, 271, 1986.
18. **Bourgois, A., Delzay, M., and Fert, V.,** Process for obtaining complex factor VIII/vWf of therapeutical use and resulting products. U.S. Patent 4,670,543, 1987.
19. **Hornsey, V. S., Griffin, B. D., Pepper, D. S., Micklem, L. P., and Prowse, C. V.,** Immunoaffinity purification of factor VIII complex, *Thromb. Haemostas.,* 57, 102, 1987.
20. **Ganz, P. R., Tackaberry, E. S., Palmer, D. S., and Rock, G.,** Human factor VIII from heparinized plasma: purification and characterization of a single-chain form, *Eur. J. Biochem.,* 170, 521, 1988.
21. **Goodall, A. H., Kemble, G., O'Brien, D. P., et al.,** Preparation of factor IX-deficient human plasma by immunoaffinity chromatography using a monoclonal antibody, *Blood,* 59, 664, 1982.
22. **Bajaj, S. P., Rapaport, S. I., and Maki, S. L.,** A monoclonal antibody to factor IX that inhibits the factor VIII:Ca potentiation of factor X activation, *J. Biol. Chem.,* 260, 11574, 1985.
23. **Bessos, H., Micklem, L. R., McCann, M., et al.,** The characterisation of a panel of monoclonal antibodies to human coagulation factor IX. *Thromb. Res.,* 40, 863, 1985.
24. **Smith, K. J.,** Monoclonal antibodies to coagulation factor IX define a high frequency polymorphism by immunoassays, *Am. J. Hum. Genet.,* 37, 668, 1985.
25. **Bessos, H. and Prowse, C. V.,** Immunopurification of human coagulation factor IX using monoclonal antibodies, *Thromb. Haemostas.,* 56, 86, 1986.
26. **Smith, K. J.,** Immunoaffinity purification of factor IX from commercial concentrates and infusion studies in animals, *Blood,* 72, 1269, 1988.

27. **Church, W. R. and Mann, K. G.,** A simple purification of human factor X using a high affinity monoclonal antibody immunoadsorbant, *Thromb. Res.,* 38, 417, 1985.

28. **Sinha, D., Koshy, A., Seamen, F. S., and Walsh, P. N.,** Functional characterisation of human blood coagulation factor XIa using hybridoma antibodies, *J. Biol. Chem.,* 260, 10714, 1985.

29. **Akiyama, H., Sinha, A. D., Seamen, F. S., Kirby, E. P., and Walsh, P. N.,** Mechanism of activation of coagulation factor XI by factor XIIa studied with monoclonal antibodies, *J. Clin. Invest.,* 78, 1631, 1986.

30. **Saito, H., Ishihara, T., Suzuki, H., and Watanabe, T.,** Production and characterization of a murine monoclonal antibody against heavy chain of Hageman factor (factor XII), *Blood,* 65, 1263, 1985.

31. **Small, E. J., Katzman, J. A., Tracy, R. P., Ratnoff, O. D., Goldsmith, G. H., and Everson, B.,** A monoclonal antibody that inhibits activation of human Hageman factor (factor XII), *Blood,* 65, 202, 1985.

32. **Gniewek, R. A. and Kurosky, A.,** Immunological characterization of human plasma factor XIII including monoclonal antibodies, *Fed. Proc.,* 46, 2244, 1987.

33. **Velso, D., Silver, L. D., Hahn, S., and Colman, R. W.,** A monoclonal anti-human plasma prekallikrein antibody that inhibits activation of prekallikrein by factor XIIa on a surface, *Blood,* 70, 1053, 1987.

34. **Berrettini, M., Lammle, B., White, T., et al.,** Detection of in vitro and in vivo cleavage of high molecular weight kininogen in human plasma by immunoblotting with monoclonal antibodies, *Blood,* 68, 455, 1986.

35. **Suzuki, K., Matsuda, Y., Kasumoto, H., et al.,** Monoclonal antibodies to human protein C: effects on the biological activity of activated protein C and the thrombin-catalysed activation of protein C, *J. Biochem.,* 97, 127, 1985.

36. **Laurell, M., Ikeda, K., Lindgren, S., and Stenflo, J.,** Characterization of monoclonal antibodies against human protein C specific for the calcium ion-induced conformation or for the activation peptide region, *FEBS Lett.,* 191, 75, 1985.

37. **Yan, S. B. and Grinnel, B. W.,** Characterisation of fully functional recombinant human protein C (rHPC) expressed from the human kidney 293 cells, *Fed. Proc.,* 46, 2243, 1986.

38. **Mitchell, C. A., Keleman, S. M., and Salem, H. H.,** The anticoagulant properties of a modified form of protein S, *Thromb. Haemostas.,* 60, 298, 1988.

39. **Litwiller, R. D., Jenny, R. J., Katzmann, J. A., Miller, R. S., and Mann, K. G.,** Monoclonal antibodies to human vitamin K-dependent protein S, *Blood,* 67, 1583, 1986.

40. **Prowse, C. V.,** Monoclonal antibodies in blood coagulation, in *Biotechnology in Blood Transfusion,* Proc. 12th Annu. Symp. Groningen Blood Bank, Overby, L., Smit-Sibinga, C. Th., and Das, M. C., Eds., in press.

41. **Zimmerman, T. S. and Fulcher, C. A.,** Ultrapurification of factor VIII using monoclonal antibodies, U.S. Patent, 4,361,509, 1982.

42. **Griffith, M., Liu, S., Neslund, G., Tsang, I., Lettelier, D., and Berkebile, R.,** Preparation of high specific activity plasma AHF by anti-F VIII:C immunoaffinity chromatography (Abstr. 1123), *Thromb. Haemostas.,* 58, 307, 1987.

43. **Limentani, S. A., Furie, B. C., Poiesz, B. J., Montagna, R., Wells, K., and Furie, B.,** Separation of human factor IX from HTLV-1 or HIV by immunoaffinity chromatography using conformation specific antibodies, *Blood,* 70, 1312, 1987.

44. **Takase, T., Rotblat, F., Goodall, A. H., et al.,** Production of factor VIII-deficient plasma by immunodepletion using three monoclonal antibodies, *Br. J. Haematol.,* 66, 497, 1987.

45. **Hornsey, V. H., Waterston, Y. G., and Prowse, C. V.,** Artificial factor VIII deficient plasma: preparation using monoclonal antibodies and its use in one stage coagulation assays, *J. Clin. Pathol.,* 41, 562, 1988.

46. **Dadi, H. K., Felton, J., Pirie, V. R., and Tsang, R.,** Monoclonal antibodies to factor VIII and their use in the preparation of factor VIII depleted plasma in Abstr. Congr. Int. Soc. Blood Transfusion, London, July 1988. Poster Th.4.16.

47. **Bessos, H., Prowse, C. V., and James, K.,** Human coagulation factor IX: direct depletion and recovery from plasma, using immobilised monoclonal antibody, *Med. Lab. Sci.,* 45, 261, 1988.

48. **Dadi, H. K., Felton, J., Rosewell, I., and Tsang, R.,** Monoclonal antibodies to factor IX and their use in the preparation of factor IX depleted plasma in Abstr. Int. Congr. Soc. Blood Transfusion, London, July 1988. Poster Th.4.17.

49. **Zimmerman, T. S. and Fulcher, C. A.,** Anti-porcine von Willebrand factor, *Thromb. Haemostas.,* 54, 889, 1985.

50. **Fulcher, C. A., Houghten, R. A., de Graaf Mahoney, S., Roberts, J. R., and Zimmerman, T. S.,** Synthetic peptide probes of factor VIII immunology and function, *Thromb. Haemostas.,* 58, 536, (Abstr. 1969), 1987.

51. **Hornsey, V., Micklem, L. R., McCann, M. C., James, K., Dawes, J., McClelland, D. B. L., and Prowse, C. V.,** Enhancement of factor VIII von Willebrand factor ristocetin cofactor activity by monoclonal antibodies, *Thromb. Haemostas.,* 54, 510, 1985.

52. **Hornsey, V. S., Prowse, C. V., and Pepper, D. S.,** Reductive amination for solid phase coupling of protein: a practical alternative to cyanogen bromide, *J. Immunol. Methods,* 93, 83, 1986.

53. **Kohn, J. and Wilchek, M.,** A new approach (cyano-transfer) for cyanogen bromide activation of Sepharose at neutral pH, which yields activated resins free of interfering nitrogen derivatives, *Biochem. Biophys. Res. Commun.,* 107, 878, 1982.
54. **Griffin, B. D., Micklem, L. R., McCann, M. C., James, K., and Pepper, D. S.,** The production and characterisation of a panel of 10 murine monoclonal antibodies to human procoagulant factor VIII, *Thromb. Haemostas.,* 55, 50, 1986.
55. **Sandberg, A., Bergsdorf, N., Brandstrom, A., and Sundstrom, S.,** 100 cycles of immunoaffinity purification of t-PA, *Fibrinolysis,* 2 (Suppl. 1), A335, 1988.
56. **Cartwright, T.,** Isolation and purification of products from animal cells, *Trends Biotechnol.,* 5, 25, 1987.

Chapter 9

IMPROVED CYANOGEN BROMIDE ACTIVATION

Duncan S. Pepper

TABLE OF CONTENTS

I. INTRODUCTION

The cyanogen bromide activation method for the synthesis of solid phase immunosorbents has many advantages but has not received the attention it deserves because people do not like preparing their own activated gels and the product has a reputation for being inefficient, unstable, and leaky. In fact, by using the modifications proposed by Kohn and Wilchek[1,2] the synthesis is fast, painless, efficient, easily quality controlled, and stable for greater than 1 year. Large (or small) quantities can be prepared at low cost and stored ready in the wet state for immediate use, and no skill in organic chemical synthesis or expensive apparatus is required. Perhaps most surprisingly, when experience is gained and confidence found, a fume cupboard is not even necessary. The whole process including quality control can be performed in 1 h.

II. EQUIPMENT REQUIRED

1. $-20°C$ Freezer *or* 6-1 stainless steel Dewar flask (e.g., Statebourne OD6) *or* polystyrene foam box
2. ''Jumbo'' size magnetic stirrer (e.g., Stuart SM8) and polytetrafluoroethylene (PTFE) magnets $\simeq 10 \times 60$-mm bar and 17 mm-diameter disk
3. Digital thermocouple meter and probe, range $-50°$ to $+50°C$, (e.g., Kane May 3000)
4. Pyrex screw-cap bottle 500 ml (e.g., Duran 21-801-445)
5. Sintered glass or plastic (#1 or #2 porosity frit available from most major laboratory suppliers) gas bubbler tube 15-mm diameter bulb, with 30 cm $\times$ 8 mm-OD tube length (Figure 1)
6. pH meter
7. 15-ml Calibrated conical glass centrifuge tubes
8. 10-ml PTFE lined screw-cap sample tubes
9. Venturi water pump vacuum line and trap

III. REAGENTS FOR ACTIVATION

1. Cyanogen bromide 25 g or 100-g bottles (e.g., Aldrich C9,149-2)
2. Store cyanogen bromide at $< -20°C$:
    - Triethylamine analytical grade (e.g. BDH 10409-5W)
    - Acetone analytical grade 2.5 l
    - Acetic acid glacial, analytical grade 1 l
    - Solid carbon dioxide $\simeq 250$ g
    - Suitable gel, e.g., Sepharose, Sephacryl (Pharmacia), or Fractogel, etc. (Merck, BDH, Pierce) 100-ml settled volume

IV. REAGENTS FOR COUPLING

- 1 mM HCl in H_2O
- 1.0 M NaHCO$_3$ pH 9.0 (or 1.0 M NaH$_2$PO$_4$ pH 6.5; see discussion)
- 1.0 M ethanolamine pH 9.0
- 0.1 M citric acid pH 2.0
- 0.1 M Na$_2$HPO$_4$ pH 8.0

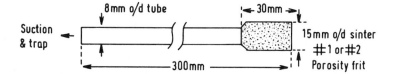

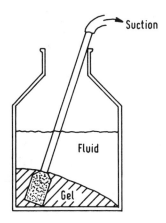

FIGURE 1.   A typical gas bubbler or stick filter tube. Dimensions are not critical.

V. KONIG REAGENT FOR ACTIVATION ASSAY AND QUALITY CONTROL

1.   Concentrated HCl 2.5 ml + pyridine 12 ml + barbituric acid 0.5 g + H$_2$O 10 ml mixed together for 10 min at 20°C or until all solids dissolve
2.   Stable at 4°C for >3 months
3.   Dimethyl formamide GPR (250 ml)

VI. METHOD

The basis of this method is the use of molar equivalents of cyanogen bromide and triethylamine in acetone/water mixtures at −25°C and pH 7 to 9. At this temperature a stable complex of triethylamine and cyanogen bromide is formed which is very reactive at neutral pH with the hydroxyl groups of agarose. The production of active cyanate esters thus proceeds cleanly and smoothly with near quantitative yield and no significant side reactions. To facilitate the handling of the small amounts of cyanogen bromide needed and to increase accuracy in dispensing, the cyanogen bromide is dissolved *in situ* in the manufacturers bottle as supplied. Use a volume of acetone in milliliters equivalent to one half the numerical weight of cyanogen bromide in the bottle. Thus, to a 25-g bottle add 12.5 ml of acetone or to 100-g bottle add 50 ml of acetone. At the same time as adding the acetone, also add a 17-mm diameter circular PTFE magnetic stirring disk to enhance the dissolution time. With care, the lid need only be removed from the bottle for a few seconds. If the CNBr has not dissolved after 5 to 10 min stirring at +20°, dissolution can be completed by brief immersion in a water bath at +37°C. The stock solution now contains 1.00 g of CNBr in each 1.00 ml of acetone and can be dispensed in convenient aliquots (e.g., 5.5

ml) in screw-cap glass vials (10-ml size with PTFE lined caps) and stored at $< -20°C$ prior to use for >1 year. For dispensing I use disposable plastic pasteur pipettes and calibrated glass conical centrifuge tubes (10 or 15 ml). The use of mechanical air-displacement pipettes is not recommended as they can be contaminated by the volatile CNBr. When using cold-stored CNBr acetone solutions, the crystals of CNBr can be rapidly redissolved by immersion in a 37°C bath for a few seconds.

VII. PROCEDURE

A. ACTIVATION

Add 100 ml gel + 100 ml water (i.e., 50:50 slurry) to an empty 500-ml Pyrex bottle and while stirring add 300 ml acetone slowly. Suck off acetone via water pump when the gas bubbler tube is immersed in the gel slurry and when gel is "dry" disconnect suction, add 400 ml acetone while mixing and repeat suction until about 50 to 100 ml of acetone is left on the gel volume of 100 ml. The use of a simple water venturi pump is much the preferred method of sucking fluids via the sintered filter, as it avoids the need to trap or dispose of noxious fluids. If such a pump is not available, then either a peristaltic pump or oil-free vacuum pump may be used. In either case, it is essential to ensure that an efficient trap/fume venting is available to handle acetone, acetic acid, and cyanogen bromide. Cool the gel/acetone slurry to below $-20°C$ by immersion of the bottle in solid CO_2 in the vacuum flask while stirring. Measure the temperature with the thermocouple and remove from the dry ice when the temperature has fallen to $-35°$ to $-30°C$. (The gel does not have to be anhydrous. The purpose of the acetone is to lower the freezing point below $-30°C$ and enhance the solubility of CNBr.) While stirring add exactly 5.0 g of CNBr as 5.0 ml of acetone solution from a pasteur pipette and be careful that this drops onto the surface of the stirred gel slurry and does not run down the wall of the bottle. The temperature should not rise above $-25°C$. Remove a 1.0-ml sample of slurry and add to 2.0 ml H_2O in a 10-ml test tube for subsequent pH measurement (sample No. 1). While stirring, add exactly 10.0 ml of triethylamine dropwise over about 30 s, again being careful that none is lost on the walls of the vessel. After a further 90 to 120 s of stirring the temperature should have risen from $-25°$ to between $-10°$ and $-5°C$, showing that a chemical reaction has taken place. Often, this is also accompanied by the appearance of a pale yellow color. Remove a second 1-ml sample and dilute in 2.0 ml of water for subsequent pH measurement. When the reaction has run for a total of 2 to 3 min (but *not* longer), add 300 ml of glacial acetic with constant stirring to stop the reaction and stabilize the activated gel. Remove a third 1.0-ml sample and dilute in 2.0 ml of H_2O for subsequent pH measurement. Sample No. 1 should have a pH of 3 to 5. Sample No. 2 should have a pH of 7.0 to 9.0 (Note: if the pH is not above 7.00, the reaction is unlikely to have proceeded efficiently, and in the future increase the amount of triethylamine by 2 to 3%; likewise if the pH is above 9.0, reduce the amount of triethylamine). Sample No. 3 should have a pH of 2 to 3; if the pH is not below 3.00, add more acetic acid to achieve this. Once glacial acetic acid has been added and the pH has fallen below 3.0, the activated gel is very stable, and all subsequent steps can be performed at room temperature without further cooling.

B. STORAGE OR IMMEDIATE USE

At this point you can choose to use the gel immediately or store it.

For immediate use the gel is transferred from acetone/acetic acid to 1 mM HCl in water (pH 3.0) by successively sucking out the solvent via the vacuum line and gas bubbler tube and then adding more of the desired solvent. With the suction line disconnected the gas bubbler tube makes a handy stirrer rod to break up the cake of gel before turning on the magnetic stirrer. After three successive additions of water the smell of acetone and cyanogen bromide should be absent, and the moist gel cake can be used directly for coupling. Because the lachrymatory activity of cyanogen bromide is so intense even at low vapor concentrations, it virtually guarantees that exposure to significant levels of volatile cyanide will not occur. The operator would have to breathe painful levels of vapor for several hours continuously before clinical symptoms of chronic cyanide poisoning (muscle weakness) appear. Second, the efficiency of this method ensures that lower concentrations of CNBr are needed, and for much shorter periods of time and lower temperatures, these factors, together with the use of a water venturi pump, ensure that no significant cyanide exposure occurs, even when acetic acid is used.

If the gel is to be stored, the washing solvent is 90% acetone + 10% glacial acetic acid, and after three successive additions and filtrations the moist gel should not smell of cyanogen bromide. The gel should be resuspended in an equal volume (100 ml) of acetone/acetic acid (90:10) and stored tightly capped in a spark-free freezer at $< -20°C$. The activated gel is stable for at least 1 year under these conditions.

C. ASSAY OF ACTIVATION LEVEL

At this point it is possible to use sample No. 3 (acid stabilized) for investigation of the degree of cyanate ester activation. This is optional but is recommended as it is simple, quick, and can be operated either as a qualitative or quantitative assay. Proceed as follows:

1. To wash the gel free of excess reagents and yield a suspension of known percent solids volume per volume solids, transfer the 3.0-ml slurry to a 15-ml calibrated conical glass centrifuge tube and fill to the 15-ml mark with 1 mM HCl in H_2O, mix by immersion, and centrifuge at $1000 \times g$ for 1 min. Aspirate fluid supernatant by means of a pasteur pipette attached to a vacuum line and repeat a further two times until all smell of acetone and CNBr is removed. Note the packed volume of gel (it helps if the lower part of the conical tube is calibrated in 0.10-ml divisions) and add an equal volume of water such that the resulting slurry is exactly 50% v/v gel solids.
2. Into two 10-ml screw glass vials dispense either 0.200 ml of H_2O (control) or 0.200 ml of *well-mixed* gel slurry (ideally 50:50 v/v solids) using a disposable pipette tip which has been cut back so that the lumen is > 1.0-mm diameter to prevent bead clogging.
3. To each vial add 0.8 ml of the Konig assay reagent and mix and leave at room temperature for 15 min; then add 5.0 ml of dimethyl formamide and leave at room temperature for a further 15 min (this serves to separate the purple chromogen from the beads).
4. Then carefully pipette 1.00 ml of the dimethyl formamide reaction mixture (without disturbing the beads) into 15.6 ml of H_2O giving an overall dilution from the original bead volume of 1:1000.
5. Read the absorbance at 575 nm in a 1-cm cell and multiply the value by 66 to give μmoles of −CNO ester per milliliter of gel.

This assay is linear over a range of 0 to 1.5 A units. For a qualitative assay, the supernatant can be assessed visually after 15 min with the Konig reagent (or if a strong color is yielded, after further dilution in dimethyl formamide).

VIII. COUPLING

1. Dispense the desired volume of gel by using twice the volume of slurry (50% v/v as stored) into a screw-cap Pyrex glass vessel (e.g., 150-ml or 500-ml tissue culture reagent bottle) and dilute five to tenfold with 1 mM HCl in water (pH $\simeq 3.0$). Suck

out all fluid using the *in situ* filter method (gas bubbler sinter on suction line) and repeat the process a further two times or until all smell of acetone and CNBr is gone.

2.   Prepare protein solution for coupling (preferably at >1 mg/ml) in 1 *M* Na$_2$CO$_3$ pH 8.5 ± 0.5, and add this directly to the moist gel pellet. The minimum volume of protein solution required should provide an easily mixed gel slurry (say, 0.5 to 1.0 × gel volume) whereas the maximum volume can be up to ten times the gel volume or whatever fills the reaction vessel yet still can be mixed.

3.   Check the pH; between 7 and 9 is acceptable (see discussion) and roller mix for 2 to 20 h at 20°C

4.   Block unreacted –CNO ester groups with 1 *M* ethanolamine pH 9 for 2 h at 20°C and then wash by *in situ* filtration alternately with water, 0.1 *M* citric acid pH 2.5, water, and 0.1 *M* Na$_2$HPO$_4$ pH 8.0

The amount of protein actually coupled is best measured by incorporation of, e.g., 10^2 Becquerel of ^{125}I-tracer protein into the coupling solution, but if this is not possible, the Pierce BCA protein assay (No. 23225) will yield a soluble chromogen with solid phased protein which is quantitative if the same (soluble) protein is used to prepare the standards.

IX. DISCUSSION

The low temperature triethylamine –CNBr method of Kohn and Wilchek is very fast to perform and very efficient. The chemical reaction is almost quantitative, and 5 g of CNBr will activate 100 ml of Sepharose 4B to levels of about 80 μmol/ml of –CNO. This allows for either a coupling capacity of ten times that previously obtainable with commercial CNBr-activated gels, or by lowering the amount of CNBr used to 0.5 to 1.0 g/100 ml of gel, "normal" levels of activation around 10 μmol/ml can be achieved.

Interestingly, other more "difficult" hydroxylic gels can also now be CNBr activated; the author has had great success with all porosity grades of Sephacryls, Fractogels, Trisacryls, and even "impossible" gels based on dextran or cellulose will yield some useful levels of –CNO active ester.

The method could, in principle, also be used with any hydroxylic polymer in e.g., sheet, fiber, or composite form. The highly macroporous gel Sephacryl S-1000 was found to activate much more efficiently (3 μmol –CNO/ml → 15 μmol/ml) if it was autoclaved at 120°C for 15 min prior to activation. The reason for this is not known, but autoclaving may enhance the availability of free reactive hydroxyls at the gel polymer chain surface. Remember, when changing the amount of CNBr, to use an equivalent amount of triethylamine: each 1.0 g (or 1.0 ml of acetone stock CNBr solution) requires 2.0 ml triethylamine.

The stability of activated gels is considerably greater than is usually realized. At − 20°C in 90% acetone + 10% acetic acid, activated Sepharose can be kept for at least 12 to 18 months, activated Fractogel can be kept for at least 6 to 12 months, and activated Sephacryl for at least 3 to 6 months. When transferred to water at pH 3.0 (1 m*M* HCl), the gel is still remarkably stable (half-life of 15 to 20 h at + 20°C). When the pH is raised to 8.0 for coupling in aqueous solutions, the half-life is reduced to about 2 h at + 20°C.

Using gels with a high degree of activation (50 to 100 μmol/ml) allows the preparation of one large batch and its storage for immediate use over the coming year, a great convenience. It can easily be checked before use by a visible reaction with the Konig reagent. Another advantage of high levels of activation is that coupling can be carried out at much lower pH (6 to 7), where coupling is via alpha-amino groups and inactivation of biological function via multipoint attachment is less likely to occur, yet the percent efficiency of coupling yield is still high (80 to 90% with 1 mg/ml IgG). Note, however, that leakage is lower with multipoint (high pH) coupling.

Capacity can *roughly* be estimated as 1 mg of IgG per micromole of −CNO, i.e., 10 μmol −CNO per milliliter of Sepharose 4B will saturate at about 10 mg of IgG per milliliter of gel; however, this is unlikely to be achieved unless at least twofold excess of protein is used in coupling.

Activation levels achieved depend on the type of gel and the porosity grade used (cross-linking does not seem to make a large difference). In general, Sepharose gels activate better than Fractogel's which in turn are more reactive than those of Sephacryl. Within a gel type the lower molecular weight exclusion limits tend to activate to higher levels than the high molecular weight exclusion types, probably due to the reduced available "surface area" in the large-pore gel types. The practical range is thus from 100 μmol −CNO per milliliter (Sepharose 6B) to 10 μmol/ml (Sephacryl S-1,000).

Buffer choice will be dictated by the desire for α-NH$_2$ (single point) or ϵ-NH$_2$ (multipoint) attachment, i.e., pH 6.0 to 7.0 or 8.0 to 9.0, as well as the expected stability of the protein to be coupled. It is possible to use higher pH values for "unreactive" proteins or polymers and the use of, e.g., 0.1 to 0.5 M Na citrate in the buffer as a stabilizer can be useful. The high ionic strength of 1 M Na$_2$CO$_3$ (pH 8 to 9) or 1 M Na$_2$HPO$_4$ (pH 6 to 7) also helps to improve the coupling efficiency, probably by shielding charged groups and eliminating mutual repulsion at the gel surface.

This author has operated on a scale from 1 ml gel to 3000 ml. Generally the larger the volume, the easier it is to maintain the low temperature due to thermal inertia. For gel volumes above 300 ml it is quicker to replace the *in situ* filter with a one-piece bench top polypropylene Buchner (e.g., Bel-Art No. H14620).

REFERENCES

1. **Kohn, J., Wilchek, M.,** A colorimetric method for monitoring activation of Sepharose by cyanogen bromide, *Biochem. Biophys. Res. Commun.,* 84, 7, 1978.
2. **Kohn, J., Wilchek, M.,** A new approach (cyano-transfer) for cyanogen bromide activation of Sepharose at neutral pH, which yields activated resins, free of interfering nitrogen derivatives, *Biochem. Biophys. Res. Commun.,* 107, 878, 1982.

Chapter 10

A USER GUIDE TO PROTEIN A

Duncan S. Pepper

TABLE OF CONTENTS

I. BACKGROUND

Solid-phase immobilized protein A has considerable utility for the purification and characterization of monoclonal and polyclonal antibodies from a wide range of species. Both affinity and capacity can be considerably enhanced by applying the sample in a high pH/high ionic strength buffer. Elution and identification of different subclasses by a continuous, linear, decreasing pH gradient is easily achieved by simple mixing devices based on two-chamber reservoirs of buffer. Useful information on subclass identity, purity, and degradation can also be obtained simultaneously during preparative scale purification.

II. INTRODUCTION

The bacterial protein A secreted by certain strains of *Staphylococcus aureus* has been known for many years to bind to the Fc region of IgG molecules, and this has been made the basis of diverse "nonimmune" antibody and antigen assays or detection systems.[1,2] More recently, the large-scale production of protein A by continuous fermentation and in high purity at modest cost has made solid phase immobilized protein A an attractive alternative to precipitation and ion-exchange purification methods. Monoclonal antibody technology has provided an additional impetus in the need for high-purity labeled and conjugated antibodies for therapeutic and diagnostic use. Fortuitously, the species of greatest interest (mouse, rat, human) all show significant amounts of binding for the majority of subclasses in use in monoclonal form (Table 1).

It is the purpose of this chapter to show that, contrary to common belief, the low-affinity binding of many subclasses can readily be improved, that subclasses can readily be separated in a simply generated linear pH gradient, and that important information relating to subclass identity, purity, degradation, and quality control can easily be generated without extra effort during the purification process.

III. MATERIALS

Pure protein A is available in bulk from Fermentech (see Appendix) or can be obtained ready, immobilized on a suitable porous gel at 2 to 5 mg/ml from many suppliers (Bioprocessing, Bio-Rad, Fermentech, Pharmacia, Pierce). If large volumes (>100 ml) of tissue culture fluids are being used as the raw material for monoclonal antibody production, then preliminary concentration by ultrafiltration 10- to 20-fold is desirable. The author uses Amicon flat membranes or hollow fibers of PM10 material, though others would, undoubtedly, be satisfactory with molecular weight cutoff in the range 10,000 to 100,000.

Columns for packing can be of any suitable size or shape to suit the scale of operation, e.g., 1- to 2-ml disposable columns (Amicon, Bio-Rad, Pierce) are suitable for <10 ml of samples, whereas 100-ml columns (40-mm diameter × 80-mm height) are suitable for <1-l samples. More recently, protein A immobilized in shallow-bed affinity membrane capsules has become available in prepacked ready-to-use form (Domnick-Hunter) in 1-ml (Memsep 1000) and 100-ml (Memsep 2000) sizes. Gradient generators based on two-chamber equal-volume mixing devices are very convenient. For smaller columns (1 to 90 ml) the Pharmacia GM-1 gradient mixer is adequate (2 × 275 ml), whereas for larger columns (50 to 300 ml) the BRL 580-1220 GC gradient mixer is better (2 × 1000 ml). Suitable magnetic stirrer base and magnet are also required with the latter gradient mixer device. The Pharmacia GM-1 mixer contains an integral stirrer motor and paddle. A suitable pump and fraction-collection combination is also required for the chosen size of column, e.g., to operate at 1 to 10 ml/min and collect 1 to 10-ml fractions over 50 to 100 fractions. Although not essential, a very convenient on-line monitor for pH/A_{280} absorbance plus chart recorder (two-channel)

TABLE 1

The Relative Affinity of Monoclonal and Polyclonal IgG For Protein A Data Compiled From a Wide Variety of Published and Unpublished Sources.

Species	Subclass	Affinity	pH of elution	Notes	Ref.
Mouse	IgG_1	Low	6 — 7	a	4,5
	IgG_{2a}	Medium	4.5 — 5.0	b	
	IgG_{2b}	High	3.5 — 4.0		
	IgG_3	High	4.0 — 4.5		
Rat	IgG_1	Low	6.0	a	9
	IgG_{2a}	None	9.0	f	
	IgG_{2b}	Low	8.0	a	
	IgG_{2c}	High	3 — 4		
Human	IgG_1	High	3.9 — 4.6	d	
	IgG_2	High	4.3 — 5.0		
	IgG_3	None	9.0	f	
	IgG_4	High	3 — 5	c	
Goat	IgG_1	Low/none	9.0	f	10
	IgG_2	Medium	5.9		
Hamster	IgG_1	Medium	5.3		
	IgG_2	Medium	5.9		
Horse	$IgG_{a,b}$	Medium	4 — 5		
	IgG_c	Low	6 — 7	a	
	IgG_T	None	9.0	f	
Bovine	IgG_1	Low	6 — 8	a	
	IgG_2	High	3 — 4		
Sheep	IgG_1	Low	6 — 7	a	
	IgG_2	High	3 — 4		
Rhesus monkey	$IgG_{1,2,3}$	High	4.3 — 5.1	e	
Guinea pig	IgG_1	High	4.7	d,e	
	IgG_2	High	3.2 — 4.5		
Canine	$IgG_{a,b,c,d}$	High	3 — 4	d,e	
Porcine	IgG	High	3 — 4	e	
Feline	IgG	High	3 — 4	e	
Rabbit	IgG	High	3 — 4	e	
Chicken	IgG_Y	None	9.0	e	
Turkey	IgG_Y	None	9.0	e	
Pigeon	IgG_Y	None	9.0	e	

Notes: (a) Considerable improvements in binding performance can be achieved by applying the sample in 3 *M* NaCl buffered to pH 9.0. (b) Mouse allotypes IgG_{2a} (aj) or (b) elute at pH 5.0 or 4.5, respectively. (c) This subclass is polydisperse and cannot be eluted in a narrow pH range. (d) Soluble protein A forms an insoluble precipitate with IgG from human, guinea pig, and dog. (e) All subclasses elute in a similar pH range and cannot be resolved. (f) No detectable affinity for this subclass, but it may be obtained free of contamination by the other bound subclasses.

is available from Pharmacia. Otherwise off-line measurement of pH and UV absorbance can be carried out manually after completion of the run using conventional pH meter and spectrophotometer. It is convenient in this case to use a combination pH electrode with a stem diameter of <8 mm so as to enable direct insertion into fraction collector tubes. A totally automated system (MABLAB) using expert system control to good manufacturing practice (GMP) requirements is also available commercially from Oros for pharmaceutical applications of protein A.

Buffer composition — The same stock solution can be used to prepare both ends of the gradient, e.g., make up 2 l; split this into two separate 1-l amounts, add solid NaCl to

one (start buffer) and adjust pH to 9.0 while the other half is adjusted to pH 2.5.
Stock buffer (100 mM phosphate + 100 mM citrate)

e.g., $NaH_2PO_4 \cdot 2H_2O$ 31.2 g
or $Na_2HPO_4 \cdot 10H_2O$ 71.6 g
or $Na_2HPO_4 \cdot 2H_2O$ 35.6 g   plus/ Na_3 citrate$\cdot 2H_2O$ 58.8 g
or KH_2PO_4 27.2 g   or citric acid$\cdot H_2O$ 42.0 g
or K_2HPO_4 34.8 g

Make up to 2 l with H_2O but *do not* adjust pH at this stage.

Split stock buffer into two equal aliquots of 1 l. To one add 176 g of solid NaCl to provide a final concentration of 3.0 M NaCl, and adjust the pH to 9.0 with 40% w/w NaOH solution. This is the starting buffer.

The other half of the stock buffer is adjusted to pH 2.5 by the addition of concentrated HCl. For prolonged storage, 3 mM NaN_3 (0.2 g/l) can be added to the starting buffer as a bacteriostat; however, its addition to the acid buffer is neither desirable nor necessary.

IV. METHOD

A. SCREENING OF COLUMN FRACTIONS

Screening of antibodies can be by any of several basic methods, e.g., specific antibody assays will require antigen-specific reagents for local radioimmunoassay or enzyme-linked immunosorbent assay (ELISA). More general assays are particularly useful; the author finds radial immunodiffusion (Mancini) as a useful semiquantitative screening assay. Using a commercial polyclonal sheep anti-mouse (or other species required) polyclonal antibody in a 10 × 10 × 0.3 cm petri dish, 81 samples can be screened simultaneously when 3-mm holes are punched in the gel on a 1-cm^2 matrix pattern in the agarose gel. Although slow ($\approx$24 h) this is a reliable technique requiring little sample or operator attention. Another useful general technique where the IgG subclass can be anticipated is to spike the sample with an internal standard of 10^2 Becquerel ^{125}I-IgG of the anticipated subclass(es).

A simple, rapid and sensitive "dipstick" ELISA is available from Amersham (RPN.29) for qualitative identification of murine antibody class, subclass, and light chain type. For any species of IgG, a useful qualitative identification can be obtained via sodium dodecyl sulfate polyacrylamide gel electrophoresis (SDS-PAGE) with and without reduction. The characteristic single band at 150,000 (nonreduced) and double bands at 50,000 and 25,000 (reduced) are diagnostic for IgG and useful also for showing up impurities (such as albumin) and proteolysis of the heavy chain (reduced molecular weight less than 50,000). An even more rapid qualitative fingerprint of IgG microheterogeneity can be obtained by thin-layer gel isoelectric focusing (IEF) which typically gives 4 to 5 closely spaced bands for each IgG and can be used to confirm the purity and identity of a particular IgG when its pattern is known. Any changes in appearance of the pattern may be indicative of proteolysis, class switching, or contamination.

B. SAMPLE PREPARATION

In general ascites and serum and plasma will require dilution whereas tissue culture fluids will require concentration. The purpose of diluting ascites, serum, and plasma is to provide a quick method of adjusting (small) volumes of sample to a suitable pH, ionic strength, and viscosity. A two- to threefold dilution in the starting buffer (100 mM phosphate + 100 mM citrate 3 M NaCl pH 9.0) should be suitable. Check that the sample pH is still greater than 8.0 (especially if weakly bound IgG subclasses can be anticipated). Usually, the sample will contain undesirable contaminants such as fibrin clots, lipids, and pristane oil. These are best removed in a preliminary step (after dilution) of centrifugation at 20,000

× g for 20 min at 4°C. The clarified fluid can be aspirated carefully via a pasteur pipette (and pump for larger volumes) so as to avoid contamination with the floating lipids and pristane oil and the sedimenting fibrin clots. Final clarification should be via a 0.45-μm disposable 25 or 47 mm microporous membrane (e.g., hydrophilic, low protein binding) using a suitable size of luer syringe or pump. The sample can be collected directly in a calibrated tube or other suitable container (for 1- to 15-ml volumes a conical bottom centrifuge tube in plastic or glass calibrated to 1.0 ml and 0.1 ml is very useful) which allows later estimation of flow rate and required time for sample application.

With tissue culture fluid (or other low concentration sources of antibody) it is useful to concentrate the feedstock 10- or 20-fold. This has two primary benefits: (1) it reduces the time required for sample application and (2) it enhances the efficiency of binding and yield of the weaker binding subclasses. After concentration, solid NaCl should be added to the sample to achieve a final concentration of 3.0 M NaCl, and 0.1 N NaOH should be added slowly dropwise with stirring until the pH is 8.5 ± 0.5. The addition of buffer ions is not strictly necessary as the protein content is sufficient for self-buffering. The presence of the dye neutral red is not usually a problem as it passes straight through the column in the void volume. Prior to application the concentrated and adjusted sample should be clarified; as tissue culture fluid is usually "cleaner" than ascites, simple microfiltration through a 0.45-μm filter is usually sufficient.

In either case (ascites or tissue culture) it can be useful to add internal standards of ^{125}I-IgG of the expected subclass(es) immediately prior to applying the sample. This can act as a useful quick check of (1) the identity of any particular peak and (2) the presence or absence of unbound IgG subclasses in the void volume indicating possible overloading.

C. COLUMN PRECONDITIONING

If the protein-A solid phase column is new or has not been used for some time, it is recommended to run a dummy elution cycle prior to sample application. The use of a blank gradient (no sample) from pH 9 to 2.5 ensures that any leaking material is washed out and any noncovalently bound contaminating proteins are stripped off. The maximum binding capacity of any previously complexed protein A is also thus regenerated by removal of specifically bound residual IgG. If minimal levels of protein A are required in the product, e.g., for therapeutic purposes, then step elution precycling up to five times before use should reduce leakage levels to below 10 ng/ml in the majority of column eluate fractions. If fouling by repeated previous use is a problem, then additional regeneration schemes can be used, e.g., 1 M KI; 3 M KSCN; 7 M urea; or 0.1% SDS in pH 9.0 buffer. Lipid contamination may also be removed by 1% neutral detergents such as Tween 20, Triton X-100, etc. Generally, such contamination can best be prevented or minimized by suitable sample pretreatment; however, if heavy contamination is a recurrent problem (e.g., with serum or plasma), then the use of an in-line dummy column (upstream from the protein A column) and packed with the same support gel bead and activation chemistry and blocked with a bland reagent or protein is useful. Alternatively, since fouling is normally seen in the upper few millimeters of a packed gel bed, the protein A gel beads can simply be overlayered in the same column with an extra 1 to 2 cm of bland gel which can be aspirated, discarded, and replaced as it fouls.

D. SAMPLE APPLICATION

With any on-line monitors of UV absorption and pH previously calibrated the sample should be pumped onto the column (preconditioned to pH 9.0 in 3 M NaCl) at the same flow rate as intended throughout the run. With sample volumes greater than 1 to 2 ml it is convenient to read the flow rate directly from the calibrated sample tube by estimating the volume decrease during 60 s of pumping. From this flow rate the time taken to apply the rest of the sample can be estimated; hence, it is not necessary to watch the sample throughout the application process. When using the Pharmacia GM-1 or BRL 580-1220 GC gradient generators the sample is most conveniently applied by first closing the gradient outlet valve, disconnecting the tubing, and taping or clamping the tube firmly into the bottom of the sample container. Any ingress of air into the sample tubing and column should be minimized. When all the sample is aspirated, rinse the vessel walls briefly with a few milliliters of starting buffer and replace the gradient generator connection tube so as to allow continued washing with starting buffer, but DO NOT open the connecting valve between the two chambers. Continue washing the sample through the column until the absorbance is reduced to an insignificant value. With large volume samples it is convenient to collect the break-through and wash fractions separately in a large vessel rather than use the fraction collector.

E. GRADIENT GENERATION

Temporarily switch off the pump and refill the starting (alkaline buffer) outlet reservoir (i.e., the chamber with the mixing device *in situ*). The level should either be marked from previous trial or estimated. Because the 3.0 *M* NaCl makes the starting buffer more dense than the final acid buffer, it has to be lower to balance without cross-flow when opening the interconnecting valve. With the Pharmacia GM-1 gradient mixer filled to maximum volume (i.e., 2 × 275 ml), the starting buffer (pH 9.0) will need to be about 10 mm lower than the finishing buffer level (pH 2.5). It is better to underestimate this difference than overestimate it: when the valve is opened, any excess heavy pH 9.0 buffer passes through the connecting valve and underlies the lighter acid buffer. It then passes back into the mixer chamber when the pump and gradient are started.

When the gradient and pump are started, the event should be marked on the chart record and an estimate made of total gradient volume, fraction volume, number of fractions, and time taken for 95% of the gradient to flow through the column. Ideally, the fraction collector should finish and switch off the pump, at this point, to avoid the column running dry.

F. FRACTION PROCESSING

If on-line data have been generated, the pH and UV absorbance data together should yield unequivocal evidence of the presence, position, and subclass (see later) of any IgG. If there is any doubt about this, then direct γ counting of the fraction collector tubes in a multiwell, γ-isotope counter will quickly confirm where each internal standard subclass has eluted. Also, the presence of >5% of applied tracer in the void volume (usually dyed purple by the neutral red dye) indicates overloading, and the breakthrough material should be retained for reapplication. If on-line data are not available the pH, ^{125}I tracer, and A$_{280}$ (absorbance) should be determined as quickly as possible by manual means. It has been suggested that, e.g., 1 *M* Tris pH 9.0 ≃ 0.2 ml should be included in all fraction collector tubes to neutralize acid gradients. However, in the author's experience this is rarely necessary because most antibodies elute in the pH range 5 to 7, and few elute in the pH 3 to 4 region. Provided this is processed within a few hours, no significant proteolysis or precipitation is seen. Also, the addition of Tris negates the potential for manual pH measurement of the gradient off-line and may interfere with subsequent immobilization efficiency. When the desired antibody peaks have been located and confirmed, the fractions can be pooled (but leave 10 to 100 μl of each fraction if bioassays are to be run) and neutralized to pH 5 to 6 with 1.0 *M* Na$_2$CO$_3$ added dropwise with continuous stirring. Pooled fractions should be measured for volume and A$_{280}$ so as to calculate the total mass of IgG present (assume 1 mg/ml has an absorbance of 1.4). For further storage it is recommended to dialyze overnight at 4°C vs. 50 m*M* phosphate and 50 m*M* citrate buffer at pH 6.0 to remove excess NaCl. Adjust the protein content to the appropriate desired value, e.g., 1 to 2 mg/ml by dilution (or concentration if small amounts were yielded), add solid maltose to 10% w/v, and dissolve by warming to 37°C. Centrifuge or filter (0.45 μm) to clarify and check the final UV absorbance again for protein content. The fluid can then be dispensed in bulk, or aliquoted and stored liquid at +4°C or frozen at −20° to −40°C or freeze-dried. The author prefers the latter option. If long-term liquid storage is anticipated, addition of 3 mM (0.02% w/v) NaN$_3$ is desirable or sterile filtration and dispensing. The addition of azide ions will increase the UV absorbance slightly and should be done after protein measurement and calculation. The use of maltose is very convenient as it allows both long-term liquid storage and repeated freezing and thawing of IgG solutions without the problem of protein aggregation which can seriously overestimate protein content by absorbance at 280 nm. On thawing, simple warming to 37°C for a few minutes is sufficient to dissolve all protein and sugar that may have cryoprecipitated. Maltose also acts as an excellent stabilizer and vehicle for freeze-drying and allows almost instantaneous redissolving when rehydrated. The presence of phosphate, citrate, and maltose at pH 6.0, while stabilizing the pure IgG do not interfere with UV absorption measurements, iodination, or coupling reactions when the antibodies are immobilized (e.g., to CNBr-agarose) or conjugated (e.g., to enzymes by the glutaraldehyde method). Maltose should be removed by dialysis if protein is estimated by the Pierce BCA (bicinchoninic acid) assay as it generates false color and should also be removed if conjugation of antibody is contemplated by periodate oxidation *of IgG* (note the normal Nakane and Kawaoi[3] conjugation involves periodate oxidation of the peroxidase enzyme, not the IgG).

G. COLUMN REGENERATION

Generally it is good practice to run the pH gradient all the way down to pH 3.0 or below, even when running known weakly binding subclasses. This is because it removes any nonspecifically bound protein as well as other strongly bound subclasses of irrelevant IgG which may be contributed by host ascites or calf serum. Simple washing of the column back up to pH 9.0 in starting buffer is enough to prepare the column for immediate reuse. Where storage for a prolonged period is contemplated, it is better to leave the column at pH 3.0 because (1) bacterial and fungal growth is inhibited, (2) leakage of protein A from CNBr linkage is less at this pH, and (3) protein A is quite stable at pH 3.0 and less likely to be proteolyzed by residual plasma proteases.

V. DISCUSSION

A. AFFINITY

Much of the early published literature and even some of the most recent does not give adequate detail of the conditions used to study binding or the actual subclasses bound or not bound. Thus, interpretation of much published work is hazardous. An attempt to do this is summarized in Table 1 but should be interpreted with caution. In particular, the binding of many subclasses which are reported "not to bind" or to bind "weakly" can be improved considerably by application in high salt and high pH buffer[4,5] such that large-scale preparation in good yield is feasible. Murine IgG$_1$ is the best example of this. Where 3 M NaCl at pH 9.0 is insufficient, consider using even higher concentrations. Note, however, that there is no clear dividing line between salt-enhanced hydrophobic binding and simple salting out onto the column. A good general rule is to use a salt concentration that is high but does not quite cause salting out of IgG from solution. Clearly this needs to be related to the particular subclass(es) of interest. If NaCl is unable to effect binding, then consider using Na$_2$SO$_4$ or Na$_3$ citrate (the use of ammonium sulfate or glycine should be avoided as high concentrations of amino compounds will compromise subsequent solid-phase coupling or enzyme conjugation). Again, the actual concentration of salt needed will depend on the animal species and subclass, but a good starting point is 10 to 12% w/v of the salt.

B. STEP VS. GRADIENT

Much early work (reviewed in References 6 and 7) used step-wise elution with a series of discrete buffers of decreasing pH. This is a hazardous technique unless the actual elution conditions of a particular antibody and subclass is known from preliminary gradient experiments. If a pH-step interval is chosen that only partly elutes a single subclass, then the antibody can elute in two or sometimes three discrete "peaks" over subsequent steps. This result has given rise to erroneous claims of heterogeneity within an antibody. Because individual antibodies within a given subclass are slightly different, it is unwise to use a fixed pH interval to define a particular subclass for purification. Conversely, the absence of heterogeneity within a single antibody (i.e., monoclonal) means resolution and peak width are much sharper in monoclonal antibodies than in polyclonal antibodies. Except in cases where antigens have to be prepared from polyclonal sources (e.g., for the preparation of subclass-specific polyclonal reagents), the resolution between different subclasses is not strictly required, for in monoclonal antibodies only a single subclass is present whereas in polyclonal antibodies it is desirable to recover the entire spectrum of subclasses to retain all activity. However, the approximate range of pH elution (see Table 1) can be a valuable guide to subclass identification and with care and experience can be used to replace the conventional immunologic subclass identification in favorable cases (murine antibodies in particular). Obviously, this identification is less reliable if step elution is used. Where repetitive purification of the same antibody is contemplated, then, step elution can be based on the gradient region required to elute only the species of interest; however, occasional regeneration to below pH 3.0 is desirable when ascites or other sources of host polyclonal sera are present. Occasionally, particularly in high-performance liquid chromatography (HPLC) and Fast Protein Liquid Chromatography® (FPLC) the rapid concentration changes in step buffer elution produce false UV peaks due to refractive index changes. These false peaks can be discounted by manual reading of fractions at 280 nm.

C. IDIOSYNCRACIES

Occasionally, individual monoclonal antibodies are found which do not conform to the general rules. This is an inevitable consequence of the change from polyclonal to monoclonal source since the physical properties are no longer averaged out. Examples include rat IgG_{2b} antibodies; some bind quite strongly while others do not. Some secreting fusion partners give rise to partially active or inactive forms of the general form AA, AB, BB, where A and B are the two different (active and inactive) heavy chains which will be present in the ratio 1:2:1. Since different subclasses or microheterogeneity may be represented by the individual heavy chains, it follows that they may be resolved into three peaks or a broad protein peak. In the latter case bioassay of actual antibody activity by antigen binding assay will indicate which fractions to pool. The primary protein A binding site on IgG is probably present in the intact CH2-CH3 region of the heavy chains. If either chain of the pair is damaged or missing, e.g., by proteolysis, then affinity will be reduced. Conversely, if aggregates or polymers are present, then affinity may be increased. These situations are manifest by minor peaks eluting earlier or later than the main peak. Other binding sites in the IgG molecules are occasionally seen, though these are weaker and more variable in occurrence and cannot be relied upon. A similar situation pertains to IgM, which occasionally (but unpredictably) binds to protein A. This is usually insufficient to produce more than a minor part of a polyclonal IgM and cannot be relied upon for the majority of monoclonal antibodies.

Occasionally, two subclasses (peaks) may be seen; this result can arise from either inadequate cloning or, possibly, class switching during prolonged subsequent culture after adequate cloning. One interesting report of wide variation within a single species of polyclonal antibody from goats exists.[8]

D. WHAT IS THE CAPACITY OF PROTEIN A?

This question is frequently asked but cannot easily be answered. The manufacturers data usually refer to the saturation capacity of the immobilized protein A using a large excess of human polyclonal IgG. On theoretical grounds (four binding sites, molecular weight of IgG 150,000, molecular weight of protein A 40,000) 1 mg of protein A might bind 10 mg of IgG; however, steric considerations are likely to reduce this in practice, particularly when solid phased. A more realistic figure is 5 mg IgG per milligram protein A. Lower affinity subclasses (e.g., murine IgG_1 or rat IgG_{2b}) will show variable binding capacity depending on the amount of salt and pH in the application buffer, but a realistic figure of 2 mg IgG per milligram of protein A is useful. Finally, the affinity of protein A for IgG is significantly less than the best antigen-antibody reactions (10^{-7} to 10^{-8} vs. 10^{-9} to $10^{-10}M$) so that the law of mass action will operate and low concentrations of antibody (1 to 2 μg/ml) will be bound less efficiently than high concentrations (1 to 2 mg/ml), and this is one of the reasons for sample preconcentration. Taking this finite affinity together with the need not to overload the column, it is prudent to use only half the column binding capacity so as to avoid loss of valuable antibody or weaker subclasses in the breakthrough volume. This finally reduces to a rough working capacity of ≈1 mg IgG per milligram of protein A. Although a rather conservative estimate, it is the author's experience that this saves time and antibody in the long run and improves resolution and reliability. If producing a coupled protein A column, aim to couple protein A at between 2 to 10 mg/ml of gel, depending on the scale of operation. The author routinely processes 50 ml of ascites on 50-ml columns containing 2 to 5 mg protein A per milliliter of gel, assuming that ascites contains between 2 to 5 mg/ml of IgG.

E. GRADIENT MANIPULATION

Rather unexpectedly, simple mixing of two buffers of the same composition (100 mM phosphate and 100 mM citrate) but of different pH is able to generate a linear-gradient decreasing pH from 6.0 to 2.5. Since phosphate has poor buffering capacity above 7.0, the gradient shows a sudden dip in this region, which is not normally a problem with murine subclasses. However, where other species or weaker subclasses are investigated, it is possible to "smooth out" this dip by using other buffers. Phosphate has two ionizations in the range of interest at pH 2.15 and pH 7.20, whereas citrate has three ionizations in the range of interest at pH 3.13, 4.76, and 6.40. Rather fortuitously, none of these overlap, and no gaps in buffering capacity exist, either. To cover the range above phosphate, a third buffer species is added which covers the range 7 to 9, and the author has had some success with tricine (pK 8.15), piperazine (pK 5.6 and 9.80), carbonate/bicarbonate, and *N*-ethyl morpholine (pK 7.6). Ideally, these buffers should be devoid of UV adsorption and reactive amino groups and be of low cost. With most impure samples, the protein content itself is sufficient to buffer the sample at the starting pH but is generally insufficient to smooth out any abrupt changes due to absence of buffering capacity in the gradient generation. A less obvious advantage of using multivalent ions such as phosphate and citrate (rather than, e.g., glycine) is that even without their buffering range, when almost fully ionized at pH 9, they provide a powerful structure-promoting influence on the protein IgG such that prolonged storage at alkaline pH and 20°C does not noticeably denature the antibody activity. Similar considerations probably apply in the weak acid region also, where partial ionization of citrate still provides a divalent ion. For similar considerations, protein A makes an ideal second step for IgG purification after sodium sulfate or ammonium sulfate precipitation as the redissolved precipitates do not need to be dialyzed free of the excess sulfate ions, instead these will (1) enhance binding of IgG to protein A and (2) stabilize the IgG while the pH is raised to 9.0. The elimination of the dialysis step thus saves considerable time, and precipitation may also replace ultrafiltration as a preconcentration step with tissue culture supernatants for the same reasons.

F. MISCELLANEOUS POINTS

Frequently, when monitoring the UV absorption a slowly rising baseline (total optical density rise $\simeq 0.1$ A units) will appear during the elution below pH 6.0. The reason for this is uncertain, but it may be due to azide ion, the presence of impurities in the buffer(s), and/or their changing ionization and spectra with decreasing pH. Because the same phenomenon also occurs with a step change, it is unwise to rely solely on A_{280} measurements for peak identification when <1 mg of antibody is present; additional validation with bioassays, immunoassays, and/or ^{125}I tracers is highly desirable in cases where small amounts of antibody are processed.

Occasionally, one may wish to immobilize antibodies by methods which do not cause loss of functional capacity from either steric/or chemical problems. Solid phase protein A is an ideal reagent in this respect as it immobilizes antibodies only via the Fc region, leaving the antibody binding side unhindered, sterically accessible, and chemically unchanged. Simple mixing of antibody (which need not be pure) in appropriate buffer (i.e., 3 M NaCl buffered to pH 8.5 if using, e.g., murine IgG$_1$ subclass) will provide temporary noncovalent immobilization. However, for subsequent use in, e.g., immunoassays or immunopurification, it is usually desirable to fix the antibody permanently to the protein A by a crosslinking process. The author finds the procedure of MacSween and Eastwood[11] simple and reliable. The gel beads-protein A-IgG complex is washed free of unbound protein in starting buffer, e.g., by centrifugation, and then paraformaldehyde is added to a final concentration of 0.5% and incubated at 37°C for 45 min followed by washing. This procedure was found to yield excellent antibody "survival" in the case of several monoclonal antibodies that were inactivated by direct cyanogen bromide immobilization. Antibody saturation of protein A is desirable to prevent later binding of any IgG in samples. Where the antibody is to be used for immunopurification, bear in mind that the pore size of the supporting gel (i.e., molecular weight exclusion limit) has to be large enough to allow free access of antigen to the immobilized protein A-IgG complex within the pores. For the majority of antigens (up to $\simeq 10^7$ mol wt) Sephacryl S-500 and S-1000 have proved to be of large enough pore size.

Many tissue-cultured antibodies will have been derived from hybridomas grown in fetal calf serum, which should, ideally, not contribute significant amounts of bovine IgG. However, the variable quality of such serum will occasionally give rise to unexpected peaks of bovine IgG, which elute before murine IgG, in the pH 6 to 8 region (Table 1 and Figure 1). A simple screen of such peaks with Mancini immunodiffusion plates against bovine IgG and murine IgG (or the species under investigation) will easily distinguish the desired antibody and impurity peaks. Conversely, the same technique can be used to screen for suitable and unsuitable batches of calf serum in hybridoma production.

Since albumin is usually the major protein component in both ascites- and tissue-culture-derived material, it represents the most serious potential contaminant of the final high-purity IgG product. Since nonspecific (ion exchange) binding of albumin can easily occur to any positively charged groups in the gel (e.g., proteins or activation groups), suppression of such charge interaction by excess salt is desirable. The use of 3 M NaCl in the sample and starting buffer achieves this, and although not essential to enhance the binding of the more strongly bound subclasses of antibodies, its presence is still desirable to reduce albumin contamination of the IgG product.

Antibodies to protein A are commercially available from Sigma (rabbit) and from IMS and Sera-lab (raised in chickens). The latter have the useful property of not binding via the Fc region, so the binding is purely "immunologic". Immunization of rabbits, chickens, guinea pigs, and mice has been well tolerated when using high-purity protein A, and extracorporeal immunosorption of human plasma on immobilized protein A devices is also well tolerated clinically. However, the author has personal experience of problems with protein A toxicity in goats (and, therefore, probably sheep also) which could only be

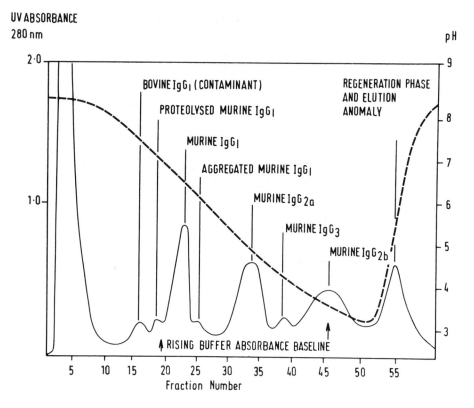

FIGURE 1. Idealized elution scheme of different murine antibodies from protein A (normally only one or two of these peaks will be seen with any one sample).

controlled by using formaldehyde inactivated protein A ("toxoided") and careful clinical management with i.v. fluids; even so, 10 mg of formaldehyde-treated protein A administered subcutaneously in Freund's adjuvant still showed detectable toxicity in goats.

Because protein A binds only to the Fc region, it is a useful reagent to separate Fab' or Fab'$_2$ fragments from intact IgG or Fc fragments. The former appear in the unbound breakthrough peak, whereas the latter will be retarded and elute at the pH appropriate to the subclass of their heavy chain. Precautions are desirable to prevent proteolysis of the protein A column. Ideally, digestion of the IgG should be done with solid-phased pepsin or papain. However if this is not feasible, soluble proteases should be inactivated before chromatography. Pepsin can be inactivated simply by adjustment of the pH to 8.5, while papain can be inactivated with iodoacetamide or iodoacetic acid (since protein A contains no thiols or disulfide bands and is not affected by thiol compounds or thiol-blocking compounds).

During regeneration of protein A (and other affinity systems) it is customary to raise the pH rapidly from 2.5 to 9.0, and during this phase the fluid is often diverted to waste. However, it is wise to monitor this process by UV adsorption as some "lost" antibodies may elute when the pH is rising, although they were not eluted during the falling gradient. This is not a consequence of gradient vs. step since the same phenomenon occurs when stepping the pH down and up. In extreme cases, more antibody can be eluted during the regeneration phase than during the elution phase. The reason for this is unknown, but it is prudent to collect the regeneration-phase fluid.

APPENDIX OF SUPPLIERS' ADDRESSES

Amersham International plc., Lincoln Place, Green End, Bucks HP20 2TP, U.K.
Amicon-Grace, 24 Cherryhill Drive, Danvers, MA 01923
BRL/Gibco, 3175 Staley Road, Grand Island, NY 14072
Bioprocessing Ltd., Medomsley Road, Consett, Tyne & Wear DH8 6PJ, U.K.
Biorad, 1414 Harbor Way South, Richmond, CA 94804
Domnick Hunter Ltd., Durham Road, Birtley, Tyne & Wear DH3 2SF, U.K.
Fermentech Ltd., Research Park, Riccarton, Midlothian EH14 4AS, U.K.
IBF, 35 Avenue Jean-Jaures, 92390 Villeneuve-la-Garenne, France
IMS, AB, P.O. Box 8012, S-75008 Uppsala, Sweden
Oros Systems Ltd., 715 Banbury Avenue, Slough SL1 4LJ, U.K.
Pharmacia LKB, S-75182 Uppsala, Sweden
Pierce Chemical Co., P.O. Box 117 Rockford, IL 61105
Sera-lab Ltd., Crawley Down, Sussex RH10 4FF, U.K.
Sigma Chemical Co., P.O. Box 14508, St. Louis, MO 63178

REFERENCES

1. **Langone, J. J.,** (^{125}I) protein A: a tracer for general use in immunoassay, *J. Immunol. Methods,* 24, 269, 1978.
2. **Langone, J. J.,** ^{125}I-labelled protein A as a general tracer in immunoassay: suitability of goat and sheep antibodies, *J. Immunol. Methods,* 34, 93, 1980.
3. **Nakane, P. K. and Kawaoi, A.,** Peroxidase labelled antibody — a new method of conjugation, *J. Histochem. Cytochem.,* 22, 1084, 1974.
4. **Ey, P. L., Prowse, S. J., and Jenkin, C. R.,** Isolation of pure IgG$_1$, IgG$_{2a}$ and IgG$_{2b}$ immunoglobulins from mouse serum using protein A-Sepharose, *Immunochemistry,* 15, 429, 1978.
5. **Seppala, I., Sarvas, H., Peterfy, F., and Makela, O.,** The four subclasses of IgG can be isolated from mouse serum by using Protein A-Sepharose, *Scand. J. Immunol.,* 14, 335, 1981.
6. **Langone, J. J.,** Protein A of *Staphylococcus aureus* and related immunoglobulin receptors produced by streptococci and pneumococci, *Adv. Immunol.,* 32, 157, 1982.
7. **Lindmark, R., Thoren-Tolling, K., and Sjoqvist, J.,** Binding of immunoglobulins to Protein A and immunoglobulin levels in mammalian sera, *J. Immunol. Methods,* 62, 1, 1983.
8. **Richman, D. D., Cleveland, P. H., Oxman, M. N., and Johnson, K. M.,** The binding of staphylococcal protein A by the sera of different animal species, *J. Immunol.,* 128, 2300, 1982.
9. **Rousseaux, J., Picque, M. T., Bazin, H., and Biserte, G.,** Rat IgG subclasses: differences in affinity to protein A-Sepharose, *Mol. Immunol.,* 18, 636, 1981.
10. **Delacroix, D. and Vaerman, J. P.,** Simple purification of goat IgG$_1$ and IgG$_2$ subclasses by chromatography on protein A-Sepharose at various pH, *Mol. Immunol.,* 16, 837, 1979.
11. **MacSween, J. H. and Eastwood, S. L.,** Recovery of immunologically active antigen from staphylococcal protein A-antibody adsorbent, *J. Immunol. Methods,* 23, 259, 1978.

Chapter 11

IMMUNOAFFINITY ADSORPTION: PRACTICAL APPLICATION IN LABORATORY RESEARCH AND PRODUCTION

Andrew Lyddiatt

TABLE OF CONTENTS

I. INTRODUCTION

Immunoaffinity association between antibodies, antigens, or other immunobiochemicals (protein A, G, etc.) has been widely exploited in analytical and preparative techniques wherein one of a pair of interactants is immobilized on a solid phase in a form which permits association with the other present as a soluble component of contacted feedstocks. Such mechanisms are the basis of many analytical techniques (enzyme linked and radioimmunoassays, electrophoretic visualization, diagnostic histology, etc.) as well as immunoaffinity adsorption applied to product recovery. The latter process may be defined as the contact of a stationary phase, covalently derivatized with active immunochemicals (the ligand), and a mobile phase containing a mixture of solutes from which an appropriate product may preferentially interact through unique affinities for the modified solid. Impurities and contaminants having weak or negligible affinity may readily be displaced from surfaces and voids by appropriate washes, and product desorbed or eluted by perturbation of specific interactions in modified conditions of pH, ionic strength, or competitive binding. This technique, frequently referred to as immunoaffinity chromatography, has been spectacularly successful in laboratory-scale purifications dedicated to the molecular dissection of biological systems and the manufacture of research tools and diagnostic probes.[1-6] In contrast, limitations of material design, unpredictable operational longevity, and economics have restricted true manufacturing operations to high-value, low-volume products (diagnostic probes, immunoaffinity ligands, therapeutic agents).[7,8] In immunoaffinity purifications, it should be stressed that the requirement of fractional partition between solid and mobile phases, which characterizes classical interactive chromatography,[9] is largely absent since only specific products associate at binding strengths exceeding weak, nonspecific interactions. The term immunoaffinity adsorption will herein be employed to describe all such immunochemical interactions exploited in fixed-bed, fluidized-bed, and slurried batch contactors, although the terms liquid chromatography (LC) and high performance liquid chromatography (HPLC) will distinguish adsorption to "large" particles (>40 μm) at near ambient pressures from high performance operations.

Polyclonal antibodies (PAbs), raised and isolated in animal immunization programs, have been widely used in the analysis and purification of many antigen types.[10] The nature of the immune response means that PAb preparations exhibit a range of binding affinities against a variety of antigenic epitopes, the detailed characteristics of which may vary from batch to batch.[6] Thus, immobilized PAb will bind antigen with a range of strengths promoting dilute recoveries and/or incomplete desorption with all but the strongest eluants (urea, thiocyanate, etc.). Such operational problems may be overcome by the use of monoclonal antibodies (MAbs), raised against specific antigenic epitopes and selected in terms of their specificity and intermediate strength of binding (dissociation constants of 10^{-6} to 10^{-9} M).[11,12] The latter ensures that effective desorption, essential to the operation of efficient, repetitive purification, is achieved under mild conditions. Many diagnostic antibodies, selected for their virtual irreversible antigenic association, are unsuited as immunoaffinity ligands and as products purified by means of immobilized antigen. Practitioners should also realize that conventional immobilization of MAbs by random covalent coupling to a porous matrix of undefined external and internal topographies inevitably generates a pseudopolyclonal ligand since geometric location, single or multipoint attachment, and biochemical damage sustained at immobilization cannot be effectively controlled. However, all accumulated evidence suggests that the degree of variation of antigen binding generated through such phenomena is less than that characteristic of true PAb preparations.

Agents applicable to the purification of antibodies (PAbs or MAbs) from blood plasma, ascites fluids, or hybridoma cultures include appropriate specific antigens, and the staphylococcal proteins A and G.[11,13,14] The former are limited since antibodies are most frequently manufactured to probe or recover previously intractable antigens. However, systems involving specific MAbs against polyclonal human IgG-Fc have been successfully employed for the positive purification of one or other interactant using immobilized preparations of the second.[15,16] In contrast, protein A and protein G have been widely used for the recovery of antibodies. Disadvantages of cost and low affinities for certain immunoglobulin types and classes have been partly overcome in recent years with improved manufacturing processes and the introduction of commercial and customized alkaline, hydrophobic buffer systems to enhance binding efficiencies.[17]

In spite of differences between various components available for assembly of the immunoaffinity systems outlined above, practical considerations associated with controlled assembly and operation are common to all.[18] However, the molecular size of interactants (and associated complexes) will influence the molecular penetration of solid phases characterized by varied effective pore size. Thus, MAbs immobilized on 4% cross-linked agarose for the purification of a low molecular weight peptide (mol wt = 1000) will ultimately perform differently to a similar MAb specific for the larger protein β-galactosidase (mol wt = 460,000). In this context, the choice of 4% or 6% agarose (most commonly Sepharose CL-4B or CL-6B) characterized by 2×10^7 and 2×10^6 Da molecular exclusion limits for globular proteins may be critical to the controlled assembly and successful operation of efficient adsorbents for defined tasks.

There are many adsorbent components, derivatized solid phases, and complete chromatographic systems commercially available to equip the laboratory practitioner with a broad portfolio of immunoaffinity purifications.[14] Activated matrices such as CNBr-, carbonyldiimidazole (CDI)-, or epoxy-activated Sepharose), may be derivatized with the ligand of choice.[1] Alternatively, ready-to-use assemblies such as protein-A or protein-G Sepharose can be utilized in all types of laboratory equipment, while dedicated chromatographs equipped with specific adsorbent columns serviced by precision pumping, monitoring, and collection equipment may be purchased as complete packages.[14] Such may perform very well, but, in addition to cost disadvantages, they inevitably invite operators to regard purification systems as black boxes. This suits many workers, since laboratory purification is frequently a handle-turning means to the exciting end of exploiting novel products as research tools. However, this is ultimately a false economy of commitment since investment in understanding biochemical mechanisms and controls of purification will be amply repaid through efficient, repetitive application of materials. When commercial systems malfunction, solutions must be empirically sought since little is publicized concerning the nature of adsorbent fabricates. Such limited information includes general physical tolerances, maximum capacities for specimen products, and recommended operating conditions in selected preparations. Maximum capacities are frequently derived from operationally irrelevant, saturation studies in the presence of excess product, and purification examples may not relate to realistic feedstocks and specific products. Both characteristics are of limited value in trouble-shooting difficult preparations.

In contrast, the essentials of active ligand concentrations, potential productivity under defined operating conditions, and predicted working lifetime represent important data which are readily obtained yet remain notably absent from commercial manuals, and are only rarely discussed in the research literature.[6] However, such information is readily assembled from simple characterization studies to the benefit of the practitioner wishing to maximize the efficiency and effectiveness of purification strategies.[6,19,20] Such will enhance the design, assembly, and operation of feedstock contactors appropriate to the solution of individual purification problems.[20]

Selected recommendations for suitable procedures gathered from the literature, and from the work of colleagues in the Biochemical Recovery Group in Birmingham, are reviewed herein, using as realistic examples the controlled immobilization of MAb or polyclonal

TABLE 1

Biochemical and Physical Characteristics of the Immunoaffinity Adsorbents Selected for Fixed-Bed and Fluidized-Bed Applications

Solid phase	Activation procedure	Immunoaffinity ligand
LC-Sepharose CL-4B[1] 4% cross-linked agarose 40—160-µm pd 2×10^7 Da exclusion	Cyanogen bromide, tresyl chloride	Hu-IgG, anti-huIgG MAb, protein-A
LC-Macrosorb K-4AX[2] 1:1 composite of kieselguhr (15-µm pores) and 4% cross-linked agarose 150- to 300- or 1000-µm pd 2×10^7 Da exclusion	As Sepharose plus carbonyldiimidazole	As Sepharose
HPLC-HMPS composites[3] with hydrophilic surfaces 8-µm pd (non-porous) 10-, 20-µm pd 10-nm pores	Cyanogen bromide, tresyl chloride, periodate, carbonyldiimidazole	HuIgG, anti-huIgG, protein-A
HPLC-Spherisorb VLS[4] silica particles 20-µm pd ; 28-nm pores	Glycidoxypropylsilane capping plus activation as for HMPS	As HMPS
HPLC-Lichrosorb[5] silica particles 10-µm pd; 6, 100 and 400 nm	As Spherisorb	As HMPS

Note: Particle diameters (pd), average pore diameters, and molecular exclusion limits as given by commercial suppliers: 1, Pharmacia, Uppsala, Sweden; 2, Sterling Organics, Newcastle-upon Tyne, UK; 3, Polymer Laboratories, Church Stretton, U.K.; 4, Phase Separations, Clwyd, U.K.; 5, BDH, Poole, U.K.

huIgG antigens for the purification of appropriate interacting partners. In particular, the exploitation of novel composite adsorbent materials and contacting devices are offered as potential solutions to practical problems encountered in integrated production and purification. The activation, derivation, characterization, and utilization of anti-huIgG McAb and huIgG PAb agaroses and kieselguhr-agaroses in the respective recoveries of huIgG from blood plasma and MAb from serum-based hybridoma cultures are compared in fixed-bed operations with respect to purification efficiency, physicobiochemical stability, and operational longevity. Advantages and disadvantages of operating kieselguhr-agarose composites in high-flow regimes, particulate flows, and fluidized or slurried contactors are discussed. The ease of assembly of affinity HPLC matrices is reviewed, and an immediate role for miniaturized preparative systems as near on-line biosensors is contrasted with current markets for productive HPLC systems.

II. EXPERIMENTAL

A. SOURCE MATERIALS

A selection of commercial liquid chromatographic (LC) and high performance liquid chromatographic (HPLC) solid phases were activated by different chemical methods (Table 1), derivatized with selected ligands, and critically compared in systems utilizing immobilized MAb (anti-huIgG MAb), protein-A, or antigen (HuIgG PAb) to purify specific antigens from blood plasma, or MAb from serum-based animal cell cultures (Table 2).[21] MAb was affinity purified, protein A obtained commercially and repurified as necessary on huIgG-Sepharose, and huIgG purified from outdated blood supplied by the National Blood Transfusion Centre Birmingham using subtractive adsorption of impurities from plasma on DEAE-52 cellulose or QA Zeta-Prep (Pharmacia-LKB).[21,22] Ligands were routinely analyzed for

TABLE 2
Summary of Experimental Immunoaffinity Systems Used in Adsorbent
Characterization and Practical Assessment

Immunoaffinity ligands and	A1	HuIgG; LC (1—20 mg/ml) or HPLC (1—50 mg/g)
adsorbents	A2	Anti-huIgG MAb; LC or HPLC as above
	A3	Protein-A LC (1—20 mg/ml) or HPLC (1—120 mg/g)
Feedstocks		Anti-huIgG MAb (30—600 μg/ml) in cell culture fluid containing bovine albumin, transferrin, and cell lysates (3—10 mg/ml total protein) (A1)
		Human blood plasma rich in albumin (30—50 mg/ml), huIgG (10—20 mg/ml), and other blood protein (A2,3)
		HuIgG (10 mg/ml) partially purified on anion-exchange adsorption (A3)
Working buffer		Phosphate buffered saline (PBS) pH 7.4 ± bacteriostat (A1-3)
		Tris-glycine (0.1 M) containing 0.3 M NaCl ± bacteriostat (pH 8.0) (A3)
Eluant		Potassium thiocyanate (3 M) in PBS pH 7.4 (A1,2)
		Glycine-HCl (0.1 M) pH 2.8 (A3)
Product polish		Rapid gel permeation on Sephadex G-25 in PBS pH 7.4 and concentration by ultrafiltration (10^4 molecular weight cut-off membrane) (A1-3)
Product QC		ELISA, affinity HPLC, reduced/unreduced SDS and native PAGE protein assays (A1-3)
Adsorbent reuse		Reequilibrate in working buffer (A1-3)

spectrophotometric characteristics, protein content, molecular homogeneity on sodium dodecyl sulfate polyacrylamide gel electrophoresis (SDS-PAGE), and specific enzyme-linked immunosorbent assay (ELISA) activity before immobilization.[22] Anti-huIgG-Fc MAb was produced from batch or continuous suspension cultures (0.05 to 5 l) of TB/C3 murine hybridomas in serum-based media and purified directly or after three- to fivefold concentration by ultrafiltration against membranes characterized by 3×10^4 Da mol wt exclusion.[21,22] TB/C3 cell lines were donated by the Department of Immunology, University of Birmingham.

B. SOLID-PHASE ACTIVATION AND DERIVATIZATION

Homogeneous agaroses and kieselguhr-agarose composites were activated by cyanogen bromide (CNBr)[22-24] and tresyl chloride procedures[22,25] with particular care taken to minimize attrition during mechanical agitation. Degrees of CNBr activation were estimated by the method of Kohn and Wilchek.[26] HPLC-silicas were capped by glycidoxypropyl silane[27] and chemically activated with cyanogen bromide and tresyl chloride as above, and by methods utilizing CDI[28] or periodate oxidation.[27,29] HPLC hydrophilic macroporous supports (HMPS) were activated directly by one of four methods described above. Activated LC and HPLC solid phases were reacted with pure protein ligands (MAb, PAb, protein A; 1 to 20 mg/ml) in phosphate buffered saline (PBS) at pH 7.4. Reactions of 1 to 16 h at 4° and 25°C yielded ligand concentrations of 0.1 to 20 mg/ml settled, swollen adsorbent. Uncoupled protein was displaced in PBS by decantation or filtration washing and unreacted groups blocked by treatment with 1 M ethanolamine in PBS at pH 7.4. Adsorbents were extensively washed in appropriate buffers at pH 4.0 and 8.0, and in 1 M NaCl, and purged with 10 volumes each of PBS, 3 M KSCN in PBS, and PBS in preparation for productive operations.[22,29] Ligand concentrations were estimated by spectrophotometric difference analysis of start material and protein contaminated washes using published extinction coefficients for individual purified ligands.

C. SOLID-PHASE CHARACTERIZATION

Estimated maximum capacities of immunoadsorbents (qm) and dissociation constants

(Kd) were established for all solid phases in equilibrium batch experiments[19] and fixed-bed saturation studies[20-22,29] using products in complex feedstocks rich in impurities (MAb in serum-based culture fluid; PAb in blood plasma). Flow characteristics and pressure resistances of adsorbents were determined in fixed-bed contactors equipped with mercury manometers or pressure transducers.[15,22] Fresh and used particles were characterized before and after repetitive utilization using a Malvern laser densitometer.

Immunoadsorbents were tested in preparative fixed-bed applications (LC and HPLC) using realistic feedstocks of clarified cell culture fluids or blood plasma rich in MAb and PAb, respectively. Feedstocks were used at native pH and ionic strength (pH 6 to 7.5; 150 mM chloride equivalents) except where blood plasma was diluted to reduce adverse rheological phenomena. After loading, immunoglobulin adsorbents were washed in 2 to 10 volumes of PBS and eluted in a front of 3 M KSCN in PBS at pH 7.[15,22] Product was quantified by on-line UV spectrophotometry and immediately desalted by gel permeation on beds ($>5 \times$ sample volume) of Sephadex G-25 (Medium) equilibrated and eluted in PBS. Protein-A adsorbents were eluted in 0.1 M glycine-HCL at pH 2.8,[11] and product was treated as above. All adsorbents equilibrated were rapidly re-equilibrated in PBS and maintained at low flow rates in sterile PBS for further use or stored at 4°C in the presence of 0.02% sodium azide or thimerosal. Protein products were concentrated by ultrafiltration (10,000 molecular mass exclusion; Amicon stirred cell or Millipore Minitan) as necessary and stored at 4°C, or once frozen at -20°C, in PBS pH 7.4 containing bacteriostat. Quality controls of SDS-PAGE with Coomassie blue and silver staining,[22,30] protein analysis by the Bradford method,[31] protease evaluation,[22,32] and biological activity in ELISA or affinity HPLC[29] were routinely applied. Repetitive evaluation of adsorbent performance was undertaken with fixed beds (1 to 5 ml) arranged in Pharmacia FPLC apparatus equipped and programmed to control up to 150 consecutive minipreparative cycles. Productivity of immobilized ligand was monitored throughout as above and used adsorbents analyzed physically and biochemically by a combination of electron microscopy, immunohistology and molecular analysis.[16]

D. INTEGRATED PRODUCTION AND AFFINITY RECOVERY

HuIgG-adsorbents (agarose, kieselguhr-agarose) were tested in on-line integrated recovery from continuous suspension cultures of TB/C3 murine hybridomas.[21] A 700-ml vessel, maintained at 36°C under pH and dO$_2$ control was maintained in steady state (5×10^6 cells per milliliter) for up to 60 d by dilution (2 to 5% volume per hour) with culture medium enriched with 5% fetal calf serum (FCS). Vessel effluent was pumped over a weir, through a sterile trap and glass wool filter to a cell settler at 4°C sized for residence times of 2 to 24 h at defined dilution rates. A second pump transferred clarified liquor to a fixed bed of adsorbent (5 to 50 ml) and subsequently to waste (see Figure 1). Particles were monitored in flow streams by microscopy, MAb by retrospective ELISA or near on-line affinity HPLC, and acid proteases by biological assay.[22,29] Saturated adsorbents were manually replaced with fresh material, stripped of product, and reequilibrated for further use as above.

E. ALTERNATIVE CONTACTORS

Kieselguhr-agarose composites, derivatized with diethyl-amino-ethane (DEAE)-ion exchange groups or immunoaffinity ligands, were operated in single stage, liquid fluidized beds (Figure 2) with upward recirculating suspensions of hybridoma cells in media containing MAb, red blood cells in plasma, and synthetic mixes of bovine albumin and yeast (10^9 cells per milliliter).[33] Loading was monitored by continuous UV spectrophotometry or retrospective biological analysis. Charged adsorbents were collapsed by flow reversal, washed free of entrapped debris by backflushing with PBS and eluted and reequilibrated in conventional fixed-bed protocols.[33]

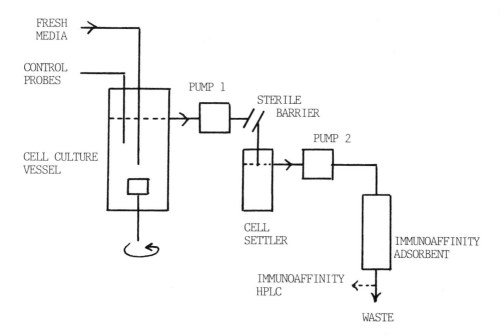

FIGURE 1. Schematic representation of continuous culture of murine hybridomas serviced by integrated product adsorption. The culture vessel (700 ml) was supplied with, fresh serum-based media at 36°C and gassed as necessary. Excess broth was pumped over a weir, through a sterile barrier to the gravity settler at 4°C. Clarified broth was subsequently pumped through a fixed-bed contactor containing immobilized antigen (huIgG) specific to the monoclonal antibody product. Anti-huIgG was monitored by analysis of samples taken from the culture vessel, cell settler, and waste outlet using ELISA or affinity HPLC methods.

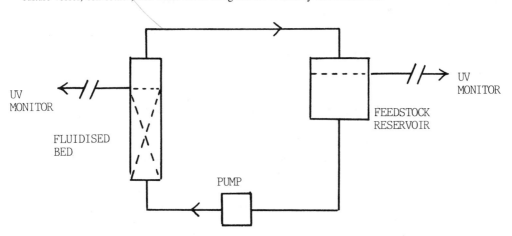

FIGURE 2. Schematic representation of a recirculating liquid fluidized-bed contactor. Fluidized beds (20 to 500 ml) were charged with kieselguhr-agarose composites and fluidized by recirculation of feedstock (0.2 to 0.6 cm/ s) to yield bed expansions of two- to fourfold. Products were desorbed by flow reversal under conditions described for fixed beds.

Immunoaffinity HPLC matrices[29] were tested in fixed-bed preparative application (1 × 5 cm), operated manually in protocols described for LC materials, or in analytical columns (3 × 5 mm) serviced by on-line gel permeation columns (4.4 to 7.5 × 200 to 300 mm; Sephadex G-15 or TSK-HG2000 SW) (see Figure 3).

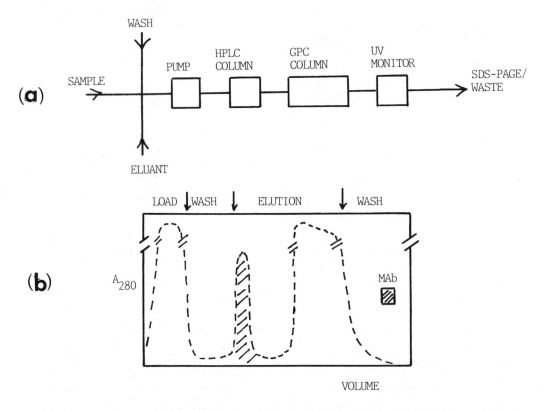

FIGURE 3.   Bioquantitation by immunoaffinity HPLC. (a) Miniaturized preparative HPLC equipment was utilized featuring contactors (0.05 to 0.3 ml) serviced upstream by a standard high performance pump and controller, and downstream by on-line gel permeation and a UV-monitoring integrator. (b) A representation of a typical analysis of crude anti-huIgG produced in serum-based hybridoma culture and quantified using huIgG-silicas or HMPS arranged in mini-HPLC contactors. Adsorbents equilibrated in PBS, pH 7.4, were loaded with clarified sample, washed in PBS, and eluted in 3 M KSCN in PBS pH 7.4. On-line gel permeation of eluate permitted integration of UV peaks in the gel permeation chromatography (GPC) void volumes.

III. IMMUNOAFFINITY LC ADSORPTION

A. COMMERCIAL MATERIALS

There is a wide range of porous solid phases commercially available for use as components in the custom assembly of immunoaffinity adsorbents or ready derivatized for application in defined purification systems.[34] Materials, fabricated from agarose, cellulose, acrylamide, hydrophilic coated polymers, or composites thereof, have proved qualities of ready activation, physical, and biochemical resistance, adequate capacities (expressed through porosites and surface areas), and minimal nonspecific binding (promoted by charged or hydrophobic impurities). Recent trends in manufacture have been toward smaller macroporous spheroids of uniform particle size characterized by improved chromatographic resolution in fixed beds operated at low to medium pressures.[20,34,35] Ironically, the uniformity of solid-phase geometry required of reversed-phase, gel permeation, or ion-exchange fractionation is not mandatory for the most specific of affinity interactions, although such qualities will influence efficiency of adsorption and subsequent eluate volumes. It is interesting to note that novel fixed-bed contactors characterized by annular designs,[36] or the use of membranes as solid phases,[37] have recently been claimed to equal or better the performance of conventional fixed beds without penalty of flow resistance, fouling, or peak trailing.

B. FAST-FLOW MATERIALS

A further design trend in adsorbent manufacture has been the establishment of rigid

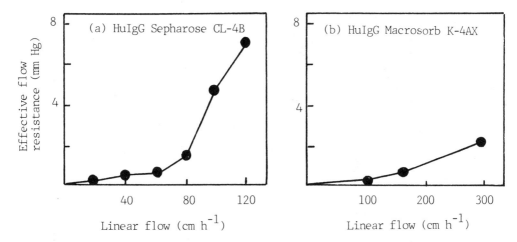

FIGURE 4. Comparison of the resistance to flow of homogeneous and composite agarose adsorbents arranged in fixed beds. Fixed beds (1.5 × 5 cm) containing (a) huIgG Sepharose CL-4B or (b) huIgG Macrosorb K-4AX were fitted with mercury manometers or pressure transducers via the head space. Each bed was challenged with varied flow velocities in the range 1 to 300 cm/h. At each velocity, steady-state pressure readings were taken and expressed in millimeters of mercury. Agarose beds exhibited significant compression at the highest velocities while composites remained unaffected.

materials with minimal flow resistance suited to operation in deep fixed beds at high linear velocities.[20,34] These reduce requirements for expensive investment in stacked modular contactors individually limited to depths of 15 to 25 cm for elastic materials (agaroses, celluloses, etc.) in larger-scale applications.[8,38] However, opportunities for high volumetric throughput should be viewed with caution in the context of immunoaffinity adsorption at both laboratory and production scales. As noted earlier, immunoaffinity processes require intermediate strengths of binding to effect specific association at the expense of impurities, but not to resist dissociation under mild conditions selected to promote limited molecular perturbation. Operating conditions for immunoaffinity adsorption have been widely discussed,[6,20,39] and the consensus of practical and theoretical opinion indicates that near-quantitative adsorption and desorption is most efficiently achieved at low flow rates (5 to 50 cm/h) where approximations to local equilibrium of product-ligand interactions are expected.[19] This is essential to efficient and cost-effective productivity from expensive immunoaffinity ligands in both laboratory and production plant. High flow rates may be beneficial in reducing washing, cleaning, and reequilibration times and but will only enhance operations with low-cost buffers.[20,40] It is conceivable that an immunoaffinity adsorbent operated to subtract impurities (e.g., albumin contaminants of animal cell products) may benefit from high-velocity operation, but only if the dissociation constant of the interaction is low and the solid phase shows little resistance to rapid molecular diffusion (small particle, large pore size). It is significant that these characteristics do not feature in the advertising literature for such materials.[41]

C. INFLUENCE OF LC SOLID-PHASE DESIGN

The two adsorbents considered herein represent a homogeneous agarose product (Sepharose CL-4B), chemically cross-linked to reduce natural elasticity, and a larger kieselguhr-agarose composite (Macrosorb K-4AX). The latter comprises a macroporous kieselguhr particle (140 to 300 or 1000 μm) filled with 4% cross-linked agarose which serves to sacrifice effective surface area for characteristics of minimal compressibility.[15,20,40] Figure 4 indicates that huIgG agarose suffers resistance to flow at high velocities (>60 cm/h) where beds are visibly compressed, while the composite is largely unaffected. Any benefits to affinity adsorption will be marginal, particularly as Figure 5 demonstrates that the composite material suffers greater diffusion resistance than homogeneous agarose at flow rates above 20 cm/h

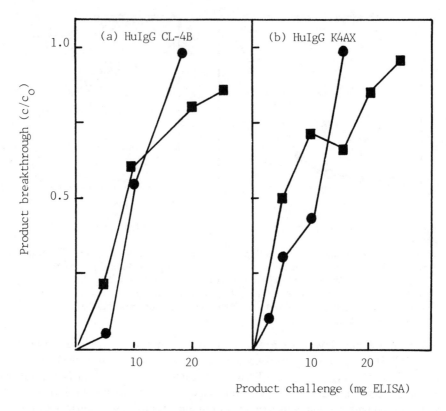

FIGURE 5. Comparison of the fixed-bed performance of homogeneous and composite immunoaffinity adsorbents. Fixed beds (1.5 × 5 cm) containing (a) huIgG Sepharose CL-4B (40 to 160
μm) or (b) Macrosorb K-4AX (150 to 1000 μm) were challenged with cell-free hybridoma media
rich in serum proteins (3 to 8 mg/ml) and anti-huIgG MAb product (0.13 mg/ml by ELISA = c_o).
The changing ratio of MAb concentration in the bed effluent (c) relative to the feedstock concentration
(c_o) represents the efficiency of adsorption and was recorded in terms of increasing product challenge.
Each material was tested at flow velocities of 23 cm/h (●—●) and 93 cm/h (■—■), equivalent
in these beds to 13 and 53 ml/h, respectively, to illustrate the greater diffusion resistance of larger
composite particles.

during adsorptive stages of operational cycles. Effective adsorption is reduced at high flow
rates in large particles. In this example the particle size range for agarose is 40 to 140 μm,
while that for composite is considerably broader (140 to 1000 μm). Narrower range composites (140 to 300 μm), which have recently become available, are less affected by such
diffusion resistance.[42]

D.   CHOICE OF ACTIVATION CHEMISTRY

The performance of huIgG Sepharose and HuIgG Macrosorb in batch and column
experiments indicates that the nature of the solid phase or methods of chemical activation
do not greatly influence the maximal capacity or effective dissociation constants estimated
for the various assemblies (see Table 3). This is a general conclusion which assumes possible
10% and tenfold errors for qm and Kd inherent in the fine analytical methods, but is drawn
from a wide study of activation chemistries, solid phases, and immunoaffinity systems.[22,29]

However, cyanogen bromide activation can be safely undertaken in methods that enable
degrees of reaction to be monitored and controlled.[23,24,26] Excessive activation of matrices,
seen with many commercial products requiring extended shelf life, may promote rapid
reaction at protein coupling which yields high concentrations of ligand located at particle
surfaces.[16] Preliminary experiments suggest that 50% maximum CNBr activation of agarose

TABLE 3
Adsorption and Desorption Characteristics of LC and
HPLC Adsorbents Exploiting huIgG as the Immunoaffinity
Ligand

Solid phase	[L]	[qm]	[Kd (M)]	[P]
1. Sepharose CL-4B	4.5	7.8	3.6×10^{-7}	1.0
Sepharose CL-4B	11.5	13.2	6.1×10^{-7}	1.0
Macrosorb K-4AX	1.4	3.5	8.1×10^{-7}	1.4
Macrosorb K-4AX	3.0	6.4	4.3×10^{-7}	1.2
Macrosorb K-4AX	6.0	8.7	3.5×10^{-7}	0.5
2. Sepharose CL-4B	2.9	3.1[a]	3.1×10^{-7}	1.0
Macrosorb K-4AX	2.3	2.2[a]	3.5×10^{-7}	1.0
3. Lichrosorb Li-60	6.4[b]	1.1[c]	3.0×10^{-7}	1.7
Lichrosorb Li-4000	20.6[b]	11.2[c]	6.0×10^{-8}	0.64
Polymer HMPS (10 µm)	5.5[b]	1.5[c]	5.0×10^{-8}	0.7

Note: LC solid phases were activated by cyanogen bromide (1) or tresyl chloride (2), and HPLC solids by periodate oxidation methods (3) (see Section II.A). Ligand concentrations [L] were estimated by difference analysis at coupling and expressed in mg/ml settled LC adsorbent or[b] mg/g dry HPLC solids. Maximum capacities (qm), expressed as mg/ml settled LC adsorbent or[c] mg/g dry HPLC solid were estimated by batch equilibrium or[a] fixed-bed saturation studies. Dissociation constants (Kd) were estimated from batch equilibrium experiments. Productivities (P) are expressed as anti-huIgG MAb recovered per unit immobilized ligand determined as mg ELISA activity (postgel permeation) per mg huIgG.

in homogeneous (CL-4B) or composite assemblies (K-4AX) yield uniform distributions of ligand at high and low loadings (1 to 12 mg/ml agarose).[16,22] By comparison with CNBr, all other chemical methods are more expensive and may require organic solvents.[34] In terms of ease of manipulation and readiness of quantitation, CDI activation will increasingly rival CNBr in popularity.[28,29,43] In contrast to other work,[44] there were no indications that any of the systems or activation methods (CNBr, CDI, periodate, tresyl chloride) were uniquely characterized by unacceptable levels of ligand leakage.[45] Although the ELISA methods employed here are insufficiently sensitive to absolutely rule it out, circumstantial evidence from SDS-PAGE analysis and the maintenance of ligand productivity in controlled repetitive testing supports this view. Care and attention during chemical derivation, assembly, and operation of materials suggests itself as the generic antidote against ligand leakage.[45,46]

E. BIOCHEMICAL PERFORMANCE

The quality of purification is not greatly influenced by choice of solid phase or activation chemistries. Both huIgG CL-4B and K-4AX yield pure anti-huIgG MAb in a single step from crude feedstock containing less than 1% total protein as product.[22] Common impurities (Table 2) include albumin, transferrin, and low molecular weight additives remaining from media used in serum-based hybridoma culture. Also present are undefined intracellular components released at death and cell breakage during upstream processing. Similar purification performance has been noted for MAb and protein A adsorbents applied to the purification of huIgG from blood plasma.[29] Proteases, present in animal cell cultures and blood plasma, or cell culture additives such as Pluronic or antifoams do not copurify, or interfere with the performance of these systems.[22]

As with all similar materials, there are physical limits to the maximum ligand concentrations which can be usefully applied to the productive recovery of macromolecules. In the system exploiting interactions between specific MAb and one or two molecules of huIgG antigen, immobilized IgG concentrations greater than 15 to 20 mg/ml settled Sepharose are readily attainable but are not reflected in productivity figures (i.e., MAb recovered per unit immobilized ligand). The presence of 50% solid kieselguhr in K-4AX composites means that the attainable maximum immobilization per unit adsorbent volume is limited to half that value by the availability of reactive agarose. However, for both materials, the molecular dimensions of ligand and products, and the physical limitations of the "pore" dimensions of 4% agarose, account for experimental observations that efficient applications (defined in terms of effective repetitive productivity) are limited to ligand concentrations in the range 3 to 6 mg/ml settled adsorbent.[22] It is important to recognize that as ligand concentrations increase, initial adsorption and subsequent desorption recoveries decrease. Optimal immobilized ligand concentrations appropriate to high repetitive productivity from defined solid-phase geometries will vary with the sizes of products, ligands, and expected stoichiometries of binding. The design limitations of protein-A adsorbents (4×10^4 Da binding a maximum three or four IgG molecules) will therefore differ from those for anti-huIgG MAb (1.4×10^5 Da binding maximum two molecules of huIgG).

It should be emphasized that requirements to double volumes of K-4AX fixed beds to achieve capacities comparable with homogeneous agaroses incur penalties of only hardware design and cost. The rigidity of composites withstands operation in deep fixed beds.[15,40] In addition, it appears that composite K-4AX better tolerates physical shock, resulting from the use of strong chaotropic agents (e.g., 3 M KSCN) as frontal eluants, than homogeneous CL-4B agarose. Fixed beds of the latter may exhibit near total resistance to fluid flow in 3 M KSCN, and physical deterioration of particle structure in repetitive usage not seen with K-4AX composites.[16] Such phenomena, possibly related to chaotropic interference with agarose-water interactions which are offset by kieselguhr structures in composites, can compromise the repetitive operation of packed bed depths greater than 10 cm.[15,22]

F. REPETITIVE OPERATION

A general conclusion from study of a number of solid phases in various purification systems is that disciplined technique, applied at assembly, operation, and storage of materials, ultimately governs the likelihood of extended productive longevity for immunoaffinity adsorbents. By comparison, the selection of solid phases and chemical routes to activation and derivation seems less important. The use of pure active ligand undoubtedly minimizes nonspecific interaction, and this represents a challenge to the view that the purity specification of affinity ligands need not be as strict as diagnostic or therapeutic products.[22,47] Purified ligands also enable accurate estimates of degrees of immobilization. The use of ethanolamine as blocking agent, following coupling to CNBr-activated (and other) matrices, may introduce charged groups through reaction of the primary amine, and glycine capping (yielding a balanced, local charge) or natural hydrolysis may be better strategies.[48] Careful study of albumin and lysozyme binding to ethanolamine-capped materials in PBS indicates negligible interaction[22,29] which is confirmed in silver-stained SDS-PAGE of MAb and PAb affinity purified from cell culture or blood plasma. The ionic strength of PBS buffers may be critical here, but can readily be duplicated in many immunoaffinity systems where pH and ionic strength are not sharply optimal. It should be emphasized that steps taken to minimize nonspecific interaction in fabricated adsorbents, such as attention to solid phase selection, ligand purification, and post-immobilization capping, may frequently be partially negated at ligand coupling. Immobilized protein inevitably possesses the characteristics of a weak, mixed-charge ion exchanger, the inherent electrostatic interactions of which are best countered by manipulation of operating pH and ionic strength of process feedstocks.

Studies of repetitive operation of CL-4B and K-4AX materials emphasize the need for good operational protocols.[45,46] The influences of physical forces, proteases in feedstocks, and bacterial contamination require control if solid phases, immobilized ligands, and products are to maintain biochemical and structural integrity. Proteases in blood and animal cell culture fluids mostly have acidic pH optima.[21,22] They may be restricted by added inhibitors or careful control of protein content and pH. Serum-based media contain sufficient carrier proteins to counter product and ligand attack. Problems may arise with low-protein media, or with adsorbents carrying protein-A as ligand requiring acid conditions for product recovery.[11,13] Post-load washing must be sufficiently prolonged to displace all proteases before the pH is reduced since acidic proteases common in animal cell cultures were proved active against IgG, protein A, and albumin below pH 4.0.[22] Bacterial contamination may be minimized by the presence of bacteriostat (thimerosal or azide) or the use of sterile solutions and affinity contactors protected by sterile filters (0.2 μm). Such precautions can yield uniform performance over 50 to 70 cycles, while inattention may promote serious diminution or collapse.[45,46,49] Clearly, a careful balance is required between overzealous protection and practical simplicity, since all such measures increase system and operational complexity.

IV. INTEGRATED PRODUCTION AND RECOVERY OF MONOCLONAL ANTIBODIES

The trend toward increased flexibility and scaleup of animal cell culture has stimulated the design and assembly of many types of sophisticated devices for mass culture and easy product recovery.[50] The study herein of controlled assembly and operation of immunoadsorbents required a regular, defined supply of anti-huIgG MAb for construction of huIgG purification systems. This was readily attained in magnetically stirred, suspension cultures (0.1 to 5 l) of TB/C3 hybridomas maintained in defined environments of dissolved CO_2, dO_2 and controlled temperature. Batches were clarified of cells by centrifugation or microfiltration (Millipore Minitan; 0.45-μm surface-modified PVDF membranes) after 5 to 7 d growth and production, and MAb purified in a single step by immunoaffinity purification as above (see Section III). Labor-intensive batch culture could be avoided, and product quality better standardized, by adaption of simple batch-culture vessels to run continuously under conditions prescribed for productive, sterile growth.[21] Such a system (Figure 1), established merely as a pragmatic service for local requirements, generated 20 to 30 ml/h effluent containing spent media components (principally albumin and transferrin), MAb product (30 to 50 mg/l), and levels of acidic protease enhanced by cell death and lysis associated with the on-line gravity settler.[21,22] Insertion of an anti-huIgG immunoaffinity adsorbent downstream of the settler enabled continuous on-line recovery of MAb product. Fixed beds could be sized with regard to maximum capacity (qm) and expected binding performance at the representative flowrates and product concentrations. Working lifetimes of 1 to 7 d could be predicted and achieved without significant loss of product in terms of either biological activity or total protein. Manual interchange of loaded and fresh beds, with subsequent desorption of product by 3 *M* KSCN, generated a regular supply of antibody over extended periods (3 to 60 d; see Figure 6). The quality of MAb product, proved by SDS-PAGE, ELISA, and negative protease assays, the productivity (20 to 75 mg/d), and the low running costs recommend this simple device (or scaled alternatives) as a solution to repetitive requirements for standardized, pure animal cell products in research laboratory and production plant.

The use of the gravity cell settler is a simple and inexpensive alternative to solid-liquid separation by mechanical devices such as crossflow microfiltration, minicyclones, or centrifuges.[51] However, passive clarification only approaches completion with long residence times which may compromise product stability and integrity.[45] In normal operation (residence times of 2 to 4 h), some particulates wash over and hazard the fixed-bed adsorbent. Beds of homogeneous agaroses (huIgG-CL-4B; 1 to 2 $\times$ 5 to 10 cm) serve as depth filters and may block within 1 to 3 h at particular cell numbers and flow rates. In contrast, composite

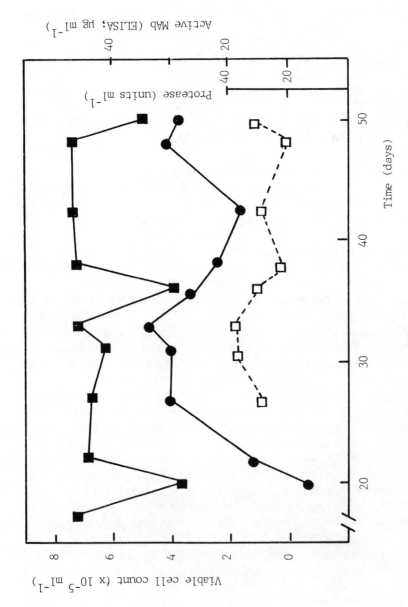

FIGURE 6.   Continuous culture of murine hybridomas. Continuous culture in apparatus depicted in Figure 1 was run for 50 d and concentrations of viable cells (■——■), anti-huIgG MAb (●——●), and acidic protease (□---□) determined in samples taken from the culture vessel exploiting methods described in Section II. Protease concentrations are expressed in units of pespin equivalents. Decrease in cell viability on days 17 to 20 and 34 to 38 reflect temporary problems with gas supply, but illustrate the robustness of systems. Monoclonal antibody was recovered on-line as described in the text.

materials (huIgG-K-4AX; 140 to 1000 μm), readily tolerate equivalent solids loading during the complete adsorptive lifetime of integrated fixed beds (2 to 5 d), and can be freed of entrained particles by backwashing with PBS prior to chaotropic elution. The suspension tolerance of composite fixed beds invites a broader application in productive biotechnology.[20,35] There is additional scope for the automation of such simple cell culture devices. Systematic affinity HPLC bioquantitation (see Section VII) of MAb products, albumin wastes, and protease by-products could be harnessed to control vessel dilution and switches from saturated beds to fresh immunoadsorbents (with concomitant elution and reequilibration) in continuous processes.

V. LIQUID FLUIDIZED-BED ADSORPTION

Many feedstocks destined for immunoaffinity adsorption are initially contaminated with biological particles (e.g., fermentation broths, homogenized cell preparations). Solid-liquid separation, whether conducted by centrifugation or microfiltration in laboratory or production plant,[52,53] imposes time penalties and promotes associated biological deterioration of products and requires investment in specialized equipment. Suspensions may be contacted with immunoaffinity adsorbents in slurried contactors, where particles are subsequently displaced by decantation washing, or in a single or multistage liquid fluidized bed where continuous washing can be applied as part of the operational cycle[33,54,55] (see Figure 2). Solid phases suited to these operations must have significant densities such that particle settling rates are counteracted by upward fluid velocities. In general, homogeneous particles of agarose, cellulose, or acrylamide are unsuited because of low density, but kieselguhr-agarose composites have proved especially amenable to fluidization. When appropriately derivatized with ion exchange or immunoaffinity ligands (protein A, anti-huIgG MAb, or huIgG PAb) as for fixed-bed applications, they have proved suited to the recovery of specific products from biological suspensions such as hybridoma cultures or whole blood preparations without detriment to final product yield or quality.[33,56] In the laboratory, single-stage fluidized beds (10 to 500 ml expanded volume) are best operated in recirculating mode until equilibrium binding is attained as judged by appropriate UV spectrophotometry. Larger-scale operations can exploit single pass, multistage fluidized contactors which attain the residence times necessary for efficient adsorption at high volumetric throughput (200 to 700 cm/min). Preliminary studies have indicated that the output from continuous hybridoma culture (see above) could be fed and bled through a fluidized-bed adsorption loop, thereby removing requirements for gravity or mechanical clarifiers (see Figure 7). As with the integrated fixed-bed contactors, the key to the operation is the physical and biochemical nature of the composite solid phases used for the controlled implementation of immunoaffinity adsorption.

VI. PREPARATIVE IMMUNOAFFINITY HPLC

Recent developments in HPLC solid phases has seen increased availability of "wide-pore" particles (5 to 10 μm; >30-nm pore diameter) suited to the purification of macro-molecules exploiting bioaffinity and ion-exchange interaction in addition to the conventional gel permeation and reversed phase processes.[34,57,58] Small HPLC particles reduce observed resistances to molecular diffusion, but require expensive investment in precision pumping and control equipment. In addition, such materials have a true macroporous structure which contrasts with the polymer network of agaroses. Hence, the molecular dimensions of coupling chemistries, immobilized ligands, and associated products will critically effect overall performance. As with LC systems (see Section III), many commercial solid phases, ready activated and/or derivatized with useful ligands such as protein A, G, or specific antibodies, are commercially available as loose material or prepacked in columns and cartridges. They

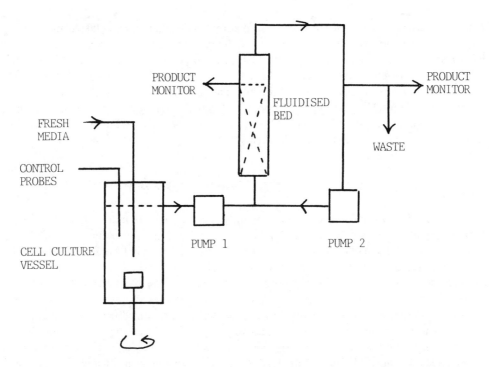

FIGURE 7.   Integration of a recirculating fluidized-bed contactor with continuous culture of murine hybridomas. Whole broth effluent from the continuous cell culture apparatus described in Figure 1 is fed at 15 to 30 ml/h into (and bled out of) a recirculating loop flowing at 15 to 30 l/h. The latter is used to fluidize huIgG composite particles and enable adsorption of specific MAb directly from the whole broth.

are expensive and details of predicted productivity, active ligand concentration, and handling protocols are, as with LC materials, rarely discussed in the technical literature. Most common solid phases are fabricated from silica which can be readily capped with hydrophilic agents such as glycidoxypropylsilane to yield a surface readily activated by conventional chemical routes.[27,29] Complete capping is essential for custom-assembled silicas if nonspecific interaction is to be avoided. Patches of raw silica in commercial materials represent a common source of poor performance, but can be readily evaluated by monitoring albumin and lysozyme binding.[29] In contrast, the recently introduced hydrophilic polymer composites (e.g., HMPS) offer materials with the distinct advantage of susceptibility to direct activation in chemical protocols essentially identical to those applied to LC materials.[29,59]

In preparative tests, diol-silicas, and HMPS polymers derivatized with protein-A, anti-huIgG MAb, and huIgG PAb (see Table 3) achieved uniformly excellent purification of their interacting partners as products from blood plasma (huIgG) and animal cell culture fluid (MAb). Controlled ligand coupling and ethanolamine blocking, as identified for LC materials (see Section III), were important here. The specific activity of products matched those of homogeneous preparations from other methods of purification. However, selection of particle diameter and pore size is far more critical than for LC materials. Surface area decreases as molecular accessibility increases with pore size. In preparative purifications of anti-huIgG MAb from serum-based hybridoma culture,[29,59] intermediate pore sizes (30 or 100 nm) yielded the best productive capacities. In contrast, 6-nm pore material was characterized by limited penetration of IgG ligands and products, and 400-nm material lacked sufficient surface area for effective production.

Immunoaffinity HPLC, whether with customized matrices or commercial material, offers a rapid means of purification and related study of immunochemical interactions. However, material costs are high and the dispensing, monitoring, and control hardware is expensive with integrated systems exceeding $30,000. Such systems are ideal in clinical situations where multiple discrete purifications are required from a range of small, valuable samples. Larger-scale HPLC is expanding, particularly with reversed-phase fractionation of small molecules and certain affinity purifications.[57] However, it is likely that increased application of affinity HPLC will require significant market expansion generated by wider application of conventional affinity LC systems.

VII. IMMUNOAFFINITY HPLC IN BIOQUANTITATION

One immediate application of affinity HPLC materials is in miniaturized preparative systems characterized by quantitative adsorption/desorption performance.[29,59] Such systems, when challenged with a complex mixture containing the product as analyte, will yield on elution a single species which may be quantified by UV spectrophotometry or other detection system. Thus, immobilized protein-A could be used to quantify certain classes of IgG in blood samples in time scales a fraction of appropriate ELISA or other biological assays. In addition, the use of an immobilized antigen to evaluate specific antibodies has the key advantage of quantifying biologically active analytes in the manner of conventional ELISA determinations. Such applications have clear advantages of minimal sample preparation, high specificity, and absolute resolution not seen with analyses based upon reversed-phase or ion-exchange interactions,[51,60] which are compromised by difficulties of peak and baseline selection.

Preliminary work has demonstrated that huIgG silicas or polymers (e.g., HMPS) characterized by ligand concentrations of 2 to 20 mg/g can be successfully applied to the bioquantitation of specific anti-huIgG MAb produced in TB/C3 hybridoma suspension cultures[21,29,59] (see Figure 8). As with preparative matrices, care must be taken to assemble materials without susceptibilities to nonspecific binding using solid phases distinquished by minimal diffusion resistance. A study of commercially available materials indicates that solid polymeric composites (HMPS; 8 μm) or derivatized diol silicas (10 μm; 6 or 400 nm) perform best. It is thought that in the IgG system, the 6-nm pore material behaves as a solid particle since IgG has been variously estimated to have a largest dimension of 5 to 9 nm.[45] The wide pores of 40-nm material benefit a system where unrestrained molecular diffusion, rather than high capacity, is at a premium. Analytical time scales of 10 to 15 min represent a fraction of those for conventional ELISA in this system, but are extended because of the requirement to fractionate desorbed materials from UV-dense eluant. A system based upon protein-A using glycine-HCL as eluant requires no such gel permeation, and analysis can be achieved in 5 min or less.

Commercial materials have not been used in these tests although there are indications that protein-A silicas may have applications in IgG quantitation.[29] However, experience with commercial preparative materials suggests that custom-built materials with prescribed maximum capacities and predictable working lifetime will best serve the analytical applications. It is essential to size columns and choose operational flow rates in the context of expected sample volumes and analyte concentrations. Elution volumes must be chosen to achieve near-complete desorption, and equilibration times to maximize working lifetime. In practice, correct choice of materials and conditions yields a linearity of response from nanogram to milligram of analyte, the lower sensitivity being limited by applied detection.[29]

Miniaturized preparative immunoaffinity HPLC systems have potential for quantifying analytes in bioprocesses (cell culture, purification, waste disposal) and ultimately gathering information that can control the processes as preset limits are met. Thus, it is easy to envisage such devices monitoring breakthrough of anti-huIgG MAb from fixed-bed recovery integrated downstream of continuous hybridoma cultures (see Section IV). Detectable analyte and product levels would trigger insertion of fresh adsorbent onstream and the subsequent de-

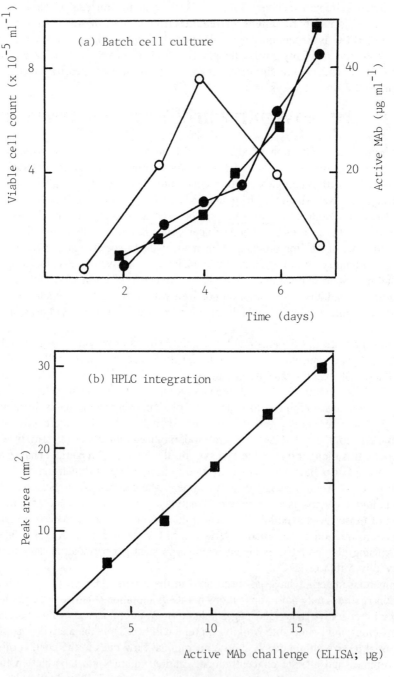

FIGURE 8.   Bioquantitation of anti-huIgG by immunoaffinity HPLC. (a) Batch cultures of murine hybridomas producing anti-huIgG MAb were grown in serum-based media over 7 d. Viable cell counts (O—O) were determined in sterile samples by histological staining and microscopic hemocytometry. Concentrations of monoclonal antibody were determined by conventional ELISA techniques (■—■) and miniaturized immunoaffinity HPLC (see Figure 3) (●—●). (b) Dose response curve of successive challenges of anti-huIgG MAb to a miniaturized immunoaffinity HPLC exploiting immobilized huIgG on silica particles. The area of UV peaks eluting in the void volume of gel permeation columns downstream of the HPLC mini-contactor (see Figure 5) were estimated and related to the ELISA activity of standard samples of affinity-purified MAb (■—■).

sorptive processing of the saturated bed. MAb levels (or other analytes) in the cell culture vessel could be monitored by appropriate specific HPLC analysis of sidestreams taken from the permeate of a tangential flow sampler.[51] Such devices offer the nearest approximation yet possible to an on-line biosensor capable of repetitive molecular quantitation without compromise to bioprocesses.

VIII. CONCLUDING DISCUSSION

A wide range of commercial products are available to the laboratory practitioner for purposes of assembling working purification systems exploiting immunoaffinity interactions. Such materials range from activated solid phases which can be derivatized with a ligand of choice, through immunoaffinity fabricates characterized by common ligand types such as protein-A, to integrated assemblies of media, pumping hardware, and control computation. All offer real prospects for the adequate accomplishment of a range of purification tasks, but troubleshooting malfunctions can be a frustratingly empirical business with the invocation of magic never far away. Much of the difficulty arises from the dearth of real practical data available concerning commercial materials. This precludes rational fine-tuning of systems to cope with individual purification problems. It is recommended that the controlled assembly and operation of immunoadsorbents specifically designed for particular purifications are the best guarantee of success. To this end, many types of LC and HPLC solid phases can be purchased in raw states, and chemical activations can readily be undertaken on laboratory scales in a safe and controlled manner.

Prescribed degrees of material activation will influence the eventual distribution of immobilized ligands and subsequent adsorption and desorption performance of matrices. Global ligand concentrations will show a broad optimum of productivity performance dependent upon the molecular size of ligands and products, the stoichiometry of interaction, and the pore dimensions of solid phases. In general, the selection of pore size is more critical in HPLC materials (6 to 400 nm) than in polymeric LC materials (molecular exclusion 2 to 20×10^6 Da). Ligands should be purified to high specific activity and molecular homogeneity before immobilization since impurities will increase nonspecific interactions. Care should be taken to cap all unreacted groups, and minimal nonspecific association validated in tests with albumin and lysozyme. Operational conditions should be chosen as a compromise between specific interaction and minimal nonspecific binding. Eluants are best chosen for their elution efficiency, and minimal effects upon products and immobilized ligands.

Materials assembled for a purification should be validated in small-scale representative tests before undertaking preparative scale purifications with valuable materials. Batch determinations of maximal capacity and dissociation will indicate likely operating conditions in terms of product concentration in feedstocks and sizing of adsorbent contactors. Where possible, materials should be run through a number of small-scale, repetitive tests to ensure that problems of ligand inactivation, incomplete desorption, or solid phase biofouling are not likely to compromise preparative applications. Where possible, fixed-bed matrices should be protected with bacterial filters, sterile operating solutions, and sanitary additives at shutdown. Where repetitive usage of materials is envisaged, a detailed operational log recording operating conditions, yields, and physical performance will pay dividends in the controlled design and fabrication of additional materials.

High-performance matrices can be as readily activated and derivatized as LC materials, although special attention is required to guarantee quantitative recovery of materials in decantation or filtration washing procedures with small-diameter particles. Whether the necessary investment in HPLC pumping and control hardware to handle these materials is practically worthwhile is an open question which will depend on specific circumstance. Many laboratories are conventionally equipped with analytical HPLC equipment which can be adapted for most preparative applications. One consequence of the use of any high-performance material, or refined chromatographic matrix, is the sample preparation required for the protection of materials against fouling and nonspecific interaction. It is ironic that many of the fractionation and clarification procedures recommended for upstream preparation of samples for immunoaffinity or conventional HPLC purification contribute to diminished exploitation of the full fractionation power of the technique and increased cumulative loss of product. This must cast doubt over the role of productive immunoaffinity HPLC, but miniaturized preparative systems offer exciting possibilities for near on-line quantitative monitoring and control of bioprocesses.

A recent review of affinity chromatography[62] concluded that common laboratory and production plant applications at a late stage of a purification sequence did not fully exploit the potential of this selective technique. As with the HPLC sample preparation referred to above, late usage reflects both upon the anxieties of operators concerned to expose valuable adsorbent materials to dirty feedstock, and the dearth of commercial materials and adsorbent contactors suited to such early use. In that respect, the suitability of kieselguhr-agarose composites for operation in fixed beds or fluidized beds without impairment of selective adsorption from suspensions is encouraging. It indicates possibilities for immunoaffinity adsorption from whole fermentation broths, cell homogenates, and other biological suspensions, and invites solid phase manufacturers to consider the production of materials specifically designed for operation in such contactors. Such will enable effective exploitation of immunoaffinity adsorption to the full limit of its inherent purification power.

IX. ABBREVIATIONS

c	Product concentration
c_o	Initial product concentration
CL-4B	4% Cross-linked agarose
CL-6B	6% Cross-linked agarose
CDI	Carbonyldiimidazole
CNBr	Cyanogen bromide
DEAE	Diethyl-amino-ethane
ELISA	Enzyme-linked immunosorbentassay
FCS	Fetal calf serum
GPC	Gel permeation chromatography
HMPS	Hydrophilic macroporous polymer support
HPLC	High performance liquid chromatography
HuIgG	Human immunoglobulin G
IgG-Fc	Constant region of IgG
K-4AX	Kieselguhr-4% agarose composite
Kd	Dissociation constant
KSCN	Potassium thiocyanate
LC	Liquid chromatography
MAb	Monoclonal antibody
MWCO	Molecular weight cutoff
PAb	Polyclonal antibody
PAGE	Polyacrylamide gel electrophoresis
PBS	Phosphate-buffered saline
QA	Quanternary amine
qm	Maximum capacity
SDS	Sodium dodecylsulfate
UV	Ultraviolet

REFERENCES

1. **Dean, P. D. G., Johnson, W. S., and Middle, F. A.,** *Affinity Chromatography,* IRL Press, Oxford, 1985.
2. **Schneider, C., Newman, R. A., Sutherland, D. R., Asser, U., and Greaves, M. F.,** A one-step purification of membrane proteins using a high efficiency immunomatrix, *J. Biol. Chem.,* 257, 10766, 1982.
3. **Jack, G. W. and Wade, H. E.,** Immunoaffinity chromatography of clinical products, *Trends Biotechnol.,* 5, 91, 1987.
4. **Secher, D. and Burke, D. C.,** A monoclonal antibody for larger-scale purification of human leukocyte interferon, *Nature (London),* 285, 446, 1980.
5. **Dunnill, P.,** Trends in downstream processing of proteins and enzymes, *Process Biochem.,* October 9, 1983.
6. **Chase, H. A.,** Affinity separations utilising monoclonal antibodies: a new tool for the biochemical engineer, *Chem. Eng. Sci.,* 39, 1099, 1984.
7. **Hill, E. A. and Hirtenstein, M. D.,** Affinity chromatography: its application to industrial scale processes, in *Advances in Biotechnological Processes,* Vol 1, Alan R. Liss, New York, 1983, 32.
8. **Moks, T., Abrahmsen, L., Osterlof, B., Josephson, S., Ostling, M., Enfors, S.-F., Persson, I., Nilsson, B., and Uhlen, M.,** Large-scale affinity purification of human insulin-like growth factor from culture medium of *Escherichia coli, Bio/Technology,* 5, 379, 1987.
9. **Sada, E., Katoh, S., Sukai, K., Tohma, M., and Kondo, A.,** Adsorption equilibrium in immuno-affinity chromatography with polyclonal and monoclonal antibodies, *Biotechnol. Bioeng.,* 28, 1497, 1986.
10. **Morris, C. J. O. and Morris, P.,** in *Separation Methods in Biochemistry,* 2nd ed., Pitman, London, 1986, chap 6.
11. **Goding, J. W.,** in *Monoclonal Antibodies: Principles and Practice,* 2nd ed., Academic Press, London, 1986, chap 6.
12. **Vetterlein, D. and Calton, G.,** The selection of monoclonal antibodies suitable for immunoadsorption, in *Affinity Chromatography and Biological Recognition,* Chaiken, I. R., Wilchek, M., and Parikh, I., Eds., Academic Press, London, 1983, 393.
13. **Goding, J. W.,** Use of Staphylococcal protein A as an immunological reagent, *J. Immunol. Methods,* 20, 241, 1978.
14. Data Sheet: Pharmacia LKB, *Protein G and Protein G Sepharose Fast Flow,* PGDS 50-012-396, Uppsala, Sweden, 1988.
15. **Desai, M. A., Huddleston, J. G., Lyddiatt, A., Rudge, J., and Stevens, A. B.,** Biochemical and physical characterisation of a composite solid phase developed for large-scale biochemical adsorption, in *Separations for Biotechnology,* Verall, M. S. and Hudson, M. J., Eds., Ellis-Horwood, Chichester, U.K., 1987, chap 14.
16. **Rudge, J. and Lyddiatt, A.,** unpublished data, 1988.
17. Data sheet: Bio-Rad MAPS buffers, BioRad Laboratories, Richmond, California, 1987.
18. **Lowe, C. R. and Dean, P. D. G.,** *Affinity Chromatography,* John Wiley & Sons, London, 1974, chap 2.
19. **Chase, H. A.,** Prediction of the performance of preparative affinity chromatography, *J. Chromatogr.,* 297, 179, 1984.
20. **Lyddiatt, A.,** Solid phases and product contactors: new options for bioselective adsorption, in *World Biotech Report,* Online Publications, London, 1988, 176.
21. **Rudge, J., Desai, M. A., Shojaosadaty, S. A., and Lyddiatt, A.,** Continuous culture of murine hybridomas with integrated product recovery, in *Modern Approaches to Animal Cell Technology,* Spier, R. E. and Griffiths, J. B., Eds., Butterworth, London, 1987, 556.
22. **Desai, M. A.,** Immunoaffinity Adsorption: Applications in the Recovery of Monoclonal Antibodies from Mammalian Cell Culture, Ph.D. thesis, University of Birmingham, England, 1988.
23. **March, S. C., Parikh, I., and Cuatrecasus, P.,** A simplified method for cyanogen bromide activation of agarose for affinity chromatography, *Anal. Biochem.,* 60, 149, 1974.
24. **Kohn, K. and Wilchek, M.,** A new approach (cyano-transfer) for cyanogen bromide activation of Sepharose at neutral pH which yields activated resins, free of interfering nitrogen derivatives, *Biochem. Biophys. Res. Commun.,* 107, 878, 1982.
25. **Nilsson, K. and Mosbach, K.,** Immobilisation of enzymes and affinity ligands to various hydroxyl supports using highly reactive sulphonyl chlorides, *Biochem. Biophys. Res. Commun.,* 102, 449, 1981.
26. **Kohn, J. and Wilchek, K.,** A colorimetric method for monitoring activation of Sepharose by cyanogen bromide, *Biochem. Biophys. Res. Commun.,* 84, 7, 1978.
27. **Ohlson, S., Glad, M., and Larsson, P.-O.,** Recent advances in high performance liquid chromatography (HPLC), in *Affinity Chromatography and Biological Recognition,* Chaiken, I. M., Wilchek, M., and Parikh, I., Eds., Academic Press, London, 1983, 241.

28. **Bethell, G. S., Ayers, J. S., Hearn, M. T. W., and Hancock, W. S.,** Investigation of the activation of various insoluble polysaccharides with 1,1'-carbonyldiimidazole and of the properties of activated matrices, *J. Chromatogr.,* 219, 361, 1981.

29. **Shojaosadaty, S. A.,** Development of High Performance Liquid Affinity Chromatography (HPLAC) Methodologies for Biochemical Recovery and Quantitation, Ph.D. thesis, University of Birmingham, England, 1988.

30. **Laemmli, U. K.,** Cleavage of structural proteins during the assembly of the head of bacteriophage T4, *Nature (London),* 227, 680, 1970.

31. **Bradford, M. M.,** A rapid and sensitive method for the quantitation of microgram quantities of protein utilising the principle of protein-dye binding, *Anal. Biochem.,* 72, 248, 1976.

32. **Kassel, B. and Meitner, P. A.,** Bovine pepsinogen and pepsin, in *Methods in Enzymology,* Vol 19, Perlmann, G. E. and Lorand, L., Eds., Academic Press, New York, 1972, 337.

33. **Wells, C. M., Lyddiatt, A., and Patel, K.,** Liquid fluidised bed adsorption in biochemical recovery from biological suspensions, in *Separations for Biotechnology,* Verrall, M. S. and Hudson, M. J., Eds., Ellis-Horwood, Chichester, U.K., 1987, chap. 16.

34. **Clonis, Y. D.,** Large-scale affinity chromatography, *Bio/Technology,* 5, 1290, 1987.

35. **Lyddiatt, A.,** New materials for selective protein recovery, in *Conf. Proc. Analyticon 87 (Use of enzymes in biotechnology),* Olympia, London, U.K., October 1987.

36. Tech. Bull. No. 101, Superflo columns, Sepragen, San Leandro, California, 1987.

37. Tech. Bull., Memsep molecular separations, Domnick Hunter, Birtley, U.K., 1988.

38. **Katoh, S.,** Scaling-up affinity chromatography, *Trends Biotechnol.,* 5, 328, 1987.

39. **Ostlund, C.,** Large-scale purification of monoclonal antibodies, *Trends Biotechnol.,* 4, 288, 1986.

40. **Bite, M. G., Berezenko, S., and Reed, F. J. S.,** Macrosorb kieselguhr-agarose composite adsorbents: new tools for downstream process and scale-up, in *Separations for Biotechnology,* Verrall, M. S. and Hudson, M. J., Eds., Ellis-Horwood, Chichester, U.K., 1987, chap 13.

41. Tech. Bull. Reselute-BSA, Celltech, Slough, U.K., 1987.

42. **Huddleston, J. G. and Lyddiatt, A.,** unpublished data, 1988.

43. **Shojaosadaty, S. A., Malhotra, J. M., and Lyddiatt, A.,** unpublished data, 1988.

44. **Peng, L., Calton, G. J., and Burnett, J. W.,** Stability of antibody attachment in immunosorbent chromatography, *Enzyme Microb. Technol.,* 8, 681, 1986.

45. **Rudge, J., Desai, M. A., and Lyddiatt, A.,** unpublished data, 1988.

46. **Lyddiatt, A., Desai, M. A., Huddleston, J. G., Rudge, J., and Shojaosadaty, S. A.,** Controlled assembly and operation of immunoadsorbents, *J. Chem. Tech. Biotech.,* in press.

47. **Fowell, S. L. and Chase, H. A.,** Variation of immunosorbent performance with the amount of immobilised antibody, *J. Biotechnol.,* 4, 1, 1986.

48. **Kohn, J. and Wilchek, M.,** Mechanism of activation of Sepharose and Sephadex by cyanogen bromide, *Enzyme Microb. Technol.,* 4, 161, 1982.

49. **Bazin, H. and Malache, J.-M.,** Rat (and mouse) monoclonal antibodies. V. A simple automated technique of antigen purification by immunoaffinity chromatography, *J. Immunol. Methods,* 88, 19, 1986.

50. **Handa-Corrigan, A.,** Large-scale in vitro hybridoma culture: current status, *Bio/Technology,* 6, 784, 1988.

51. **Kalyanpur, M. N., Skea, W., and Siwak, M.,** Isolation of cephalosporin C from fermentation broths using membrane systems and high performance liquid chromatography, *Dev. Ind. Microbiol.,* 26, 455, 1985.

52. **Axelsson, H. A. C.,** Centrifugation, in *Comprehensive Biotechnology,* Vol. 2, Moo-Young, M., Ed., Pergamon Press, Oxford, 1985, chap 21.

53. **Gabler, F. R.,** Cell processing using tangential flow filtration, in *Comprehensive Biotechnology,* Vol. 2, Moo-Young, M., Ed., Pergamon Press, Oxford, 1985, chap 23.

54. **Wesselingh, J. A. and van der Wiel, J. P.,** Adsorption of proteins: process design, in *Conf. Proc. 4th European Congr. on Biotechnology,* Vol. 2, Elsevier, Amsterdam, 1987, 546.

55. **Burns, M. A. and Graves, D. J.,** Continuous affinity chromatography using a magnetically stabilised fluidised bed, *Biotechnol. Prog.,* 1, 95, 1985.

56. **Wells, C. M., Lyddiatt, A.,** unpublished data, 1988.

57. **Clonis, Y., Jones, K., and Lowe, C. R.,** Process-scale high performance liquid affinity chromatography, *J. Chromatogr.,* 363, 31, 1986.

58. **Regnier, F. R., Pfannkoch, E., and Lin, N. T.,** in *Conf. Proc. Biotech '84 U.S.A.,* Online publications, London, 1984, 573.

59. **Shojaosadaty, S. A. and Lyddiatt, A.,** Application of affinity HPLC to recovery and monitoring operations in biotechnology, *Separations for Biotechnology,* Verrall, M. S. and Hudson, M. J., Eds., Ellis-Horwood, Chichester, U. K., 1987, chap 35.

60. **Gustafsson, J.-G., Frei, A.-K., and Hedman, P.,** Monitoring of protein product formation during fermentation with fast protein liquid chromatography, *Biotechnol. Bioeng.,* 28, 16, 1986.

61. **Low, D. K. R.,** The use of the FPLC System in method development and process monitoring for industrial chromatography, *J. Chem. Tech. Biotechnol.,* 36, 345, 1986.
62. **Bonnerjea, J., Oh, S., Hoare, M., and Dunnill, P.,** Protein purification: the right step at the right time, *Bio/Technology,* 4, 954, 1986.

Chapter 12

PRODUCTION AND CHARACTERIZATION OF SOLID-PHASE IMMUNOADSORBENTS

Duncan S. Pepper

TABLE OF CONTENTS

I. BACKGROUND

Affinity chromatography achieves its peak of performance specificity in the form of immunoaffinity (particularly with monoclonal antibodies), and the crucial step is the preparation of the solid-phased antibody (or antigen) in a useful form. Here will be summarized a series of practical approaches to the design, production, characterization, and use of such material. Because of major differences in technique, however, cell affinity chromatography will be excluded, although some of the points covered will be common to both particulate and soluble target species. The special application of depletion is also covered. All of the techniques discussed have been evaluated by the author or colleagues and arrived at over years of trial and error as appropriate techniques to be used by immunologists for a purpose rather than as an end in themselves or as a tour de force in synthetic chemistry. In short, these techniques are proved to work. However, exclusion of others does not imply criticism; rather, there has not been the opportunity to evaluate everything in parallel.

II. INTRODUCTION

It is convenient to consider the problems of producing and characterizing solid phase immunoadsorbents under the two broad headings of system components (Section III) and system performance (Section IV), that is, the individual parts and the whole.

The system components cover such things as the actual solid-phase supports (''solid'' here includes gels), the chemistry used for activation coupling and blocking, the ligand (binding molecule) itself, and the physical enclosure.

The system performance covers such things as content of ligand per unit volume of support, the specific capacity of the ligand for the target molecule (specific activity per milliliter or per milligram of ligand), the elution and regeneration solutes required, the control and measurement of leakage of the solid phased components, and the sterilization and cleaning of the solid phase.

Some of the general problems that have to be overcome in the production and characterization of solid-phased immunoadsorbents are summarized in Table 1.

III. SYSTEM COMPONENTS

A. THE SOLID PHASE

The traditional supports used for immobilizing antigens and antibodies have been the soft porous gels originally developed for gel filtration such as Sephadex (dextran), Biogel P (acrylamide), and Sepharose (agarose). However, as the need for greater pore size (mol wt exclusion limits) and physical strength (resistance to compression) became apparent, a second generation of hybrid polymers, cross-linked supports, and ceramic (inorganic polymers) were developed such as Ultrogel (acrylamide agarose), Sephacryl (acrylamide-dextran), and Fractogel, Trisacryl, etc. (cross-linked hydroxyethylmethacrylates) and porous glass/silica. More recently, hybrid inorganic supports have appeared where the strength and rigidity of ceramics are married to the biocompatibility and ease of derivatization of organic polymers, e.g., glycol-silica or magnetic iron oxide with a polystyrene shell. Some commercially available supports are given in Table 2. For a more extensive review of published work see Reference 1.

The general requirements of the ideal porous solid phase are summarized in Table 3 from the point of view of using spherical particles. For reasons of efficient packing and flow in columns the particles should be spherical and of similar size, and because biological macromolecules diffuse rather slowly, these particles should be as small as possible. However, there is a low limit to feasible bead size dictated by the pressure limit of the pump,

TABLE 1
Problems to Be Solved

Immobilize ligand on particle and retain activity
Bind product with high capacity and minimal (nonspecific binding)
Elute product with maximum yield, minimum volume, and leakage
Regenerate adsorbent with minimal loss, volume, and time
Operate system sterile and pyrogen free
Maximize repetitive use cycles

TABLE 2
Chemistry of Solid Phases

	Supplier
Sepharose, Biogel Ultrogel A - agarose ± cross-linking	Various*
Ultrogel AcA — acrylamide-agarose composite	IBF
Trisacryl — tris hydroxyethyl cross-linked acrylamide	IBF
Fractogel — pentaerythritol cross-linked glycidylmethacrylate	Merck/Pierce
Sephacryl — bisacrylamide cross-linked allyl dextran	Pharmacia
Cellufine - cellulose	Amicon
Spheron — poly-HEMA	Tessek
Composites - inorganic skeleton + organic polymer coating	Various*
Inorganics — silica, zirconia, titania, hydroxyapatite, carbon	Various*
Magnetic — iron & nickel oxides	Various*

Note: * Addresses of individual suppliers and branded products are given in the Appendix.

TABLE 3
The "Ideal" Solid Phase

1. Narrow size distribution, smallest possible diameter
2. Macroporous/microporous, maximum available surface area
3. Low cost; disposable for biomedical applications or large scale
4. Nontoxic, low leachables
5. Rigid (no deformation) and strong (no fragmentation)
6. Easily derivatized (surface chemistry)
7. Low nonspecific adsorption
8. Dense (packed and fluid bed operation)
9. Facile sterilization and sanitization

column, etc. In the case of ordinary liquid chromatography (e.g., ≤10 psi) this translates to a lower limit of about 40 μm. However, to achieve this the mean particle size of the distribution will need to be about 60 μm and an upper bound of 80 μm; thus, a lower limit of 40 μm means 40 to 80 μm; similarly, a lower limit of 75 μm means 75 to 150 μm. Most manufacturers of porous beads supply them in several particle size ranges, described as fine, medium, and course. Particles larger than 150 μm should be avoided as efficiency will be low, whereas particles below 40 μm will require higher operating pressures, i.e., high-performance liquid chromatography (HPLC) systems.

The internal pore size (and its corresponding internal surface area) is a very important parameter (see later) because it controls not only the absolute capacity for activation and coupling of ligand (i.e., the available internal surface area), but also the efficiency or freedom of access to the immobilized ligand by the target molecule (i.e., the mean pore size within the beads controls access by diffusion). These two parameters are intimately (and inversely) related, but only in the case of rigid ceramic and inorganic beads are accurate data available. Because "soft" organic gels are destroyed by drying, the conventional methods of pore

TABLE 4
Membrane Chemical Composition

Cellulose esters
Nylon (polyamide)
Polysulfone
Polypropylene
Metal sinter (SS)
Ceramic (ZrO_2, Al_2O_3)
Polycarbonate
Acrylic
PTFE
PVDF Polyvinylidinedifluoride

diameter and area measurement are not applicable; instead, indirect inference can be made from the ability of the gel beads to resolve molecules of different sizes when operated in the "gel filtration" mode. Such data are supplied by the manufacturers in the form of "mol wt exclusion limits" which vary according to the shape of the molecules used (compact proteins give higher mol wt exclusion limits than extended polysaccharides, DNA, etc.).

Unfortunately, such mol wt data cannot easily be related to absolute pore diameter as measured, e.g., in nanometers for a silica particle; instead, one is forced to rely on empirical comparisons of performance in which different particles are used for the same application. One point to bear in mind is that molecules are in constant vibratory motion in solution, and their ability to "fit" or otherwise move freely within a pore is not a simple yes or no situation. Rather, as the pore size increases, the statistical likelihood that any proportion of its volume could be occupied by the diffusing molecule increases. Thus, to achieve 95% efficiency of access for an average protein molecule, the pore size needs to be 10 to 20 times greater than the target molecule. In contrast, a pore size of 1 to 2 times the size of the target molecule will allow an efficiency of access of only 5%.[2] Thus, the optimum pore size for affinity chromatography will be *about ten times* greater than that for gel filtation. (Since pore size is a linear dimension and mol wt is effectively cubic, this means mol wt exclusion limit of some 10^3 times greater.)

More recently, recognizing that porous gels are still far from ideal for immunoaffinity supports, several manufacturers have developed porous membrane based preactivated and coupled devices (e.g., Cuno "Zetaffinity" and Domnick Hunter Memsep 1000 and Memsep 2000 devices) which combine high flow rates, low diffusion restriction, and are packed into shallow bed and wide format for optimum productivity. Many different chemical compositions of membrane are commercially available (Table 4). However, in immunoaffinity applications cellulose and polyvinylidinedifluoride (PVDF) have received the most attention (both Pall and Millipore manufacture and supply blank activated membranes in sheet form). Other novel materials (see Table 5) have also been used in addition to membranes to provide "monolithic" supports. They are relatively less well developed than porous beads but have a number of potential advantages (listed in Table 5), particularly when considering large-scale laboratory or industrial applications. Since membranes are also available in hollow fiber cartridge format, several manufacturers are also exploring this type of device for immunoaffinity production scale operations.[3]

In summary, the ideal solid phase does not yet exist, but a sensible choice can be made depending on the priorities involved, e.g., scale, cost, performance, and reuse. In the author's laboratory Sepharose 4B is used for small-scale work and Sephacryl S-1000 for large-scale work.

Although silica is cheap and rigid, its application outside of HPLC with biological macromolecules has been restricted because of problems with small pore size, high nonspecific binding, and difficulty in chemical activation. A series of porous glasses with large

TABLE 5
Novel "Monolithic" Supports

Wool felt slab
Cellulose + graft acrylic polymer, e.g., CUNO "Zetaprep"
Nylon fiber
Cotton wool, rayon (viscose) fiber
Regenerated cellulose sponge (Spontex)
Regenerated cellulose mixed ester membranes (MEMSEP)
   No packing problems or voids, air-tolerant
   Easy fabrication in shallow/wide beds or novel shapes, e.g., cylindrical
   Easy scaleup, predictable capacity/pressure/flow rate
   Nonporous filamentous supports obviate slow diffusion of macromolecules
   Nonsolvated gel allows anhydrous organic solvent changes

TABLE 6
Noncovalent and Hybrid Adsorption Chemistry

Noncovalent adsorption to	±	Cross-linking	±	Sacrificial layer
Porous glass		Glutaraldehyde		Proteins (IgG, BSA, fibrinogen)
Silica		Bis-epoxides		Poly (PHE-LYS)
Polystyrene		Tris-aziridines		DEAE-Dextran
Nitrocellulose				Polyethyleneimine
PVDF				
Charge modified nylon				

pore sizes and chemically derivatized to produce glycol and diol surface groups is available from Pierce (See Appendix) and these can be used as the basis for periodate or CNBr activation. Alternatively, at least one company provides a custom immobilization service ("Prosep" Protein Separations Ltd.) using proprietary activation chemistry on porous glass.

For batch adsorption and elution or batch adsorption and column elution a high-density solid phase is most desirable. This is where glass, ceramic, or magnetic oxide porous supports are most useful.

B. THE COUPLING CHEMISTRY, ASSAY OF ACTIVE GROUPS, AND BLOCKING

A wide choice of coupling chemistries is possible including noncovalent adsorption (Table 6) as well as covalent bonding (Table 7), and hybrids of both approaches are also known (Table 6).

Generally, covalent bonding (Table 7) is more reliable and less "leaky". However, it requires favorable reactivity of both the solid phase and the ligand, which is not always feasible. Covalent bonding also demands some skill in organic synthesis (unless commercial preactivated gels are bought) and tends to cost more. In contrast, noncovalent adsorption (Table 6) is much easier and cheaper to operate. However, it is more "leaky" and works best only with the more strongly adsorbing solid supports (e.g., silica, glass, ceramics, metal oxides) and is not feasible with hydroxylic polymeric gels. The leakiness of noncovalent adsorption can be reduced to some extent by using *in situ* cross-linking of the adsorbed layer (protein or polymer) with glutaraldehyde or similar divalent cross-linkers. A disadvantage is that some inactivation and loss of nonadsorbed material is inevitable, so a low-cost ligand (antigen or polyclonal antibody) is preferred. Another possibility is to coat the adsorbing particle with a sacrificial layer of protein or polymer (which is irrelevant or bland in the process operated), cross-link this *in situ*, and use the residual free cross-link residues (e.g., aldehyde groups if glutaraldehyde was used) to couple a second layer of active protein

TABLE 7
Choices for Activation Chemistry of Hydroxylic Supports[a]

Aqueous solvent:	Aldehyde — reductive amination	
	Oxidizers	Reducers
	Periodate	Ascorbic acid
	Bromine	Borohydride
	Acid hydrolysis	Cyanoborohydride
	Cyanogen bromide	
	Divinylsulfone	
	bis-Epoxides, epichlorhydrin	
	N-hydroxysuccinimide activated carboxyl	
Mixed solvent	Benzoquinone	
	Cyanogen bromide — triethylamine	
	Trimethylsilane derivatives[26]	
Organic (anhydrous) solvent	Carbonyldiimidazole	
	2-Fluoro-1-methyl pyridinium toluene-4-sulfonate (FMP)	
	Tresyl chloride, tosyl chloride	
	Triazine chloride	

[a] Detailed recipes for individual methods can be found in the handbooks provided on affinity chromatography by Pharmacia and IBF (see Appendix).

(antigen or antibody). An advantage of this approach is that it effectively blocks the non-specific binding and produces large spacer groups; however, a disadvantage is that capacity for binding of the second layer is reduced compared to the first layer capacity.

With covalent bonding chemistry (Table 7) a wide choice of reagents is possible, and most of the more reliable methods are also available from commercial suppliers (see Appendix); however, the choice of combinations of chemistry of activation and bead type and porosity are inevitably restricted. Nevertheless, commercial suppliers represent the best first choice for immunologists who work on a small scale or who do not wish to undertake chemical synthesis.

Broadly, chemical activation methods can be divided into three categories (Table 7), namely, aqueous solvent, mixed solvents, and anhydrous organic solvents. These represent increasingly aggressive reaction conditions but also increasing toxicity and cost. Unfortunately, the reactivity of hydroxylic gels is not great in water, and so the choice is often limited to the second or third categories. Of those reagents which work in totally aqueous solvents, aldehyde reductive amination, cyanogen bromide, and divinyl sulfone (DVS) are the most useful.

Reductive amination requires the generation of aldehyde groups in the gel, and these can be of very variable amount and reactivity. This type of gel (aldehyde) is not widely available commercially, and reaction conditions are difficult to predict. Also, different gels require different methods of producing aldehyde groups. Sepharose (agarose) gels can be oxidized with bromine[4] or acid hydrolyzed,[5] while *cross-linked* Sepharose gels can be oxidized with periodate.[6] Wright and Hunter[7] have shown that Sephacryl gels are easily oxidized with periodate to generate aldehyde groups. In the author's experience the latter method is most convenient and reliable, and it has subsequently been optimized[8] to the point where 1 to 100 mg of antibody can be coupled to 1 ml of gel with coupling efficiencies of 95 to 60%.

The cyanogen bromide method is still one of the most popular and reliable coupling methods for immunoaffinity work, probably because it is commercially available and has more user experience than any other method. Its disadvantages include toxicity and leakiness in the alkaline pH operating range. A recent improvement[9] in the method of synthesis now makes activation a simple, reliable, and painless method, albeit, by working in mixed aqueous-acetone solvent, and the coupled gels still show slight leakage in the alkaline region. This improved method is discussed at length in a separate article in this volume (Chapter 9).

Another activation chemistry which works well in totally aqueous solvents is the divinylsulfone reagent originally used by Porath and co-workers as a cross-linking reagent[10] and more recently for coupling proteins to porous gels.[11] Initial concern about leakage in the alkaline region (with low mol wt ligands) was subsequently found not to be a problem, and proteins are easily and efficiently coupled.[11,12] Divinylsulfone (DVS) -agarose gel is commercially available (see Appendix) and is stable in aqueous suspension for long periods. Although the reagent is toxic and more expensive than cyanogen bromide, its enhanced reactivity and ease of use make it an attractive alternative which deserves more attention.

As already mentioned elsewhere CNBr activation is more efficient in mixed-solvent systems. Likewise, benzoquinone can be used to activate agarose or other hydroxylic polymer gels in an ethanol-water mixture. The reagent is cheap and relatively nontoxic, but it produces colored gels with significant nonspecific binding of protein. The method has not received much attention,[13] but a recent comparison of different activation methods with a monoclonal antibody to FVIII-von Willebrand factor (vWF) showed that benzoquinone performed well,[14] so this method also is deserving of further exploitation.

Totally (anhydrous) organic solvent systems offer the most reactive systems for derivation of hydroxylic gels. However, they also present problems in solvent drying, solvent exchange, cost, flammability, and toxicity. Fortunately, several of the more useful activation chemistries are commercially available on selected gel particles.

Originally, triazine chloride was the most popular reagent of this group, for it is cheap and used in bulk for textile dying. However, it is difficult to control side reactions, and some reactive groups may remain after blocking. More recently, the "cleaner" reagent carbonyldiimidazole has been optimized[15,16] and is commercially available as "Reactigel" from Pierce or Merck (see Appendix) on a choice of three different porous supports (Sepharose 6B, Trisacryl GF-2000, and Fractogel TSK65). This eliminates the need for in-house synthesis and makes trial evaluation of this type of chemistry very convenient. However, the gels are expensive for bulk use. The carbonyldiimidazole (CDI) type of chemistry is very efficient, reactive, produces low leakage, and does not impart a charged group to the linkage.

Tresyl chloride,[17] tosyl chloride, and 2-fluoro-1-methylpyridinium toluene-4-sulfonate (FMP) (Table 7) all represent a similar chemical approach in that a labile leaving group is temporarily introduced into the gel and is subsequently displaced by a nucleophilic group (e.g., $-NH_2$ or $-SH$) on the ligand, thus, covalently bonding the gel and ligand without any intervening groups. Both tresyl-activated and FMP-activated gels are commercially available (see Appendix) and are similar to CDI in general operational use, cost, and advantages. As a group their main advantage is that they do not exhibit the alkaline leakage of CNBr-coupled proteins.

When producing "in-house" activated gels, it is very desirable to check that newly activated or stored material is adequately functional prior to coupling expensive antibodies or rare antigens. Simple qualitative color tests will often suffice; however, quantitative assays are also feasible in many cases. The author finds the Konig reagent method of Kohn and Wilchek[18] ideal for checking CNBr activation. The details are given elsewhere in this volume (Chapter 9).

For aldehyde (e.g., periodate or glutaraldehyde activation) the commercially available Schiff's stain is quite simple to use and gives a rapid and strong visible pink stain in the activated gel particles.

Vinyl sulfone can be estimated by the thiosulfate-acid titration method of Porath et al.[10] except that the reaction time should be extended from 2 to 20 h.

TABLE 8
Monoclonal Antibody Panel Characterization

Growth rate of hybridoma, in culture and ascites
Cell line survives freezing, thawing, and recloning
Immunoglobulin class and subclass
Proteolytic susceptibility
Isoelectric point
Protein-A pH gradient behavior
Iodination damage?
Solid phasing damage?
Epitope mapping: chemical, immunological, bioassay

The reactive gels such as CDI-, FMP-, Tresyl, etc. are best quantitated by elemental analysis for N or S; however, such techniques are rarely available outside chemical laboratories, and the author has found a simple indirect approach which gives qualitative results with *any* activated gel. Mix a small volume (about 0.1 ml) of active gel with 1 M hydrazine at pH 10 for 15 min at 37°C and wash well in acetic acid (1% v/v) three times. Stain the gel with 1 ml of 1% trinitrobenzene sulfonic acid freshly prepared in 0.05 M Na$_2$B$_4$O$_7$, buffer pH 9.2. An immediate red-pink-brown color appears which should *remain* in the gel after washing to neutrality. Acid washing will produce a yellow-pink slightly paler color.

C. THE LIGAND: ANTIGEN OR ANTIBODY

Although immunoaffinity methods are powerful, they are not best suited to be the initial operational step in a process starting with crude material or large volumes. Problems can arise with fouling, proteolysis, and slow flow rate and processing time. Thus, prior to immunoaffinity some preliminary treatment such as precipitation or ion-exchange is more suitable. Consider also a second upstream step in which a group-specific reagent selects a class of similar molecules, e.g., protein A for immunoglobulins (see Chapter 10). This reduces the chances of contamination by nonspecific binding of irrelevant antigens. A wide variety of such general affinity reagents are available from Pharmacia, Pierce, etc. (see Appendix).

Generally, however, immunologists will wish to prepare their own immobilized antigen, protein, or antibody. The coupling of carbohydrates as antigens is a rather specialized topic and will not be further discussed here except to mention that reversed periodate-reductive amination, vinyl sulfone, or epoxides have all been successfully used.

Most commonly, nowadays, the material to be coupled will be a monoclonal antibody, less commonly a polyclonal antibody, or "home-made" antigen of local interest. General principles apply to all these, and it will be assumed the case applies to monoclonal antibodies; where major differences apply these will be highlighted.

A necessary first step is to select the appropriate monoclonal antibody for the job required; an antibody suitable for enzyme-linked immunosorbent assay (ELISA) or histology is not necessarily suitable for immunopurification or depletion. Some points to consider are summarized in Table 8. It is much easier to select these desirable features if the fusion is planned or under the control of the end user. Preexisting or commercial antibodies may have been screened by inappropriate methods or be more suitable for nonpurification applications. In planning the fusion, it is desirable to start with as pure an antigen as possible, to use a long interval between immune boosting, and to design a primary screening assay (ELISA or radioimmunoassay [RIA]) which specifically detects the desired antibody and can handle several hundred samples of hybridoma supernatants per 24 h. The author prefers RIA screening because of its precision, sensitivity, wide dynamic range, and low nonspecific binding. However, this does presuppose the availability of pure ($\geq 50\%$) antigen. The specific positive clones should be discarded if they show slower growth rates or IgM class; neither of these is likely to be successful in immunopurification if faster growing and/or IgG secreting clones are available. Within the IgG subclasses, some will be more susceptible to proteolysis (particularly murine IgG 2b) either when produced in low-serum tissue culture medium which can expose the antibody to proteases secreted by the hybridoma or in use when coupled where crude feedstocks may contain proteolytic contaminants. The resulting cleaved antibody fragments can both contaminate the product and reduce column capacity and lifetime. Thus, where a choice is possible, avoid murine IgG 2b subclasses and concentrate on IgG 1 and 2a. Fortunately, these are the commonest ones, except with bacterial antigens where IgG3 appears.

It sometimes happens that when cloning a large number of positive hybridomas, one can get two clones which are identical, and, therefore, both do not need to be carried through. Unfortunately, it is not easy to predict from an immunopurification point of view at such an early stage whether two antibodies recognize the same epitope and are identical. Isoelectric focusing of purified antibody is a simple and rapid way of deciding if two antibodies are identical. The four or five closely spaced bands of each antibody can be checked visually, and if an identical pattern is seen with two clones, it is likely that they are identical and only one need be retained.

At this point in the fusion, several hundred positive clones will have been screened and reduced to perhaps 20 to 30 promising candidates. At this point some form of secondary screening is highly desirable to reduce the number to 10 antibodies in a useful panel. Ideally, such secondary screening should represent closely the intended use, i.e., immunopurification. It is rather difficult to arrange this easily in every case, but some form of ELISA or RIA is usually appropriate. If pure antigen is available, this can be used to coat ELISA plate wells and each exposed well used to capture putative antibodies. High titer and/or high-affinity binding antibodies can then be detected by enzyme conjugated second antibodies (e.g., rabbit anti-mouse Ig etc.). Duplicate plates are then set up with all strongly binding antibodies and titered. The exercise is repeated a third time but without adding second antibody; instead, the plates are washed with putative elution solvents, e.g., 0.1 M glycine HCl pH 3.0 (a restricted choice of typical eluates is given later in Section IV.C of this chapter), and after washing to remove eluted antibody, the residual bound (noneluted) antibody is quantitated by a second antibody enzyme conjugate. The "ideal" antibodies are those which show strong color in the primary binding screen but low color in the eluted screen. Antibodies which continue to show color after the elution screen will either be difficult to elute or need a further evaluation of eluates. Unlike polyclonal antibodies, monoclonal antibodies can show both high affinity and easy elutability, sometimes referred to as "switching."

Another way to screen a limited number of antibodies ($\leqslant$10) is to couple 1 mg of each to an activated gel in 1 ml aliquots, e.g., in centrifuge tubes of about 10-ml capacity. The antibody need not be pure, but variations in growth and rate and hybridoma antibody concentration will influence the absolute amount of binding. Each 1-ml aliquot of gel and antibody can be subdivided into, e.g., 100-μl aliquots and each antibody exposed to, e.g., 10^2 to 10^3 Becquerel of ^{125}I-tracer antigen. After suitable mixing times the bound tracer can be separated, the gel washed, and the tubes counted directly in a gamma counter. With good antibody-tracer combinations the percentage of tracer bound should vary between 50 and 90%. This assay tests affinity more than capacity as the tracer is used at low concentration. The same samples of gel plus antibody showing good tracer binding can then be evaluated in "elutability" screening with suitable solvents, e.g., acid, alkali, chaotropes, etc. (a suitable list is given in Table 13). Desirable antibodies are those which show good tracer binding before elution and low binding after elution.

Both the ELISA and RIA test formats can be used (see Section IV.C) to screen in more detail for suitable eluants and combinations of eluants.

When a panel of $\leqslant$10 antibodies is chosen, the antibodies should be further characterized by chromatography on protein A (see this volume, Chapter 10) to establish or confirm subclass, purity, etc. and as a convenient source of pure antibody. Although not strictly necessary, it is useful to study antibody behavior after iodination and/or solid phase coupling to ELISA plates. Both these manipulations can inactivate certain antibodies. Unless particularly valuable, such "unstable" antibodies should be avoided.

Epitope mapping can be approached in a number of ways. From the point of view of immunopurification, however, it is mainly of interest to know if two (or more) antibodies bind to the same, similar, close, or distant sites. This information is particularly useful when designing depletion systems (see Section V this chapter). The simplest approach is to set up a series of coated ELISA well rows with purified antigen and in the solution phase compete soluble enzyme conjugated antibody (antibody A) with a second unlabeled antibody (antibody B etc.). A checkerboard pattern can be set up in which every possible combination of labeled and unlabeled antibody pairs is examined. The diagonal (identical pairs) should show maximum inhibition of binding (low color) while symmetric pairs should give identical results (if enzyme conjugation does not damage one of the pair). Any other antibody pairs giving low binding are either to identical or close epitopes, and any giving no inhibition are presumably on unrelated epitopes. It frequently happens that certain epitopes are stronger antigens, and from a panel of ten antibodies perhaps only four or five really independent epitope groups will be evident.

Where polyclonal antibodies are to be coupled, it is desirable but not essential to use an immunoglobulin fraction. A limited number of activated gels will couple efficiently at up to 50 mg of protein per milliliter so are feasible to use with raw serum (e.g., periodate-Sephacryl or low temperature CNBr method); however, the functional capacity for antigen binding per milliliter of gel will be low, and the chances of leaking of bound animal serum proteins are greatly increased. One advantage of this approach is that inactivation and losses are minimal, useful when only small amounts of rare antisera are available. Generally, an IgG fraction can be coupled as such at ≤ 10 mg/ml to most activated gels (though again, efficiency of coupling and specific activity will be low). Such fractions are readily derived by Na_2SO_4 precipitation and/or protein A chromatography. Ideally, an immunopurified polyclonal antibody should be used as this minimizes the nonspecific leakage problem and enhances the specific activity per milliliter of gel. One possible disadvantage is that prior immunopurification of the antibody will remove the highest affinity fraction (because it is difficult to elute), and this will not be represented in the coupled-purified antibody. However, except in depletion work, this is unlikely to be a practical disadvantage. A serious problem with polyclonal antibodies compared to monoclonal antibodies is that it is difficult to regenerate fully the free antibody after each elution cycle, and so repetitive use cycles tend to diminish working life more quickly than with monoclonals. This is more of a problem with high mol wt (protein) antigens than with low mol wt antigens (drugs, etc.).

D. PHYSICAL CONTAINMENT

This is not a trivial problem. Traditional chromatography columns are long and thin whereas on theoretical grounds an affinity support should be wide and shallow. It is difficult to buy commercial columns which fit this specification and difficult to pack, operate, and maintain porous gel beads in such a format. The best that can be expected with most bead and column systems is a packed bed which is as high as it is wide, achieved, for instance, by the use of upper and lower column adaptors. More recently, annular columns have become available in a hollow cylinder format for scaleup of soft gels (see Appendix). An interesting recent development is the availability of shallow prepacked beds of preactivated affinity membranes (see Appendix) which conform to the ideal specification noted above. Hollow fiber devices[3] are also attractive in this respect. Housings should be transparent to allow visual assessment of fouling and nonlinear elution (channeling), and resistant to all elution and cleaning solvents. At the present time autoclavable column-packing combinations are not feasible, but some column assemblies can be autoclaved in their empty state. Batch operation or batch adsorption/column elution is also feasible where large volumes of dilute and/or particulate feedstocks are unavoidable. However, they will generally suffer from poorer capacity, loss of solid phase beads, and lower purity and concentration product. Both filter and centrifugal modes of operation are equally feasible with batch operation, but filtration is more easily scaled up.

IV. SYSTEM PERFORMANCE

A. CAPACITY FOR LIGAND

It can be surprisingly difficult to measure the amount of protein actually bound within a bead. The commonest method (also the most unreliable) is to measure the absorbance of a solution at 280 nm before and after coupling and assume the difference is due to coupled protein. Several things can go wrong. The optical density changes can be small and dilution (by gel fluid) can be large, making a correction difficult. IgG can aggregate during mixing at alkaline pH and give an apparent rise in absorbance due to light scattering. Some gels (especially Fractogel) can release UV-absorbing materials which increase the apparent absorbance after coupling. This method also cannot be used to measure the residual protein content after prolonged storage or use.

The author routinely incorporates 10^2 to 10^3 Becquerel of tracer IgG (it need not necessarily be the same antibody or even species) into coupling reactions to monitor the efficiency of coupling. Since actual gel samples can be counted directly in a gamma-counting well, nondestructive and repeated measurements of gel-protein content are feasible, and leakage can be sensitively computed over a period of several months. For situations where rapid relative ranking of capacity and leakage performance is needed in a large number of samples, then ^{125}I tracer is ideal, but it cannot readily be used to assign absolute levels of, e.g., leaked murine IgG into a potential product. In this case absolute levels down to ≤ 1 ng/ml of murine IgG in product can better be estimated by two-site sandwich ELISA (see under Section IV.D this chapter). Where isotopes are not acceptable, the author has found colorimetric assays such as the Pierce BCA protein assay to be very convenient (see Appendix). Simply mix 0.1 ml of gel phase with 2 ml of BCA reagent at 37°C for 30 min and, after centrifuging down the beads, read the absorbance of the purple chromogen at 562 nm. Standardization is simply achieved with a series of soluble protein standards (preferably the same protein as used in coupling) covering the range 10 to 100 μg.

B. CAPACITY FOR PRODUCT

We should distinguish here between the capacity to bind the primary liquid (antigen or antibody) and its subsequent operation as a functional binding ligand (which is not the same thing) which may or may not remain active after coupling. Three primary factors affect ligand capacity for product: the efficiency of chemical coupling, the steric accessibility of the "active site" on the coupled molecule, and the retention of such material after successive washing and elution cycles. First, consider the choice of solid phase porosity on coupling capacity, e.g., for IgG (Table 9). Where supports are available in a wide range of pore size, it is useful to see how the coupling of the same protein with the same activation level and chemistry will compare as a function of pore size. This is best determined empirically since increasing pore size ($\equiv$ accessibility) will be counterbalanced by decreasing internal surface area (available sites for coupling), and it is never clear where the optimum might be. Examination of the data in Table 9 show that coupling capacity for polyclonal human IgG with CNBr activation shows increasing capacity with increasing pore size until capacity plateaus and then falls slightly. The best figure of 13.8 mg IgG per milliliter of gel occurs

TABLE 9
Effect of Support Porosity on Capacity for IgG (Human IgG Saturated CNBr Support)

Support		mg IgG/ml	^{125}I-protein A bound %	Exclusion limit (mol wt × 10^{-6})
Fractogel	TSK 40	8.3	47.55	0.01
	TSK 55	11.9	65.18	1.00
	TSK 65	13.8	78.32	5.00
	TSK 75	12.5	84.65	50.00
Sephacryl	S-200	14.2	51.16	0.25
	S-300	13.5	63.91	1.50
	S-400	24.5	58.34	8.00
	S-500	16.7	72.21	50.00
	S-1000	16.3	66.08	150

within the Fractogel series with TSK 65 (MW exclusion limit of 5.0×10^6 while the best figure with Sephacryl gels (24.5 mg/ml) occurs in the S-400 grade (mol wt limit 8.0×10^6). Since both gels were activated to a large excess with CNBr, this must represent accessible surface for protein binding, and it is interesting that Sephacryl S-400 binds nearly twice as much protein IgG per milliliter of gel as Fractogel TSK 65. However, a different picture emerges if we look at the *functional* behavior of this IgG in terms of the ability of its F_c region to bind ^{125}I-protein A tracer (this is a useful general technique for assessing antibody "function" independent of antigen binding) which measures mainly affinity since the tracer is used at a low concentration (nanograms per milliliter). Here we see that the best figure for tracer binding (84.7%) occurs in the Fractogel TSK 75 gel (mol wt limit = 50.0×10^6), and this is better than the 72.7% binding which occurs with Sephacryl S-500. Of interest is the fact that in both cases functional binding is optimal in higher porosity grades of gels than total protein binding, and a better performance figure is obtained with Fractogel than Sephacryl which is the reverse of the total protein binding merit.

It is important, therefore, in choosing a system to empirically optimize *functional* capacity from a choice of gel types and porosity grades. Since the affinity of IgG for protein A is essentially constant in solution, the variation in binding must represent the influence of steric accessibility on affinity in the solid phase. Similar effects are likely with other more specific antigen-antibody reactions.

Also of note is the considerably greater mol wt exclusion limit in the empirically optimized gels above compared to the usual Sepharose 4B exclusion limit (mol wt 20×10^6), all the more so since neither IgG (150,000) or protein A (40,000) are anywhere near the exclusion limit for the former. It would seem that most workers are using gel supports of less than adequate pore size, and a general move to higher mol wt exclusion limit gels would benefit. Since we are also interested in specific antigen affinity, it is important to determine this as well. Table 10 summarizes the data of Hornsey et al.[8] in which different grades of Sephacryl were activated by periodate oxidation and coupled to ^{125}I-tracer labeled monoclonal antibody to von Willebrand Factor (vWF-Ag), and subsequent affinity was determined by ^{125}I-vWF-Ag tracer binding. Since antibody retention after coupling and chaotrope washing and cycles can be much lower than primary coupling, these data were also included. In contrast to the data in Table 9 where the gel was saturated with an excess ($\simeq$ 50 mg per milliliter of gel) of human IgG, in Table 10 the gels were coupled with only 1.5 mg of murine IgG per milliliter of activated gel. Here it can be seen that the *efficiency* of coupling improves dramatically with increasing porosity and plateaus in Sephacryl S-500 and S-1000 at 94% efficient uptake. Since antibodies are generally expensive, this improved performance relative to the less porous grades is of considerable utility. Following chaotrope

TABLE 10
Affinity of Immunosorbent for FVIII Using McAb-vWF on
Sephacryl Gels 1.5 mg Murine IgG Coupled by Reductive
Amination per ml of Gel[8]

Gel type	% Ab coupled	% Ab retained	% Tracer bound
Sephacryl S-1000	94	84	62
Sephacryl S-500	94	84	53
Sephacryl S-400	90	74	42
Sephacryl S-300	72	55	23
Sephacryl S-200	30	22	18

TABLE 11
Factors Found to Increase Ligand Activity

Increased pore size (up to 20 times molecular size)
Coupling to "pillar" rather than "spherical" structure within bead
Correct orientation (antibodies via carbohydrate or Fc to protein A)
Flexibility of ligand — correct steric presentation
Ligand density of coupling low - multiple valency interactions limited to 2—3 per molecule e.g., low coupling pH (6—8) and/or high protein content (2—10 mg/ml)
Small bead size — reduced surface skin effect
Nonporous "monolithic" supports
Reduced flow rate, e.g., 1—2 column volumes/h
Higher temperatures increase diffusion (37°C, 20°C better than 4°C)

washing and cycling (to mimic several use cycles) the degree of antibody retention still exceeded 80% relative to starting material in both S-500 and S-1000. When the affinity of the bound antibody was assessed by functional binding of ^{125}I-vWF-Ag, a significant benefit of larger pore size was again seen. However, in this case Sephacryl S-1000 was superior to S-500 (62% vs. 53% tracer binding), indicating that even with the same amount of antibody (1.26 mg/ml of gel) the more porous gel has a better functional performance. Since the mol wt of vWF-Ag varies between 1×10^6 and 10×10^6 whereas the exclusion limit of Sepharcryl S-1000 is 150×10^6, this again underlines the conclusion that gel support beads show best operational efficiency with several orders of magnitude higher porosity than the target molecule size.

In Table 11 are also listed a number of other variables which may influence the functional capacity of a coupled ligand. In addition to pore size ($\equiv$ mol wt exclusion limit) the actual physical structure of the gel may also influence performance. Linear stranded molecules ("Pillar" structures) such as agarose provide two dimensions for freedom of access by target molecules, whereas the newer synthetic polymers and ceramics are aggregates of microscopic spheres ("Berry" structures) which provide only a one-dimensional freedom of access. This may partly explain the good general capacity seen with Sepharose gels as supports.

Orientation of antibodies is important with large target antigens, and *a priori* we would expect random orientation of coupled antibodies so that perhaps only one half to one third of molecules could actually function effectively because of steric accessibility. To overcome this problem one can either immobilize antibodies by their carbohydrate (mainly present in the hinge region) using periodate activation of the antibody and reductive amination onto hydrazino gels or, more elegantly, immobilize the antibody via its F_c region to previously immobilized protein A. Permanent cross-linking *in situ* to protein A is feasible and is described in detail elsewhere (Chapter 10).

A second problem arises where small amounts of antibody or protein are coupled to heavily activated gels (e.g., 1 mg/1 ml of antibody is coupled to a gel with a potential

TABLE 12
Elution Strategy

List possible eluants (N ≤100)
Check product for survival in eluants
   Short list (N ≤10)
Check antibody for survival in eluants?
Check solid phase for survival in eluants?
Screen antibody/antigen complex for "elutability"; ideally, secondary McAb screen should be "elutability"
   Antibody immobilized on bead and tracer labeled partner or
   Antigen immobilized on plate and antibody measured by ELISA
Screen short list of eluants in minicolumn format

capacity of 100 mg/ml). Since antibody molecules contain 10 to 20 reactive available ε-amino groups, the antibody can be fixed in a rigid and inflexible state or even bent and distorted in a permanent stressed state. Neither of these situations is conducive to efficient binding and so a judicious choice has to made between level of activation and efficiency of coupling. If inactivation still occurs, reactivity can be reduced by lowering the coupling pH in the range 6.0 to 9.0 and/or increasing the amount of protein to be coupled either specific Ag/Ab or high mol wt irrelevant polymers (albumin, gelatin, IgG). As alluded to above, protein A is also a useful alternative in controlling the valency (effectively 2) with which IgG is immobilized, and this is particularly useful where the antigen binding site contains reactive amino groups. Also, because protein A is an extended linear molecule, it produces a degree of flexibility and accessibility that is absent when antibodies are coupled directly to a polymer support bead. Another problem arises where many antibodies (or other macromolecules) are coupled close together, effectively reducing the specific activity of each and, in extreme cases, even the absolute capacity. Operationally, this means that although functional capacity increases as antibody loading is increased from 1 to 10 mg/ml, it does not increase linearly, and at even higher levels, 40 to 50 mg/ml, the functional capacity may even begin to decrease. For maximum utilization of expensive antibody work in the range 1 to 2 mg/ml of coupled antibody/antigen. However, in depletion work it may be necessary to exceed this value up to, e.g., 5 to 10 mg/ml.

Finally, in dynamic flowing applications, access by target molecules to the inside of the bead is slow (diffusion of macromolecules may take several seconds per bead!), and so in practice only the surface layers of antigen and antibody are used. The core of the bead is effectively wasted. This situation can be alleviated by working with smaller beads (40 to 70 μm in the case of low pressure liquid chromatography or 5 to 30 μm in the case of HPLC). However, there is a tendency recently to move to nonporous beads (or membranes with large pores) which effectively means no diffusion limitation of capacity. Unfortunately, these "nonporous" structures have a reduced total surface area and, hence, reduced capacity, but this may be more than counterbalanced by the ability to complete absorption/washing/elution cycles in ≤ 10 minutes. Thus, productivity in milligram per device per day may be better if repetitive cycling is contemplated, e.g., the Memsep devices (see Appendix).

C. ELUTION SCHEMES

A general strategy for elution schemes (Table 12) may be outlined roughly as follows. During preliminary screening of, e.g., monoclonal or polyclonal antibodies some idea will be gained of typical "good" eluants for the various antibodies and antigens available, e.g., some antibodies elute best at low pH (e.g., 0.1 *M* glycine HCl, pH 3.0) whereas others elute best at high pH (e.g., 0.1 *M* triethylamine, pH 10); yet others elute best at neutrality with chaotropes such as urea, KI, KSCN, or organic solvents such as ethylene glycol. A short list of such eluants can usually be assembled (Table 13), and when promising antibodies are found, the list can be added to if necessary (more aggressive eluants) or modified to

TABLE 13
Possible Eluants to Consider

Miscellaneous	Extremes of pH, temperature, and salt concentration both high and low; electrophoresis/electrofocusing
Salts	KI, KBr, KSCN, $MgCl_2$, $MgBr_2$, $Ca(ClO_4)_2$, $Ca(SCN)_2$;
Glycols	50—100% v/v Ethylene glycol, propylene glycol, PEG400
Chaotropes	5—8 M urea, 3—5 M guanidine hydrochloride, 0.2 M lithium diiodosalicylate (LIS)
Amines	Ethanolamine, triethanolomine, triethylamine — 1M and pH 7—10 ethylenediamine, dimethylaminopropylamine, diaminohexane
Solvents	50% v/v dimethylsulfoxide, acetonitrile; 10% dioxane
Detergents	0.1—2% w/v sodium dodecylsulfate

stabilize or synergize activity. The "elutability" screens described in Section III.C based on ELISA or RIA are particularly convenient for screening up to 100 different combinations of eluant. Since some antigens or antibodies will be destroyed by certain eluants, it is easy to eliminate some (e.g., extremes of pH) on paper. Others will need to be tested for innocuity against the soluble product, e.g., by bioassay or other appropriate means. Finally, it should not be forgotten that some eluants are not innocuous to the solid phase bead itself. Silica and porous glass are well known not to be suitable above pH 8.5 because of leaching. Noncross-linked agarose is prone to enhanced breakdown with multiple cycles of reuse in chaotropic solvents because the agarose chains are only held together by hydrogen bonds.[19] Fractogels seem particularly prone to leak coupled antibodies when exposed to 1 M KI or 3 M KSCN. This is an inherent property of the bead polymer as it does not occur with any other CNBr-activated gel beads.

Usually, it is possible to synergize the action of one eluant by another. For instance, the combination of alkaline pH and organic amine[14] or glycol[20] is very effective with monoclonal and polyclonal antibodies amounting to a very efficient switch. The most useful single eluant for polyclonal antibodies is acid buffer. Often glycine HCl pH 2.5 to 3.0 is used; however, citric acid or acetic acid are also equally effective and have the respective virtues of being a stabilizer and volatile. Several alkyl amines are also volatile in the alkaline range. This is useful if you have to do freeze-drying, handle very small masses of protein, or need to carry out bioassays, e.g., tissue culture when the buffer and eluants are toxic. In addition to synergy, one may also need to add stabilizers, preservatives, etc., for instance, Mg^{2+}, Ca^{2+}, ethylenediaminetetraacetic acid (EDTA), amino acids to stabilize enzymes, antigens, etc. Caution should be exercised in using thiols or strong reducing agents ($NaBH_4$) as reduction of intrachain or interchain disulfide bonds may either dissociate subunit chains causing leakage or inactivation or may render the solid phase reactive as a solid phase thiol capable of binding other thiol proteins. Eluants which might appear unusable with proteins in solution (acetonitrile, isopropanol, methanol, dimethylsulfoxide) due to denaturation and/ or precipitation can be used with considerable impunity with solid phase covalently immobilized proteins (e.g., antibodies) because their stability is considerably enhanced by immobilization. The stability of the antigen (soluble partner) then becomes the limiting factor. One of the most effective eluants known is 0.2 M lithium diiodosalicylate (LIS); however, it is likely to denature most proteins in soluble form and is probably more useful for glycoprotein or polysaccharide antigens. More specific elution is occasionally appropriate, e.g., with competing biospecific eluants (antigen or antibody), e.g., when antigen is required for immunization and the presence of excess homologous antibody is acceptable or for ^{125}I-tracer where the presence of unlabeled carrier antigen is acceptable. Temperature change (to lower or higher than room temperature) is also sometimes helpful but is difficult to predict or screen for. With difficult elutions, especially with polyclonal antibodies and/or deep packed beds, some advantage may be gained by eluting the column in reverse flow, i.e., so that free antigen does not have to traverse through regions of free antibody. In

TABLE 14
Leakage Mechanisms

Within solid support
    Polymer chain dissociation of hydrogen bonds
    Hydrolysis of polymer chain subunits
    Mechanical fragmentation of solid support
Within linkage
    Hydrolysis of bead link
    Hydrolysis of link ligand
Within ligand
    Proteolysis of antibody (Fab-Fc)
    Reduction of disulfide bonds
    Chaotrope dissociation of hydrogen bonds

extremis, where elution is not possible, a cleavable (chemical or enzymic) link can be incorporated between solid phase and ligand. Examples include disulfides (thiol cleavable), diols (periodate cleavable), or gelatin (collagenase cleavable). Since the eluant will contain antibody complexed to antigen and since the column is essentially "single use", this is restricted to small amounts, valuable samples, or antigen for immunization.

D. LEAKAGE CONTROL AND ASSAY

Probably the most important single issue currently limiting large-scale application of immunoaffinity ligands is the concern surrounding leakage of the protein ligand, solid phase components, or activation chemicals. Part of the problem lies in the difficulty of measuring easily and accurately such leakage. Most attention has been focused on the protein ligand (especially murine IgG) which can be monitored by ^{125}I-tracer leakage, by ELISA in the effluent, or indirectly by Coulter counting of microparticles. Occasionally, indirect evidence of leakage is provided by the appearance of anti-murine IgG antibodies in recipients of infused clinical products. By contrast almost no evidence is available on leakage of solid phase polymer, probably because this is difficult to measure. For example, the polymers will not be very antigenic, will not adsorb in the ultraviolet, and will not be capable of sensitive chemical or biological assay. However, indirect evidence, at least in the case of noncross-linked agarose[19] clearly indicates that soluble agarose chains can be stripped from beads during chaotropic cycling or working. The various possible mechanisms of leakage are summarized in Table 14; it should be stressed that in reality all three classes of leakage mechanism will occur to varying extents simultaneously. It is, thus, short-sighted to focus on improving only one of them (e.g., chemical linkage) without addressing all three. The proportion of each mechanism — support, linkage, ligand — will depend on the chemical composition of each and the cycle of operation. Within the solid support, leakage may occur, e.g., with noncross-linked agarose, by dissociation of hydrogen bonds during chaotropic elution (KI, urea, etc.) such that whole agarose chains with intact ligand attached may appear in the eluted product. Alternatively, in acid or hydrolytic (enzymic?) environments the intrachain glycosidic bonds may be cleaved releasing fragments of short polygalactose chain with ligand attached. In batch operation simple mechanical fragmentation of beads may produce microscopic fragments of beads (e.g., $\simeq 0.1$ to 1.0 μm) which pass through bed support filters, do not sediment, and remain in the product.

Much attention has been focused on the cleavage of the covalent chemical activation and linkage of the ligand to the solid phase. However, relatistic numerical comparative data are few and far between.[21,22] Theoretically, cleavage can occur in two sites: either at the bead-link side or the link-ligand side of the coupling molecule(s). One interesting observation is that covalent linkage to, e.g., a large protein can itself produce mechanical "strain" in the penultimate peptide bond of a protein such that it is particularly susceptible to cleavage

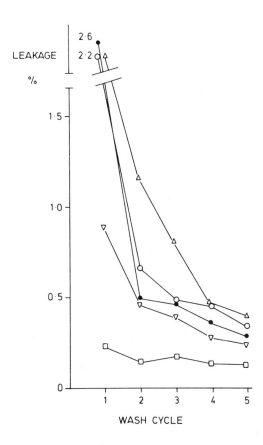

FIGURE 1. Leakage (% of total loading) of murine monoclonal antibody ESN 3 against human factor IX coupled to different gels by different chemistries: △ carbonyldiimidazole activated Trisacryl GF-2000; ○ nitrophenylchloroformate Trisacryl GF-2000; ● Divinylsulfone activated Fractogel TSK-65; ▽ carbonyl-diimidazole activated Sepharose 6B; □ periodate activated Sephacryl S-1000. Each 1-ml gel was coupled to 1 mg of antibody with 10^4 Becquerel of ^{125}I-IgG tracer; efficiency of coupling after preliminary blocking and washing was respectively, 63, 40, 57, 79, and 87%. Leakage was expressed as % normalized to total counts residing in the gel. Each cycle of elution represents 30 min mixing at 20°C in 5 volumes of 0.2 M glycine - HC1 pH 10.0 followed by PBS washing back to neutrality. Minimal detectable leakage is 0.02%. (Unpublished data of H. Bessos.)

by the provision of "activation energy". Cleavage within the ligand molecule (e.g., antibody) can occur by proteolysis (either due to proteases within crude feedstock or accidentally provided by microbial contamination of the column) or by reduction of disulfide bands, or by chaotrope dissociation of heavy and light chains which have been proteolyzed and reduced prior to elution. Wherever possible, reducing agents such as $NaBH_4$, $NaCNBH_3$, and R–SH should be avoided, especially in conjunction with chaotropic eluants. However, the problem of cleavage of the actual linkage region is not such a great problem with macromolecules as it is with low mol wt ligands such as amino acids or drugs. The coupling of antigen/antibody will usually be via multiple side chain ε-amino groups numbering 2 to 20 in most cases. The statistical likelihood of all these hydrolyzing is considerably smaller than for a single bond. To take an extreme example, divinyl sulfone is much more leaky[10] at pH 8.5 to 9.0 than CNBr coupling to *monovalent* ligands, but when used to couple proteins it is significantly less leaky[11] and is, in fact, better than some other systems for protein coupling (Figure 1).

TABLE 15
Particulate Leakage During Elution Cycle

Cumulative effluent volume (ml)	Eluant	Coulter 2 μm channel counts/ml
100	PBS pH 7.4	23,775
300	PBS pH 7.4	765
400	Saline BP	257
1000	Saline BP	701
1100	0.1 *M* GLY:HCl pH 3.0	15,507
1400	0.1 *M* GLY:HCl pH 3.0	2,869
1500	PBS pH 7.4	6,897
2000	PBS pH 7.4	939

Note: Column consisted of 40 ml of Sepharose 4B coupled to polyclonal human IgG at 15 mg/ml via CNBr-linkage. Incoming eluants were filtered on-line via a 0.22-μm filter and control duplicate particle counts were also run in a buffer designed to eliminate all protein based aggregates (1% SDS, 8 M urea, 0.1 *M* Na_2CO_3, 10 m*M* DTT), counts in this control never exceeded 1000/ml. Estimates of sensitivity suggest 10^3 2-μm particles per milliliter correspond to 0.5 ng of leaked protein aggregates.

A more complex situation is evident when actual comparative data are examined for different gel and activation chemistries (Figure 1). This shows that "leakage" is strongly time dependent, being much greater in the early cycles of use and rapidly decreasing in asymptotic fashion to approach a value which is constant for a particular gel, activation, solvent combination. It is clear that any comparison of chemistries is meaningless unless carried out in parallel with similar gel particles, solvents, and times as well as histories of washing.

In all probability the initial leakage seen in Figure 1 is noncovalently bound (hydrophobic?) material which was not removed by the primary washing steps but dissociates slowly over a period of weeks in Tris saline buffer at 4°C or over a period of hours in acid and alkali buffers. Detailed study of the actual point when leakage is maximal (Table 15) shows that bursts of leakage occurs during the transition from one pH to another (e.g., 7→3 or 7→10) or from one solvent to another (e.g., H_2O→ *M* KI or 8 *M* urea →H_2O). Once a steady state of equilibrium was achieved in any given solvent, leakage decreased markedly. It follows that rapid cycling of buffers, e.g., acid to alkali to acid etc. is by far the best way to remove this initial burst of "noncovalent" leakage. This initial noncovalent leakage might also be avoided in the first place by incorporating a nonionic detergent (e.g., 1% v/v Tween 20) into the protein solution to be coupled prior to contact with the activated resin. Other factors which have been found to reduce leakage in various situations are summarized in Table 16. Two points in particular are worth noting: heat provocation at 60°C for 1 to 10 h can be used as an alternative to prolonged storage and washing and also represents a valuable "pasteurization" step for microbiological control. This is possible because solid phase coupled proteins are usually much more stable to aqueous heating than their soluble counterparts.

Glutaraldehyde cross-linking *in situ* after coupling[21,22] with concentrations from 0.02 to 1% v/v may improve leakage severalfold. However, a variable loss in functional activity is likely, and blocking with amines is desirable.

To summarize, leakage is always detectable if a sufficiently sensitive assay is used; however, it may be reduced by several orders of magnitude (from microgram per milliliter to nanogram per milliliter) by judicious combination of several approaches listed in Table 16. Attention to the whole system (not just activation chemistry) is desirable to minimize leakage problems. The actual levels of leaked ligand (e.g., murine IgG) are best determined

TABLE 16
Factors Found to Decrease Leakage

1. Multiple site coupling at, e.g., pH 9.3 not pH 7.0
2. Storage temperature, e.g., 4°C and pH 4 not 20°C/pH7
3. Increase primary wash volume from $2^1/_2$ to 10 volumes
4. Change from saline to PBS for primary wash
5. Fit 0.22-μm inline filter immediately upstream
6. Store gels 6 to 8 weeks before use
7. Dummy cycling with pH 2 to 9 cycling or chaotropes
8. Heat provocation after coupling, e.g., 60°C for 1—10 h and/or autoclave gel prior to use
   (i.e., *before* activation or coupling)
9. Postcoupling cross-linking with 0.04—1% v/v glutaraldehyde pH7
10. Immobilize predigested antibody — Fab'2 or Fab
11. Couple protein in presence of 1% v/v Tween 20

TABLE 17
Sterilization and/or Sanitization
of Solid Phases

Autoclaving
60_{Co} - γ irradiation (wet)
Ethylene oxide
Sodium hydroxide
Sodium hypochlorite
Ethanol, isopropanol
Quaternary amines
Pasteurization
Detergents
Formaldehyde, glutaraldehyde

by a sensitive ELISA in the final product. Commercial ELISA reagents should be capable of measuring down to 0.1 to 1.0 ng/ml of murine IgG in human protein samples.

E. STERILIZATION AND CLEANING

There are no absolutely reliable methods that can be guaranteed to work in sterilization and cleaning of immunopurification affinity reagents. Rather, a judicious choice has to be made based on the known stability of the solid phase, the coupling chemistry, and the ligand. Prevention is the first step. All fluid entering the device should be filtered off-line to remove gross contamination and should contain bacteriostats or other preservatives. In addition, a final 0.22-μm in-line filter should always be present upstream of the column or device. This not only prevents bacterial colonization of the column, but also reduces particulate contamination and clogging of the upper surface of the bed. Table 17 lists some of the basic approaches which can be used alone or in combination. Usually, the protein part of an immunosorbent will not resist autoclaving or γ-irradiation. However, many polysaccharides, glycoproteins, and gels will survive quite well, especially when coupled to the solid phase. Ethylene oxide works wells in a moist environment but is prone to inactivate proteins via alkylating reactions, and it is very difficult to remove completely all residues. The most effective universal reagent is probably sodium hydroxide which in concentrations between 0.10 M and 0.5 M (2 h 20°C) is very effective[23] both as a sterilizer and for cleaning solid phases of "irreversibly" bound macromolecules. Obviously, compatibility with individual antigen or antibody is a critical requirement for this reagent, and, in general, it cannot be assured that it is compatible with antigens or antibodies. Sodium hypochlorite (effectively, chlorine dissolved in sodium hydroxide) is a very effective sterilizer even when diluted, but most proteins are unlikely to survive significant contact; however, polysaccharides and glycoproteins may. Concentrations of ethanol or isopropanol in the range 15 to 20% v/v are probably the best choice for effective sterilization and compatibility with immobilized proteins etc. They are relatively mild, of low toxicity, and easily washed out before use. Again, proteins are much less prone to denaturation by alcohols when solid phase coupled. A variety of quaternary amines are available in water-soluble form for sanitization (''Vantocil'' and ''Cetavlon'' ICI are both active against yeasts and bacteria at 10 to 20 μg/ml concentrations). Since they are quite inert to proteins at these concentrations, it is also possible to consider mixed systems with, e.g., ethanol and quaternary amines. As alluded to earlier ''wet heat'' or pasteurization by heating in an aqueous system to 60 ± 1°C for several hours may be feasible with antibodies which would otherwise aggregate in solution but are effectively prevented from doing so when immobilized. Denaturation is much slower than aggregation at these temperatures and most antibodies will survive at least 30 min at 60°. With careful choice of pH (e.g., 5 to 6) and buffer stabilizer (e.g., 0.5 M Na$_3$ citrate) it should be possible to heat for several hours with only modest loss ($\simeq$ 25%) of antibody functional activity. In addition to any sterilizing action, such heating prior to use will accelerate the removal of any weakly bound molecules that might otherwise leak out during later processing. Detergents specifically active against phospholipid-bound viruses (both neutral, e.g., Tween 20, Tween 80, and anionic, e.g., sodium cholate, sodium deoxycholate) are known to be well tolerated by both soluble and immobilized proteins. In addition to virucidal activity (which may be useful while product is temporarily bound to an immobilized purification system) detergents may also function usefully in removing nonspecifically bound proteins which might coelute with product and/or will remove lipid, lipoprotein, etc. which may tend to reduce sorbent capacity by physical blocking of pores. Finally, one should not neglect the traditional aldehyde sterilants (formaldehyde, glutaraldehyde) which are active in aqueous systems at pH 7.0 in the concentration range 0.1 to 1.0%. Although some antigen or antibody function may be lost ($\simeq$25%) on first use,[21] subsequent reuse of aldehyde may be without further loss. Some experimentation is desirable in terms of time (1 to 3 d) and temperature (20 to 40°C). The aldehyde treatment can also be usefully considered as a cross-linking process that will reduce leakage as well. It may be prudent to follow aldehyde treatment with a blocking amine (ethanolamine, glucosamine, glycine, lysine) to eliminate any free residual aldehyde groups which might otherwise remain to bind irreversibly product and contaminants. Where simple cleaning and regeneration of a solid phase is required rather than sterilization, a combination of 0.1% w/v sodium dodecylsulfate and 8 M urea or 0.1 M Na$_2$CO$_3$ is very effective, but should be perfused as quickly as possible and followed by extensive washing to neutrality.

V. DEPLETION

At first sight, it might seem that depletion of a single species from a feedstock fluid by immunoaffinity would be identical in operation to conventional immunopurification; however, experiences teaches that this is not so. The first major differences is that whereas 90% yield in a standard purification process would be very acceptable, a residue of 10% remaining in a depleted fluid is unlikely to be acceptable in most cases. Rather, the aim is for 0.1 to 1% residual biological activity depending on the sensitivity of the bioassay. Second, since complex biological mixtures are involved, any nonspecific decay or loss of other components in a mixture may be undesirable such that one may require 90% yield of all other active constituents. Furthermore, their concentration may be important, so dilution is not acceptable either. Inactivation may occur simply by prolonged manipulation at room temperature, so low flow rates through soft gel supports may be unacceptable with complex, viscous biological fluids. Unfortunately, the benefit of 4°C operation may be cancelled out by an increase in viscosity which further prolongs processing time. A better choice is to use a more rigid, more porous gel with a higher loading of antibody. Thus, whereas in purification one may aim for maximum specific capacity and use 1 to 2 mg/ml of antibody, in a depletion system both low final concentration of analyte and high volumetric throughput may predicate use of antibody concentration in the 3 to 10 mg/ml range of concentration. Such increased antibody loadings also increase the need to ensure that leakage of immobilized antibody is minimized by the procedures described earlier, for leakage of antibody may seriously hamper any subsequent use of depleted fluids as substrates in bioassays. Another significant difference relates to (non)recovery of the adsorbed product. Since removal is more important, the use of highest affinity and/or polyclonal antibodies is preferable, provided they can be regenerated. Even the most active chaotropes (LIS, KSCN, etc.) are useful since active product is not required, only regenerated antibody, which is generally more resistant. Thus, very high affinity antibodies which are unsuitable for immunopurification may be ideal for immunodepletion. Unfortunately, since concentrations of the target molecule may be low (e.g., growth factors and clotting factors are active in the 1 to 100 ng/ml range), problems with insufficient affinity of antibodies may still persist. In the case of artificial (immuno-depleted) hemophilic plasma,[24,25] the concentration of FVIII has to be reduced from about 200 ng/ml (normal plasma level) to ≤ 1 ng/ml ($\equiv 0.5\%$ of normal plasma) without significant loss of other labile coagulation factors. Since any one antibody alone is not capable of such performance (either due to epitope loss on some antigens or to insufficient antibody affinity), empirical optimization showed that two or three different antibodies directed to different epitopes on the molecule were a more practical approach (see Chapter 8). One is then faced with several possible alternatives for immobilization: (1) mix all three antibodies and immobilize on the same gel bead; (2) immobilize each antibody separately to three batches of gel beads, but mix these into one column; (3) immobilize each antibody onto separate gel beads and operate these in three separate columns in sequential flow. After empirical evaluation it turns out the third option (the three different sequential columns) is the most efficient adsorbent for the desired analyte, FVIII. In retrospect, this can be explained on the assumption that each antibody-epitope pair operates to its maximum efficiency without saturation and is then followed by the second antibody-epitope pair and so on. This successive efficiency is lost if each antibody is immobilized in the same bead or within the same column. Provided a suitable regeneration buffer can be found, this three-column system is no more difficult to operate than a single-column system. In addition to depletion of specific analytes, exactly the same format can be used for "negative" immunoaffinity chromatography of a persistent contaminant (e.g., albumin) which is otherwise difficult to remove from products requiring high purity.

VI. DISCUSSION

Immunoaffinity purification is still in its infancy and has yet to realize its full-scale potential, e.g., in production of therapeutic materials. Partly, this is because many practical, economic, and regulatory problems have to be solved for each product. The practical advice in this chapter is not meant to be exclusive but to encourage other workers to experiment with new combinations of solid phase-coupling-ligand options. It is only by such empirical data generation that a coherent set of theoretical rules will evolve to enable reliable prediction of performance and scaleup to be obtained. If this can be achieved by breaking the rules, so much the better.

VII. APPENDIX OF MATERIALS AND SUPPLIERS

A. ACTIVATED AND NONACTIVATED POROUS BEADED SUPPORTS

Agarose, Sepharose, CNBr- and Tresyl-activated agarose, Sephacryls N-hydroxysuccinimide-activated carboxyl-agarose from Pharmacia, P. O. Box 175, S-75104 Uppsala, Sweden or from Pharmacia, 800 Centennial Avenue, Piscataway, NJ 08854.

Fractogels (Toyopearl) epoxy-CDI-activated Fractogel, agarose, Trisacryl, porous glass, glycol-glass. BCA reagent from Merck, Frankfurterstrasse 250, D-1600 Darmstadt, West Germany or from Pierce, P. O. Box 117, Rockford, IL 61105.

Trisacryl GF-2000, Ultrogels, glutaraldehyde and nitrophenyl-chloroformate (NPC) -activated Ultrogels and Trisacryl, Spherodex (dextran-coated silica), magnetic Ultrogel from IBF, 35 Ave. Jean-Jaures, 92390, Villeneuve-la-Garrene, France.

Poly-hydroxyethylmethacrylate (HEMA) and divinysulfone (DVS) -activated HEMA from Tessek, 444 Castro St., Suite 710, Mountain View, CA 94041, or from Tessek, Voldbjergvej 14A, DK8240 Risskov, Denmark.

DVS-activated agarose 4B from Kem-En-Tec, Lemchesvej II, DK2900 Hellerup, Denmark; Agarose, N-hydroxysuccinimide activated with different spacer arms from Biorad Labs Ltd. Caxton Way, Watford Business Park, Watford, WD1 8RP, U.K. Also from Biorad 1414 Harbor Way South, Richmond, CA 94804.

FMP-activated Fractogel and Trisacryl from Bioprobe International Inc., 2842C Walnut Ave., Tustin, CA 92680.

Nugel glutaraldehyde and poly-N-hydroxysuccinimide-activated silicas from Separation Industries Inc., 4 Leonard St. Metuchen, NJ 08840 or from Camlab, Nuffield Road, Cambridge, CB4 1TH, U.K.

Prosep-activated porous glass-proprietory chemistry — custom immobilization from Protein Separations Ltd., Medomsley Road, Consett, Tyne & Wear, DH8 6PJ, U.K.

Preactivated core and shell microparticles (including magnetic) from Dyno Particles AS, P. O. Box 160, N-2001, Lillestrom, Norway; also from Dynal Inc. 45 North Station Plaza, Great Neck, NY 11020.

B. ACTIVATED MEMBRANES, ASSEMBLED DEVICES, AND ANNULAR COLUMNS

Memsep 1000/1010/2000 glutaraldehyde activated membrane capsule devices, Dominick Hunter Ltd., at Durham Road, Birtley, Tyne & Wear, DH3 2SF, U.K. or DH Filters Inc., 4220 Commercial Drive, Tracy, CA 95376.

Zetaffinity-activated annular cellulose/acrylamide composite device, Cuno Inc., Molecular Separation Products, 400 Research Parkway, Meriden, CT 06450 or Anachem Ltd., 20 Charles St. Luton, Beds., LU2 OEB, U.K.

AFC-aldehyde agarose and proprietory elution buffer, also Prodisc-PVC membrane-Silica composite with PEI-glutaraldehyde activation from New Brunswick Scientific Co. Inc., P.O. Box 4005, Talmadge Road, Edison, NJ 08818 or NBS, 6 Colonial Way, Watford, Herts, WD2 4PT, U.K.

Immobilon AV-activated PVDF sheet membrane from Millipore Corp., P.O. Box 255, Bedford, MA 01730 or Millipore U.K. Ltd., 11/15 Peterborough Rd, Harrow, Middlesex, HA1 2YH U.K.

Immunodyne-activated membrane dipsticks and Silent Monitor 96-well trays from Pall

Corp, 77 Crescent Beach Road, Glen Cove, NY 11542 or Pall Biosupport Division, Europa House, Havant St., Portsmouth, Hants P01 3PD, U.K.

Annular packing columns for conventional soft gels (50 ml and larger) from Sepragen Corp. 2126 Edison Av., San Leandro, CA 94577.

REFERENCES

1. **Scouten, W. H.,** A survey of enzyme coupling techniques, in *Methods in Enzymology,* Vol. 135, Mosbach, K., Ed., Academic Press, New York, 1987, 30.
2. **Yau, W. W., Kirkland, J. J., and Bly, D. D.,** *Modern Size-Exclusion Chromatography,* Wiley-Interscience, New York, 1979, 33.
3. **Brandt, S., Goffe, R. A., Kessler, S. B., O'Conner, J. L., and Zale, S. E.,** Membrane based affinity technology for commercial scale purifications, *Biotechnology,* 6, 779, 1988.
4. **Einarsson, M., Forsberg, B., Larm, O., Riquelme, M. E., and Scholander, E.,** Coupling of proteins and other amines to Sepharose by bromine oxidation and reductive amination, *J. Chromatogr.,* 215, 45, 1981.
5. **Stults, N. L., Lin, P., Hardy, M., Lee, Y. C., Uchide, Y., Tsukada, Y., and Sugimori, T.,** Immobilization of proteins on partially hydrolysed beads, *Anal. Biochem.,* 135, 392, 1983.
6. **Liberatore, F. A., McIsaac, J. B., and Royer, G. P.,** Immobilised carboxypeptidase Y. Applications in protein chemistry. *FEBS Lett.,* 68, 45, 1976.
7. **Wright, J. F. and Hunter, W. M.,** A convenient replacement for cyanogen bromide-activated solid phases in immunoradiometric assays, *J. Immunol. Methods,* 48, 311, 1982.
8. **Hornsey, V. A., Prowse, C. V., and Pepper, D. S.,** Reductive amination for solid-phase coupling of protein. A practical alternative to cyanogen bromide, *J. Immunol. Methods,* 93, 83, 1986.
9. **Kohn, J. and Wilchek, M.,** A new approach (cyano-transfer) for cyanogen bromide activation of Sepharose at neutral pH, which yields activated resins, free of interfering nitrogen derivatives, *Biochem. Biophys. Res. Commun.,* 107, 878, 1982.
10. **Porath, J., Laas, T., and Janson, J. C.,** Agar derivatives for chromatography, electrophoresis and gel-bound enzymes. III. Rigid agarose gels, cross-linked with divinyl sulphone (DVS). *J. Chromatogr.,* 103, 49, 1975.
11. **Lihme, A., Schafernielsen, C., Larsen, K. P., Muller, K. G., and Boghansen, T. C.,** Divinyl sulphone activated agarose: formation of stable and non-leaking affinity matrices by immobilisation of immunoglobulins and other proteins, *J. Chromatogr.,* 376, 299, 1986.
12. **Jorgensen, T.,** High performance liquid affinity chromatography on hydroxyethylmethacrylate polymer, *Hoppe Seylers Biochem.,* 368 (Abstr.), 752, 1987.
13. **Brandt, J., Andersson, L.-O., and Porath, J.,** Covalent attachment of proteins to polysaccharide carriers by means of benzoquinone, *Biochim. Biophys. Acta,* 386, 196, 1975.
14. **Mejan, O., Fert, V., Delezay, M., Delaage, M., Cheballah, R., and Bourgois, A.,** Immunopurification of human factor VIII/vWF complex from plasma, *Thromb. Hemostas.,* 59, 364, 1988.
15. **Bethell, G. S., Ayers, J. S., Hancock, W. S., and Hearn, M. T. W.,** A novel method of activation of cross-linked agaroses with 1,1-carbonyldiimidazole which gives a matrix for affinity chromatography devoid of additional charged groups, *J. Biol. Chem.,* 254, 2572, 1979.
16. **Bethell, G. S., Ayers, J. S., Hearn, M. T. W., and Hancock, W. S.,** Investigation of the activation of cross-linked agarose with carbonylating reagents and the preparation of matrices for affinity chromatography purifications, *J. Chromatogr.,* 219, 353, 1981.
17. **Nilsson, K. and Mosbach, K.,** Immobilization of enzymes and affinity ligands to various hydroxyl group carrying supports using highly reactive sulfonyl chlorides. *Biochm. Biophys. Res. Commun.,* 102, 449, 1981.
18. **Kohn, J. and Wilchek, M.,** A colorimetric method for monitoring activation of Sepharose by cyanogen bromide, *Biochem. Biophys. Res. Commun.,* 84, 7, 1978.
19. **Sandberg, A., Bergsdorf, N., Brandstrom, A., and Sundstrom, S.,** 100 cycles of immunoaffinity purification of t-PA, *Fibrinolysis,* 2 (Suppl.1) (Abst.), 335, 1988.
20. **Fornstedt, N.,** Affinity chromatographic studies on antigen-antibody dissociation, *FEBS Lett.,* 177, 195, 1984.
21. **Kowal, R. and Parsons, R. G.,** Stabilization of proteins immobilised on Sepharose from leakage by glutaraldehyde cross-linking, *Anal. Biochem.,* 102, 72, 1980.

22. **Sato, H., Kidaka, T., and Hori, M.,** Leakage of immobilized IgG from therapeutic immunoadsorbents, *Appl. Biochem. Biotechnol.,* 15, 145, 1987.
23. **Berglof, J. H., Adner, N. P., and Doverstein, S. Y.,** Inactivation of microbial contamination in chromatographic separation media using sodium hydroxide, Proc. 20th Congr. Int. Soc. Blood Transfusion, London, U.K., 10-15th July Abstr. No. PM 4/14, 1988, 70.
24. **Takase, T., Rotblat, F., Goodall, A. H., Kernoff, P. B. A., Middleton, S., Chand, S., Denson, K. W., Austen, D. E. G., and Tuddenham, E. G. D.,** Production of factor VIII deficient plasma by immunodepletion using three monoclonal antibodies, *Br. J. Haematol.,* 66, 497, 1987.
25. **Hornsey, V. A., Waterston, Y. G., and Prowse, C. V.,** Artificial factor VIII deficient plasma: preparation using monoclonal antibodies and its use in one stage coagulation assays, *J. Clin. Pathol.,* 41, 562, 1988.
26. **Groff, J. L. and Cherniak, R.,** The incorporation of amino groups into cross-linked Sepharose by use of (3-aminopropyl)-triethyoxysilane, *Carbohydr. Res.,* 87, 302, 1980.

Index

INDEX